Digital Signal Processing

D. Sundararajan

Digital Signal Processing

An Introduction

 Springer

D. Sundararajan
Formerly at Concordia University
Montreal, QC, Canada

ISBN 978-3-030-62370-8   ISBN 978-3-030-62368-5   (eBook)
https://doi.org/10.1007/978-3-030-62368-5

This Springer imprint is published by the registered company Springer Nature Switzerland AG
The registered company address is: Gewerbestrasse 11, 6330 Cham, Switzerland

Preface

Due to the advances in digital system technology and fast numerical algorithms, digital signal processing is a necessity in most areas of science and technology in theory and practice. However, as the mathematical content of digital signal processing is quite high, students have to be made comfortable in their learning of this subject through a number of appropriate examples, figures, and programs. Further, it has to be pointed out how the theoretical analysis is approximated numerically to include practical considerations in applications. In conjunction with theoretical analysis and a laboratory class, programming is essential for getting a good understanding of the subject and the course should end with one or two good projects of complexity that the students can handle. Further, each student should practice basic concepts, such as convolution and discrete Fourier transform, with paper-and-pencil and programming as much as necessary to their good understanding. The essentials of digital signal processing are signals and systems, sampling, transform methods and their computational aspects, and digital filter analysis, design, and implementation.

About 20 years ago, I wrote a book on digital signal processing. After that, I wrote five books on topics related to signal processing. Further, the topics of multirate digital signal processing and discrete wavelet transform have become quite popular in practice. Based on my long experience and to include new materials with some good features, I wrote a new book. This book is primarily intended to be a textbook for an introductory course in digital signal processing for senior undergraduate and first year graduate students in electrical and electronics departments. It can also be used for self-study and as a reference. The prerequisite is a course in signal and systems or equivalent and some experience in programming. The features of this book are the detailed coverage of finite impulse response optimal filter design with MATLAB® programs (available online), clear and concise presentation of the difficult concepts using DFT of short length data or appropriate figures, physical explanation of concepts, large numbers of figures and examples, and clear, concise, and, yet, comprehensive presentation of the topics.

Answers to selected exercises marked with $*$ are given at the end of the book. A Solutions Manual and slides are available for instructors at the website of the book. I assume the responsibility for all the errors in this book and would very much appreciate receiving readers' suggestions and pointing out any errors (email:

d_sundararajan@yahoo.com). I am grateful to my Editor and the rest of the team at Springer for their help and encouragement in completing this project. I thank my family for their support during this endeavor.

D. Sundararajan

Contents

Abbreviations

BIBO	Bounded-input bounded-output
DC	Direct current, sinusoid with frequency zero, constant current or voltage
DFT	Discrete Fourier transform
DSP	Digital signal processing
DTFT	Discrete-time Fourier transform
FIR	Finite impulse response
IDFT	Inverse discrete Fourier transform
IIR	Infinite impulse response
Im	Imaginary part of a complex number or expression
LSB	Least significant bit
LTI	Linear time invariant
MSB	Most significant bit
Re	Real part of a complex number or expression
ROC	Region of convergence
SNR	Signal-to-noise ratio

Discrete-Time Signals

1

Signals are used for communication of information about some behavior or nature of some physical phenomenon, such as temperature and pressure. In the mathematical form, it is a function of one or more independent variables. Speech signal is one of the most commonly used signals by human beings. Signals usually need some processing for their effective use. Available speech signal may be too weak for audibility. It needs amplification. Signals get degraded in transmission by known or unknown processes. These signals need suitable filtering to improve their quality. After conditioning the signal, the information content of the signal has to be extracted. In applications of signal processing, we encounter a variety of signals, such as biomedical signals, audio signals, seismic signals, and video signals. In this book, we confine to the processing of signals with one independent variable. Most of the naturally occurring signals have arbitrary amplitude and are difficult to process as such. Basically, in signal processing, these signals are suitably approximated by a few basis signals so that signals are represented compactly and the processing becomes simpler and efficient. The approximation divides signal processing into two major types, processing in the time domain with the time as the independent variable and processing in the frequency domain with the frequency as the independent variable. We use the type that is more advantageous for the given signal processing task. Of course, there are lots of details, in either case, to be studied to become proficient in processing the signals in practice.

1.1 Signal Classification

Depending on the nature of sampling, quantization, and other characteristics, signals are classified. Classification helps to choose the most appropriate processing for a given signal. Unnecessary work and possible errors can be avoided, if the signal characteristics are known.

© The Author(s), under exclusive license to Springer Nature Switzerland AG 2021
D. Sundararajan, *Digital Signal Processing*,
https://doi.org/10.1007/978-3-030-62368-5_1

1

1.1.1   Continuous-Time Signals

This type, also called analog signals, is characterized by its continuous nature of the dependent and independent variables. For example, the room temperature is a continuous signal. It has a value at any instant of time. The mathematical form of a signal is $x(t)$, where t is the independent variable and $x(t)$ is the dependent variable. That is, $x(t)$ is a function of t. While the independent variable for most of the signals is time, it could be any other entity such as frequency or distance. The mathematical characterization is the same irrespective of the nature of the signal. It could be a temperature signal or a pressure signal or anything else. Figure 1.1a shows an arbitrary continuous-time signal. This is a typical example of signals occurring in practice with arbitrary amplitude profile. In digital signal processing, the first step, as mentioned earlier and presented later, is to approximate such signals in terms of basic signals so that the processing is simplified. The signal is defined at each and every instant of time over the period of its occurrence. It can assume any real or complex value at each instant. While all practical signals are real-valued, complex-valued signals are necessary and often used as intermediaries in signal processing.

1.1.2   Discrete-Time Signals

In this type of signals, also called discrete signals, the values of the signal are available only at discrete intervals. That is, we take the sample values of a

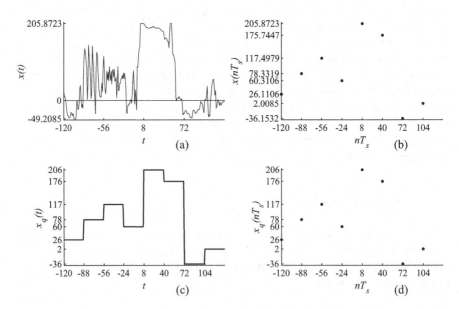

Fig. 1.1 (**a**) Continuous-time signal; (**b**) discrete-time signal; (**c**) quantized continuous-time signal; (**d**) digital signal

continuous signal at intervals of T_s, the sampling interval. The signal could also be inherently a discrete-time signal. Irrespective of the source, the discrete value $x(n)$ is called the nth sample of the signal. The sampling interval T_s is usually suppressed. Figure 1.1b shows the discrete-time signal corresponding to the signal shown in Fig. 1.1a. The mathematical form of this type of signal is $x(nT_s)$, where nT_s is the independent variable and $x(nT_s)$ is the dependent variable. That is, $x(nT_s)$ is a function of nT_s. In the expression nT_s, the sampling interval T_s is assumed to be a constant in this book and the range of the integer variable, in general, is $n = -\infty, \ldots, -1, 0, 1, \ldots, \infty$. It looks like that, we cannot recover $x(t)$ from $x(nT_s)$ as lots of the values of the signal between sampling intervals are lost in the sampling process. It is true, in this case, as the sampling interval is not short enough to represent and process it with its samples. However, any practical signal, with sufficiently short sampling interval, can be represented by a finite number of samples and reconstructed after processing with adequate accuracy.

Practical signals can be considered, with negligible error, as composed of frequency components with frequencies varying over a finite range (band-limited). The sampling theorem says that any continuous signal can be exactly reconstructed from its sampled version with a sampling interval T_s, such that more than two samples in a cycle of the highest frequency component are taken. The sampling theorem is the basis on which DSP is based. The reason we want to process the signal using its samples, rather than the continuous version itself, is that DSP includes the advantages of low-cost, higher noise immunity, and highly reliable digital components. The disadvantages are the errors accumulating due to the representation of numbers by a finite number of bits and the limitation of the frequency of operation due to sampling. In most applications, the advantages outweigh the disadvantages. Therefore, DSP is widely used in applications of science and engineering.

1.1.3 Quantized Continuous-Time Signal

The procedure in DSP is that the continuous signal is sampled, processed, and converted back to the continuous form. In reconstruction of a signal from its samples, an important component is the sample-and-hold device. This device produces the quantized continuous-time signal. Figure 1.1c shows the quantized continuous-time signal corresponding to the signal in Fig. 1.1a. The sample values are quantized to certain levels. In the figure, the sample values are rounded to the nearest integer. We could have also truncated. Now, the signal values are integer valued. By quantizing the signal, the wordlength to represent the signal is reduced. In sampling, the number of samples is reduced. These two steps, while they may introduce some errors, are unavoidable to process a signal by DSP. DSP devices can handle, due to their digital nature, only finite number of finite wordlength signals. While the DSP processing of a signal can never be exact in most cases, it can be ensured that the accuracy of representation is adequate for the given application requirements.

1.1.4   Digital Signal

In this form, both the dependent and independent variables are sampled values. The importance of this type is that practical digital devices can work only with signals of this form. Figure 1.1d shows the digital signal corresponding to the signal shown in Fig. 1.1a.

Time Domain Representation of Discrete Signals

In DSP, the processing is first carried out using discrete signals and then the effects of finite wordlength effects are analyzed. The discrete signal is a sequence of numbers with infinite precision. Even if the signal is obtained by sampling, the sampling interval is dropped and the signal is represented as $x(n)$ instead of $x(nT_s)$. While the correct representation of a sequence is $\{x(n)\}$ and $x(n)$ is single value, $x(n)$ is usually used to represent a sequence also. There are variety of ways of representing a sequence. Consider a sequence

$$\{1.1, 2.1, 3.1, 4.1\}$$

Assuming the value 1.1 corresponds to index 0, it could be written as

$$\{x(0) = 1.1, x(1) = 2.1, x(2) = 3.1, x(3) = 4.1\}$$

or

$$\{x(n), n = 0, 1, 2, 3\} = \{1.1, 2.1, 3.1, 4.1\}$$

or

$$\{\check{1}.1, 2.1, 3.1, 4.1\}$$

or

$$x(n) = \begin{cases} n + 1.1 & \text{for } 0 \leq n \leq 3 \\ 0 & \text{otherwise} \end{cases}$$

or graphically. The check symbol ˇ indicates that the index of that element is 0 and the samples to the right have positive indices and those to the left have negative indices.

1.1.5   Periodic and Aperiodic Signal

A discrete signal $x(n)$, defined for all n, is periodic, if

$$x(n) = x(n + N), \quad \text{for all } n$$

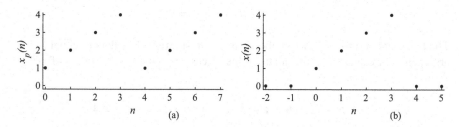

Fig. 1.2 (a) Two periods of a periodic discrete-time signal with period 4; (b) an aperiodic discrete-time signal

where N is a nonzero positive integer. The smallest N satisfying the defining condition is its period. Two cycles of a periodic signal are shown in Fig. 1.2a. The period is 4. It repeats its values over one period indefinitely. It satisfies the definition.

$$x(0) = x(4) = x(8) = \cdots \quad \text{and} \quad x(n) = x(n+4), \quad \text{for all } n$$

The sum of the samples over any interval of duration of a period of a periodic signal is the same. In typical applications, a periodic signal is generated by periodic extension of a signal defined over a finite interval. The finite number samples over one period are repeated at intervals of the period indefinitely on either side.

All the samples of an aperiodic signal have to be defined. The samples over any one period of a periodic signal are enough to define it, as the other samples are periodic repetitions. However, in signal operations, we often need the arbitrarily indexed sample values, given the values over a period. These samples can be found by periodic extension of the samples. But, it is easier to define it in terms of the given samples over one period. The problem is to relate the indices of the samples with the indices of the known values. Consider the sequence, which characterizes a periodic signal $x(n)$ with period $N = 4$.

$$\{x(0) = 2, x(1) = 1, x(2) = 4, x(3) = 3\}$$

By periodic extension, we get the values of $x(n)$ for a longer range, as shown in the table.

n	...	-8	-7	-6	-5	-4	-3	-2	-1	0	1	2	3	4	5	6	7	...
$x(n)$	...	2	1	4	3	2	1	4	3	2	1	4	3	2	1	4	3	...
$n \bmod N$	...	0	1	2	3	0	1	2	3	0	1	2	3	0	1	2	3	...

$$x(n) = x(n \bmod N) \quad \text{for all} \quad n$$

The *mod* function, $n \bmod N$, returns the remainder of dividing n by N, after truncating the quotient.

$$n - \lfloor (n/N) \rfloor N, \quad N \neq 0$$

The floor function rounds the number to the nearest integer less than or equal to its argument. For example, let us find the value of $x(6)$ from the given 4 values of $x(n)$. With $N = 4$, $n = 6$,

$$6 - \lfloor (6/4) \rfloor 4 = 6 - 1(4) = 2 \quad \text{and} \quad x(6) = x(2) = 4$$

With $N = 4$, $n = -5$,

$$-5 - \lfloor (-5/4) \rfloor 4 = -5 - (-2)(4) = 3 \quad \text{and} \quad x(-5) = x(3) = 3$$

In any operation involving the independent variable of a periodic signal, this procedure can be used.

Let $\{\check{3}, 1, 2, 5\}$ be a periodic sequence with period 4. The 33rd number in the sequence is 1. The index is obtained as $33 - (33 \bmod 4)(4) = 1$ and $x(33) = 1$. $x(-2) = x(2) = 2$. This is called circular indexing. Plot the N sample points on a circle. Usually, counterclockwise movement is taken as positive direction. Figure 1.3 shows the circular or periodic nature of periodic signal with period $N = 8$ samples. It is clear from the figure that the *mod* function adds or subtracts multiples of N until the given n is reduced to the range 0 to $N - 1$. No practical signal is periodic according to the definition. The definition is necessary for easier mathematical analysis. For practical purposes, a signal that is periodic over some required interval is adequate.

A signal that is not periodic is aperiodic. An example is shown in Fig. 1.2b and another in Fig. 1.4a. It never repeats any pattern of samples indefinitely. While most naturally occurring signals are aperiodic, the periodic signals are used to analyze them. The message signal in a communication system is aperiodic and the carrier signal is periodic. The waveforms in electrical power systems are periodic.

$$x(2) \bullet = x(2 \pm k N) =$$

$$x(3 \pm k N) = \bullet \, x(3) \qquad\qquad x(1) \bullet = x(1 \pm k N)$$

$$x(4 \pm k N) = \bullet \, x(4) \qquad\qquad x(n) \qquad\qquad x(0) \bullet = x(0 \pm k N)$$

$$x(5 \pm k N) = \bullet \, x(5) \qquad\qquad x(7) \bullet = x(7 \pm k N)$$

$$x(6) \bullet = x(6 \pm k N)$$

Fig. 1.3 Periodicity of a periodic signal with $N = 8$ samples and k an integer

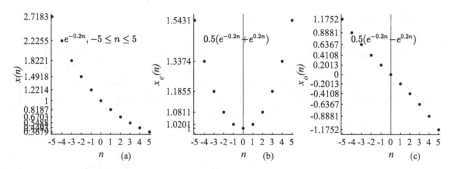

Fig. 1.4 (a) The discrete-time signal $e^{-0.2n}, -5 \leq n \leq 5$; (b) its even component; (c) its odd component

1.1.6   Even and Odd Signals

A signal is said to be even if $x(-n) = x(n)$. An even signal is symmetric with respect to the vertical axis at $n = 0$. A signal is said to be odd if $-x(-n) = x(n)$. An odd signal is antisymmetric with respect to the vertical axis at $n = 0$. Figure 1.4a–c show, respectively, the decaying exponential signal

$$x(n) = \begin{cases} e^{-0.2n} & \text{for } -5 \leq n \leq 5 \\ 0 & \text{otherwise} \end{cases}$$

and its even and odd components. Every signal can be expressed as the sum of its even and odd components.

$$x_e(n) = 0.5(x(n) + x(-n)) \quad \text{and} \quad x_o(n) = 0.5(x(n) - x(-n))$$

In other words, the sum of odd and even signals is neither odd- nor even-symmetric. The even and odd components of $x(n)$ are

$$x_e(n) = 0.5(e^{-0.2n} + e^{0.2n}) \quad \text{and} \quad 0.5(e^{-0.2n} - e^{0.2n})$$

Table 1.1 shows the samples of the signal, its time-reversed version, its even and odd components. By adding and subtracting the corresponding sample values of the signal in the first two rows and dividing by 2 yield the sample values of the components. From the figure and the table, it can be seen that the even component is symmetrical with respect to the vertical axis at $n = 0$ and the odd component is asymmetrical. Further, the original signal is equal to the sum of the sample values of the even and odd components. Due to symmetry, the sum of odd and even-symmetric signals $x_o(n)$ and $x_e(n)$, over symmetric limits N, are given by

Table 1.1 Even and odd decomposition of $e^{-0.2n}$, $-5 \le n \le 5$

n	-5	-4	-3	-2	-1	0	1	2	3	4	5
$x(n)$	2.7183	2.2255	1.8221	1.4918	1.2214	1	0.8187	0.6703	0.5488	0.4493	0.3679
$x(-n)$	0.3679	0.4493	0.5488	0.6703	0.8187	1	1.2214	1.4918	1.8221	2.2255	2.7183
$x_e(n)$	1.5431	1.3374	1.1855	1.0811	1.0201	1	1.0201	1.0811	1.1855	1.3374	1.5431
$x_o(n)$	1.1752	0.8881	0.6367	0.4108	0.2013	0	-0.2013	-0.4108	-0.6367	-0.8881	-1.1752

$$\sum_{n=-N}^{N} x_o(n) = 0, \qquad \sum_{n=-N}^{N} x_e(n) = x_e(0) + 2\sum_{n=1}^{N} x_e(n)$$

The product $z(n) = x(n)y(n)$ of an even signal $x(n)$ and an odd signal $y(n)$ is odd, since

$$z(-n) = x(-n)y(-n) = x(n)(-y(n)) = -(x(n)y(n)) = -z(n)$$

The product of two even or two odd signals is even.

A complex signal is said to be conjugate-symmetric if $x^*(-n) = x(n)$, where * indicates conjugation. A complex signal is said to be conjugate-antisymmetric if $-x^*(-n) = x(n)$. Every complex signal can be expressed as the sum of its conjugate-symmetric and conjugate-asymmetric components.

$$x(n) = x_{cs}(n) + x_{ca}(n), \ x_{cs}(n) = 0.5(x(n) + x^*(-n)) \quad \text{and}$$

$$x_{ca}(n) = 0.5(x(n) - x^*(-n))$$

As conjugation does not change real signals, these definitions reduce to the real-valued case without conjugation. Let

$$x(n) = \{1 + j2, 3 \overset{\smile}{-} j1, 2 + j3\}$$

Then,

$$x^*(n) = \{1 - j2, 3 \overset{\smile}{+} j1, 2 - j3\}$$

The time-reversed version of the conjugated signal is

$$xr^*(n) = \{2 - j3, 3 \overset{\smile}{+} j1, 1 - j2\}$$

Now,

$$x_{cs}(n) = \{1.5 - j0.5, \ 3, \ 1.5 + j0.5\}, \quad x_{ca}(n) = \{-0.5 + j2.5, \ -j1, \ 0.5 + j2.5\}$$

Circular Even and Odd Symmetries

For periodic signals, the definitions are the same as those for aperiodic signals with the difference that the indices of the sample points are defined by *mod* operation. A periodic signal $x(n)$ with period N is even-symmetric, if it satisfies the condition

$$x((-n) \bmod N) = x(n) \quad \text{for all} \quad n$$

For example, the periodic extension of the sequences, with even, $N = 10$, and odd, $N = 9$, lengths,

$$\{x(n), n = 0, 1, \ldots, 9\} = \{3, 1, 5, -3, 4, 5, 4, -3, 5, 1\}$$

and

$$\{x(n), n = 0, 1, \ldots, 8\} = \{7, 1, -5, 6, 3, 3, 6, -5, 1\}$$

are even. The samples of an even-length signal with indices 0 and $N/2$ can be arbitrary. Other samples satisfy the condition $x(n) = x(-n)$. The symmetry can also be defined by the samples of the signal over one period as

$$x(N - n) = x(n), \quad 1 \le n \le N - 1$$

For example, cosine waveforms, with period 10 and 7,

$$\cos\left(\frac{2\pi}{10}n\right) \quad \text{and} \quad \cos\left(\frac{2\pi}{7}n\right)$$

are even-symmetric. The samples over one period, respectively, are

$$\{1, 0.8090, 0.3090, -0.3090, -0.8090, -1, -0.8090, -0.3090, 0.3090, 0.8090\}$$

and

$$\{1, 0.6235, -0.2225, -0.9010, -0.9010, -0.2225, 0.6235\}$$

A periodic signal $x(n)$ with period N is odd-symmetric, if it satisfies the condition

$$-x((-n) \bmod N) = x(n) \quad \text{for all} \quad n$$

For example, the periodic extension of the sequences, with even and odd lengths,

$$\{x(n), n = 0, 1, \ldots, 9\} = \{0, -1, -5, -3, -4, 0, 4, 3, 5, 1\}$$

and

$$\{x(n), n = 0, 1, \ldots, 8\} = \{0, 1, 5, 6, 3, -3, -6, -5, -1\}$$

are odd. The samples of an even-length signal with indices 0 and $N/2$ must be zero. For odd-length signals, $x(0)$ must be zero. Other samples satisfy the condition $x(n) = -x(-n)$. The symmetry can also be defined by the samples of the signal over one period as

$$-x(N - n) = x(n), \quad 1 \leq n \leq N - 1$$

For example, sine waveforms, with period 10 and 7,

$$\sin\left(\frac{2\pi}{10}n\right) \quad \text{and} \quad \sin\left(\frac{2\pi}{7}n\right)$$

are odd-symmetric. The samples over one period, respectively, are

$$\{0, 0.5878, 0.9511, 0.9511, 0.5878, 0, -0.5878, -0.9511, -0.9511, -0.5878\}$$

and

$$\{0, 0.7818, 0.9749, 0.4339, -0.4339, -0.9749, -0.7818\}$$

are odd-symmetric.

1.1.7 Energy and Power Signals

The instantaneous power dissipated by a resistor of $1\,\Omega$ with a voltage $v(t)$ applied across it or a current $i(t)$ flowing through it is $v^2(t)$ or $i^2(t)$. The energy dissipated is the integral of power over the period power is applied. The assumption of $1\,\Omega$ makes the energy or power as indicative. The actual energy or power can be computed only with the load known.

In the discrete case, the energy of a signal is defined as the sum of the squared magnitude of the values of a real or complex-valued discrete signal $x(n)$ as

$$E = \sum_{n=-\infty}^{\infty} |x(n)|^2$$

If the summation is finite, the signal is an energy signal. For example, the energy of $x(n) = (0.6)^n, \ n \geq 0$ is

$$E = \sum_{n=0}^{\infty} (0.36)^n = \frac{1}{1 - 0.36} = 1.5625$$

The cumulative sum of the terms of the geometric series, with 4-bit precision, are

$$\{1, 1.36, 1.4896, 1.5363, 1.5531, 1.5591, 1.5613, 1.5621, 1.5623, 1.5624, 1.5625\}$$

For practical signals, the energy of the signal can be approximated using a finite number of terms of the summation.

For signals not having finite energy, average power may be finite. The average power is defined as

$$P = \lim_{N \to \infty} \frac{1}{2N+1} \sum_{n=-N}^{N} |x(n)|^2$$

For example, let $x(n) = 2u(n)$. This is a power signal since,

$$P = \lim_{N \to \infty} \frac{1}{2N+1} \sum_{n=0}^{N} (2)^2 1 = \lim_{N \to \infty} 4 \frac{N+1}{2N+1} = 2$$

The first few values of P are

$$\{4, 2.6667, 2.4000, 2.2857, 2.2222, 2.1818\}$$

For a periodic signal with period N, the average power can be determined over a period as

$$P = \frac{1}{N} \sum_{n=0}^{N-1} |x(n)|^2$$

If the power of a signal is finite, then the signal is a power signal. The average power of the sine wave $\sin(\frac{2\pi}{8}n)$ is

$$P = \frac{1}{8} \sum_{n=0}^{7} |x(n)|^2 = \frac{1}{8}(0)^2 + \left(\frac{1}{\sqrt{2}}\right)^2 + 1^2 + \left(\frac{1}{\sqrt{2}}\right)^2 + 0^2 + \left(\frac{1}{\sqrt{2}}\right)^2 + 1^2$$

$$+ \left(\frac{1}{\sqrt{2}}\right)^2 = \frac{1}{2}$$

A signal is an energy signal or a power signal or neither. Signal $x(n) = n^2$, $-\infty \le n < \infty$, is neither a power signal nor an energy signal.

1.1.8   Causal and Noncausal Signals

Signals start at some finite time, usually chosen as $n = 0$ and assumed to be zero for $n < 0$. Such signals are called causal signals. For example, the impulse, step, and ramp signals shown in Fig. 1.6 are causal. Causality condition restricts the limits of operations of signal processing. A noncausal signal is such that $x(n) \neq 0$ for $n < 0$. A periodic signal is noncausal.

1.1.9   Deterministic and Random Signals

A signal that is specified for all time is a deterministic signal. For example, the unit-step signal is deterministic since all its values are known. The future values of a random signal are not exactly known. They are characterized by some average value. For example, while the amount of rainfall at a place at a time is exactly known for past time, the future amount is not exactly known. However, the limits of the value can be predicted.

1.1.10   Bounded Signals

A discrete signal $x(n)$ is bounded if the absolute value of all its samples is less than or equal to a finite positive number. For example, $0.6^n u(n)$ is bounded, while $2^n u(n)$ is unbounded.

1.1.11   Absolutely Summable Signals

A signal $x(n)$ is absolutely summable, if

$$\sum_{n=-\infty}^{\infty} |x(n)| < \infty$$

For example, the cumulative sum of the magnitude of the samples of the signal

$$x(n) = 0.8^n u(n)$$

over a finite range is shown in Fig. 1.5a. The sum converges to 5, as we know from the closed-form formula $1/(1 - 0.8)$.

1.1.12   Absolutely Square Summable Signals

A signal $x(n)$ is absolutely square summable, if

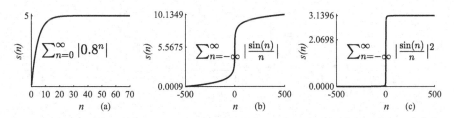

Fig. 1.5 (a) The cumulative sum of the magnitude of the samples of the signal $0.8^n u(n)$, over a finite range; (b) the cumulative sum of the magnitude of the samples of the signal $|\frac{\sin(n)}{n}|$, $-\infty < n < \infty$, over a finite range; (c) the cumulative sum of the magnitude of the samples of the signal $|(\frac{\sin(n)}{n})|^2$, $-\infty < n < \infty$, over a finite range

$$\sum_{n=-\infty}^{\infty} |x(n)|^2 < \infty$$

Consider the signal

$$x(n) = \frac{\sin(n)}{n}, \quad -\infty < n < \infty$$

At $n = 0$, $x(n) = 1$. The cumulative sum of the magnitude of the samples of the signal, over a finite range, is shown in Fig. 1.5b. It keeps increasing and will be ∞ in the limit. The cumulative sum of the magnitude of the samples of the squared-version of the signal, over a finite range, is shown in Fig. 1.5c. It converges to the limit 3.1396.

1.2 Basic Signals

As the amplitude profile of practical signals is arbitrary, it is difficult to transmit and interpret in their original form. In order to facilitate their processing, some basic and well-defined signals are used to approximate them. Further, as the response of systems is also arbitrary, systems are characterized by their response to the basic signals.

1.2.1 Unit-impulse Signal

The discrete unit-impulse signal, shown in Fig. 1.6a, is defined as

$$\delta(n) = \begin{cases} 1 \text{ for } n = 0 \\ 0 \text{ for } n \neq 0 \end{cases}$$

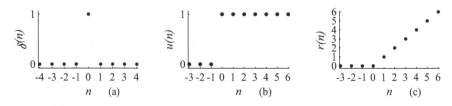

Fig. 1.6 (a) The discrete unit-impulse signal, $\delta(n)$; (b) the discrete unit-step signal, $u(n)$; (c) the discrete unit-ramp signal, $r(n)$

The impulse and sinusoidal signals are most often used to decompose an arbitrary signal in terms of components for easier processing. The unit-impulse signal has a value of 1, when its argument is zero and zero otherwise. Therefore, $\delta(n)$ is nonzero only at $n = 0$ and $\delta(n - k)$ is nonzero only at $n = k$. By using a nonzero k, we get the shifted versions of the impulse. The multiplication of a signal by the impulse and summing the product yields the value of the signal at the location of the impulse, called the sampling property of the impulse. That is,

$$\sum_{n=-\infty}^{\infty} x(n)\delta(n - k) = x(k)$$

The unit-impulse is used to decompose a signal in terms of scaled and shifted impulses in the time domain, where time is the independent variable. Similarly, sinusoidal signals with various frequencies are used to decompose a signal in terms of scaled and shifted sinusoids in the frequency domain, where frequency is the independent variable. The operation used is again the sum of products.

Consider the signal

$$x(0) = 1.1, x(1) = 2.1, x(2) = 3.1, x(3) = 4.1$$

This signal can be expressed, in terms of impulses, as

$$x(n) = 1.1\delta(n) + 2.1\delta(n - 1) + 3.1\delta(n - 2) + 4.1\delta(n - 3)$$

With $n = 3$, for example,

$$x(3) = 1.1\delta(3) + 2.1\delta(2) + 3.1\delta(1) + 4.1\delta(0) = 4.1$$

In general,

$$\sum_{k=-\infty}^{\infty} x(n)\delta(n - k) = x(n) \sum_{k=-\infty}^{\infty} \delta(n - k) = x(n) = \sum_{k=-\infty}^{\infty} x(k)\delta(n - k)$$

Decomposing a signal in terms of impulses or sinusoids is usually the major and the first operation in signal processing. Once decomposed, the processing becomes much simplified irrespective of the arbitrariness of the given signal.

1.2.2 Unit-Step Signal

The discrete unit-step signal, shown in Fig. 1.6b, is defined as

$$u(n) = \begin{cases} 1 \text{ for } n \geq 0 \\ 0 \text{ for } n < 0 \end{cases}$$

The value of the unit-step signal is 1 for positive values of its argument and zero otherwise. The value of the shifted unit-step signal $u(n + 1)$ is 1 for $n \geq -1$ and zero otherwise. To ensure that the value of a signal is zero over a range, we multiply the signal with $u(n)$. Signal $x(n) = e^{-n}$ has nonzero values from $n = -\infty$ to $n = -\infty$, whereas $e^{-n}u(n)$ has values zero from $n = -\infty$ to $n = -1$. Any pulse signal can be expressed as a combination of scaled and shifted unit-step signals. For example, the signal expressed by

$$x(n) = (n + 1)u(n) - (n + 1)u(n - 4)) = (n + 1)(u(n) - u(n - 4))$$

is

$$x(0) = 1, x(1) = 2, x(2) = 3, x(3) = 4$$

and zero otherwise. The first difference $u(n) - u(n - 1)$ is the unit-impulse signal $\delta(n)$. The running sum of the unit-impulse

$$\sum_{k=0}^{\infty} \delta(n - k)$$

is the unit-step signal $u(n)$.

1.2.3 Unit-Ramp Signal

The discrete unit-ramp signal, shown in Fig. 1.6c, is also often used in the analysis of signals and systems. It is defined as

$$r(n) = \begin{cases} n \text{ for } n \geq 0 \\ 0 \text{ for } n < 0 \end{cases}$$

Starting with value 0 at $n = 0$, it increases linearly with increment 1 for positive values of its argument and it is zero otherwise.

The shifted unit-step signal $u(n - 1)$ can be expressed in terms of difference of ramp signals as $r(n) - r(n - 1)$. The unit-ramp signal $r(n)$ can be expressed in terms of the impulse as

$$r(n) = nu(n) = \sum_{k=0}^{\infty} k\delta(n - k)$$

In terms of the unit-step signal,

$$r(n) = \sum_{k=0}^{\infty} k(u(n - k) - u(n - k - 1))$$

1.2.4 Sinusoids

The mathematical definition of the sinusoidal waveform is a linear combination of the trigonometric sine and cosine functions. The unit-circle, with a radius of 1 and the center at the origin, is characterized by the equation

$$x^2 + y^2 = 1$$

as shown in Fig. 1.7. For each real number θ, we associate a point $f(\theta)$ on the unit-circle. The arc length when we move in the counterclockwise direction starting from the point $(1, 0)$ is θ. Then, values of sine and cosine of θ are defined in terms of the coordinates (x, y) of the point $f(\theta)$ as

$$\cos(\theta) = x \quad \text{and} \quad \sin(\theta) = y$$

Fig. 1.7 The unit-circle with $\sin(\pi/3)$ and $\cos(\pi/3)$ defined

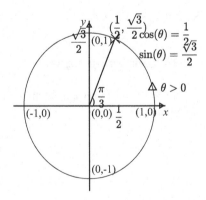

Figure 1.7 shows the unit-circle with $\sin(\pi/3)$ and $\cos(\pi/3)$ defined.

$$\cos\left(\frac{\pi}{3}\right) = \frac{1}{2} \quad \text{and} \quad \sin\left(\frac{\pi}{3}\right) = \frac{\sqrt{3}}{2}$$

As the signal is represented over a circle, the sinusoid is called a periodic or circular signal.

One radian is the measure of an angle that subtends an arc of unit length on the unit-circle. Since the circumference of the unit-circle is 2π, one revolution of the circle produces an angle of 2π rad. A degree is defined as $(1/360)$th of the angle of one complete revolution. Therefore, $2\pi = 360°$ and 1 radian $\approx 57.3°$.

The discrete sinusoidal waveform $x(nT_s)$ is expressed in two equivalent forms. Let $a \neq 0$, $b \neq 0$, and ω be real numbers. Then, for every real number nT_s (the independent variable),

$$x(nT_s) = a\cos(\omega nT_s) + b\sin(\omega nT_s) = A\cos(\omega nT_s + \theta)$$

where

$$A = \sqrt{a^2 + b^2}, \quad \theta = \tan^{-1}\left(\frac{-b}{a}\right), \quad a = A\cos(\theta) \text{ and } b = -A\sin(\theta)$$

Suppressing the sampling interval T_s, we get

$$x(n) = a\cos(\omega n) + b\sin(\omega n) = A\cos(\omega n + \theta)$$

The first and second forms are, respectively, rectangular and polar representation of the discrete sinusoidal waveform. For example,

$$4\cos\left(\frac{2\pi}{8}n + \frac{\pi}{6}\right) = 2\sqrt{3}\cos\left(\frac{2\pi}{8}n\right) - 2\sin\left(\frac{2\pi}{8}n\right)$$

$$\frac{1}{2}\cos\left(\frac{2\pi}{8}n\right) + \frac{\sqrt{3}}{2}\sin\left(\frac{2\pi}{8}n\right) = 1\cos\left(\frac{2\pi}{8}n - \frac{\pi}{3}\right)$$

The sinusoid is expressed in terms of its amplitude A and phase angle θ in polar form. In the rectangular form, the sinusoid is expressed as a linear combination of cosine and sine waveforms. The cosine and sine components are also the even and odd components of the sinusoid, respectively.

In general, the polar form of a discrete sinusoid is given by

$$x(n) = A\cos(\omega n + \theta), \quad -\infty < n < \infty$$

where A (half the distance from either of its peaks) is its amplitude, ω is its radian frequency, and θ is the phase. As the sampling interval between two adjacent samples tends to zero, the discrete sinusoid becomes a continuous sinusoid. Any continuous sinusoid can be reconstructed from more than 2 samples in a cycle. Therefore, while the continuous sinusoid appears mostly in practice, its discrete version is mostly used in the analysis of signals and systems. The period of a discrete sinusoid N has to be an integer, as the independent variable n takes only integer values. In that period, the waveform has to complete an integral number, say k, of cycles. That is, the frequency of the waveform $f = k/N$ cycles per sample has to be a ratio of integers. Then, the fundamental period is equal to N. Otherwise, the signal is aperiodic. This happens when there is an irrational factor, such as $\sqrt{2}$ or π, in the cyclic frequency f. The angular frequency $\omega = 2\pi f$ rad. At a given frequency, there exists an infinite number of sinusoids with their peaks occurring at different instants. To discriminate the innumerable number of sinusoids, we need another parameter to identify a particular sinusoid in addition to amplitude and frequency. That parameter is the phase, which indicates where its positive peak occurs with reference to the instant $n = 0$. The cosine waveform, whose peak appears at $n = 0$, is mostly (and in this book) used as the reference, and its phase is defined as $\theta = 0$ rad. The first positive peak of the sine waveform occurs at one-fourth of its cycle after $n = 0$ and its phase is $-\pi/2$ rad.

The cosine and sine waveforms are special cases of a sinusoid with phases 0 and $-\pi/2$ rad, respectively. Arbitrary sinusoids can be expressed in terms of phase-shifted cosine or sine waves as

$$A \sin(\omega n + \theta) = A \cos\left(\omega n + \left(\theta - \frac{\pi}{2}\right)\right)$$

$$A \cos(\omega n + \theta) = A \sin\left(\omega n + \left(\theta + \frac{\pi}{2}\right)\right)$$

These equations can be verified using trigonometric addition formula.

Figure 1.8a shows one cycle of the cosine waveform $\cos(\frac{2\pi}{16}n)$. The amplitude, angular frequency, cyclic frequency, the period, and the phase are

$$A = 1, \quad \omega = \frac{2\pi}{16}, \quad f = \frac{\omega}{2\pi} = \frac{1}{16}, \quad N = 16, \quad \theta = 0$$

The peak occurs at $n = 0$.

Figure 1.8b shows one cycle of the sine waveform $\sin(\frac{2\pi}{12}n)$. The amplitude, angular frequency, cyclic frequency, the period, and the phase are

$$A = 1, \quad \omega = \frac{2\pi}{12}, \quad f = \frac{\omega}{2\pi} = \frac{1}{12}, \quad N = 12, \quad \theta = -\frac{\pi}{2}$$

The nearest peak from the origin occurs at $n = 3$ after a delay of 3 sample intervals compared with the cosine waveform (one quarter of a cycle). The phase of the waveform is $\theta = 2\pi(-3)/12 = -\pi/2$ rad.

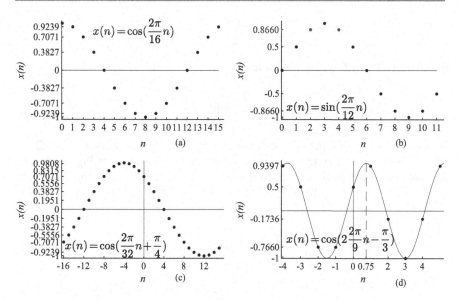

Fig. 1.8 One cycle of (a) $\cos(\frac{2\pi}{16}n)$; (b) $\sin(\frac{2\pi}{12}n)$; (c) $\cos(\frac{2\pi}{32}n + \frac{\pi}{4})$; (d) $\cos(2\frac{2\pi}{9}n - \frac{\pi}{3})$

Figure 1.8c shows one cycle of the sinusoid $\cos(\frac{2\pi}{32}n + \frac{\pi}{4})$. The amplitude, angular frequency, cyclic frequency, the period, and the phase are

$$A = 1, \quad \omega = \frac{2\pi}{32}, \quad f = \frac{\omega}{2\pi} = \frac{1}{32}, \quad N = 32, \quad \theta = \frac{\pi}{4}$$

The nearest peak from the origin occurs at $n = -4$ ahead of 4 sample intervals compared with the cosine waveform. The phase of the waveform is $\theta = 2\pi(4)/32 = \pi/4$ rad.

Figure 1.8d shows one cycle of the sinusoid $\cos(2\frac{2\pi}{9}n - \frac{\pi}{3})$. The amplitude, angular frequency, cyclic frequency, the period, and the phase are

$$A = 1, \quad \omega = 2\frac{2\pi}{9}, \quad f = \frac{\omega}{2\pi} = \frac{2}{9}, \quad N = 9, \quad \theta = -\frac{\pi}{3}$$

The sample values over one cycle are

$$\{x(n), n = -4, -3, \ldots, 4\}$$

$$= \{0.9397, 0.5, -0.766, -0.766, 0.5, 0.9397, -0.1736, -1, -0.1736\}$$

In general, the period of a discrete sinusoid is $N = 2\pi k/\omega$, where k is the smallest integer, greater than 0, that will make N an integer. For the first 3 cases, $k = 1$ and the period $N = 2\pi/\omega$. In this case, with $\omega = 2\frac{2\pi}{9}$,

$$N = \frac{2\pi k}{\omega} = \frac{9k}{2}$$

For $k = 1$ and $k = 2$, we get the values

$$\left\{ N = \frac{9}{2} \quad \text{and} \quad 9 \right\}$$

With $k = 2$, we get the period as 9 samples. The cyclic frequency is $\omega/2\pi = 2/9$. That is, the waveform completes 2 cycles in its period. As shown in Fig. 1.8d by a thin line, the nearest peak from the origin does not occur at a sample point. The peak occurs at $t = 3/4$. The phase of the waveform is $\theta = -4\pi(3/4)/9 = -\pi/3$ rad. Since

$$\cos\left(2\frac{2\pi}{9}t - \frac{\pi}{3}\right) = 1 \text{ and } \left(2\frac{2\pi}{9}t - \frac{\pi}{3}\right) = 0$$

solving for t we get $t = 3/4$.

The Importance of Sinusoids in Signal Processing

The impulse and sinusoidal signals are most often used as basis signals to decompose arbitrary signals, respectively, in the time and frequency domains, which is the first and necessary step in signal processing. The difference between these two representations is the independent variable. In the time domain, it is time and it is frequency in the frequency domain. This decomposition facilitates their processing and either representation provides the same processed signal. However, in general, the sinusoidal representation turns out to be better in terms of computational cost of signal processing and providing insight of the characteristics of signals and the operations. Therefore, most of the processing are usually carried out in the frequency domain.

The major advantages of the sinusoidal representation accrue from the three characteristics of them. When excited by a sinusoidal source signal, the signals at all parts of a linear system are sinusoidal and of the same frequency of the excitation in the steady-state. The amplitude and phase only are changed. The sum of any number of sinusoids with the same frequency is also a sinusoid of the same frequency differing only in amplitude and phase. Further, the derivative and the integral of a sinusoid is also of the same form with only changes in amplitude and phase, no matter how often the operation is carried out. These characteristics make the solution of a differential equation characterizing a system algebraic in the frequency domain. Any arbitrary practical signal can be adequately represented in terms of sinusoids. Further, fast numerical algorithms are available to get the representation of an arbitrary time domain signal in the frequency domain. Therefore, in most cases, processing of signals is more efficient using the sinusoid as the building block.

1.2.5 The Sum of Sinusoids of the Same Frequency

Let

$$a(n) = A\cos(\omega n + \theta) \quad \text{and} \quad b(n) = B\cos(\omega n + \phi)$$

The rectangular representation of the sinusoids is

$$A\cos(\theta)\cos(\omega n) - A\sin(\theta)\sin(\omega n) \text{ and } B\cos(\phi)\cos(\omega n) - B\sin(\phi)\sin(\omega n)$$

The rectangular representation of the sum of the sinusoids is

$$(A\cos(\theta) + B\cos(\phi))\cos(\omega n) - (A\sin(\theta) + B\sin(\phi))\sin(\omega n)$$

Therefore, the magnitude and the phase of the sum of $a(n)$ and $b(n)$, $c(n) = C\cos(\omega n + \alpha)$, using Pythagorean theorem, are

$$C = \sqrt{A^2 + B^2 + 2AB\cos(\theta - \phi)}$$

$$\alpha = \tan^{-1}\frac{A\sin(\theta) + B\sin(\phi)}{A\cos(\theta) + B\cos(\phi)}$$

By repeatedly applying this formula, the sum of any number of sinusoids can be found.

Example 1.1 Determine the sum of the two sinusoids $a(n) = \sin(\frac{2\pi}{8}n + \frac{\pi}{6})$ and $b(n) = -2\cos(\frac{2\pi}{8}n - \frac{\pi}{3})$.

Solution Expressing the signals in terms of cosine signal, we get

$$a(n) = \cos\left(\frac{2\pi}{8}n - \frac{\pi}{3}\right) \quad \text{and} \quad b(n) = 2\cos\left(\frac{2\pi}{8}n + \frac{2\pi}{3}\right)$$

$$A = 1, \quad B = 2, \quad \theta = -\frac{\pi}{3}, \text{ and } \phi = \frac{2\pi}{3}$$

Substituting the numerical values in the equations, we get the magnitude and the phase of the sum of $a(n)$ and $b(n)$, $c(n) = C\cos(\frac{2\pi}{8}n + \alpha)$, as

$$C = \sqrt{1^2 + 2^2 + 2(1)(2)\cos\left(\frac{-\pi}{3} - \frac{2\pi}{3}\right)} = 1$$

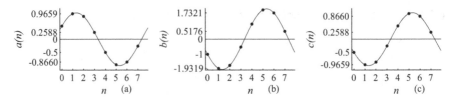

Fig. 1.9 (a) The sinusoid $a(n) = 1\cos(\frac{2\pi}{8}n - \frac{\pi}{3})$; (b) the sinusoid $b(n) = 2\cos(\frac{2\pi}{8}n + \frac{2\pi}{3})$; (c) the sum of $a(n)$ and $b(n)$, $c(n) = 1\cos(\frac{2\pi}{8}n + \frac{2\pi}{3})$

$$\alpha = \tan^{-1}\frac{1\sin\left(\frac{-\pi}{3}\right) + 2\sin\left(\frac{2\pi}{3}\right)}{1\cos\left(\frac{-\pi}{3}\right) + 2\cos\left(\frac{2\pi}{3}\right)} = 2.0944\ \text{rad} = 120°$$

The waveforms $a(n)$, $b(n)$, and $c(n)$ are shown, respectively, in Fig. 1.9a–c. By adding the corresponding samples in the first two figures must be equal to the corresponding samples in the last figure.

■

1.2.6 Exponential Signals

An exponential signal, with base a, is defined as

$$x(n) = a^n$$

where a is a constant and n is the independent variable. The independent variable is the exponent. Exponential signals generalize the signal and system analysis. Further, their compact form and ease of manipulation make them appear in most of the signal and system analysis. Figure 1.10a shows an exponentially decaying sinusoid $x(n) = (0.95)^n \sin(\frac{2\pi}{8}n)$. We get the exponentially decaying sinusoid by multiplying the constant-amplitude sinusoid $\sin(\frac{2\pi}{8}n)$ by the real exponential $(0.95)^n$. The exponential part imposes a constraint on the amplitude of the sinusoid and makes it decreasing. At the peaks of the sine signal, the signal reduces to the real exponential. Therefore, the real exponentials $(0.95)^n$ and $-(0.95)^n$ form the envelopes of the positive and negative peaks of the waveform, respectively. Figure 1.10b shows an exponentially growing sinusoid $x(n) = (1.15)^n \sin(\frac{2\pi}{8}n)$.

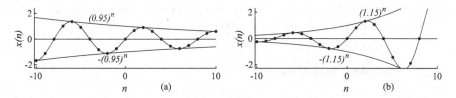

Fig. 1.10 (a) Exponentially decaying sinusoid, $(0.95)^n \sin(\frac{2\pi}{8}n)$; (b) exponentially growing sinusoid, $(1.15)^n \sin(\frac{2\pi}{8}n)$

Fig. 1.11 The unit-circle with $e^{j\frac{\pi}{3}}$ defined in the complex plane

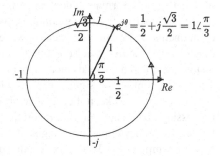

1.2.7 The Complex Representation of Sinusoids

While practical devices generate the real-valued sinusoids, an equivalent mathematical representation, in terms of complex exponentials, makes signal and system analysis much simpler. The exponent of the complex exponential is a complex number. A real sinusoid is characterized, at a given frequency, by two scalars. In the complex exponential representation, a sinusoid is represented by a single complex-valued signal with a single complex coefficient. This form is compact and is easier to manipulate and used in most of the signal and system analysis. Figure 1.11 shows the unit-circle with $e^{j\frac{\pi}{3}}$ defined in the complex plane. It carries the same information that is shown in Fig. 1.7. Two values are embedded in a complex number, as any complex number is a two element vector. With a complex coefficient and a complex exponential signal, we get an equivalent representation of a real sinusoid. That is,

$$x(n) = \frac{A}{2} \left(e^{j(\omega n + \theta)} + e^{-j(\omega n + \theta)} \right) = A \cos(\omega n + \theta)$$

due to Euler's identity. The sum of two complex conjugate exponentials $e^{j\omega n}$ and $e^{-j\omega n}$ with two conjugate coefficients $\frac{A}{2}e^{j\theta}$ and $\frac{A}{2}e^{-j\theta}$ represents a real sinusoid. The redundancy is not a problem in practical implementation. Actually, only one of the complex exponentials is used in practical analysis as they are complex conjugates. As the complex number is a two element vector, a pair of graphs is required to represent a complex signal. The plots could be for real and imaginary parts or for magnitude and phase. As the signal is represented over a circle, as in the case of real sinusoid, it is also periodic or circular.

1.2.8 Sampling and Aliasing

Although most naturally occurring signals are continuous, digital signal processing is preferred due to several advantages. The first task is to sample the signal with infinite values and represent it by a finite number of values. In sampling, it looks like that some information content of the signal are lost. A signal is composed of infinite number of frequency components. The magnitude of the higher frequency components of practical signals, however, gradually becomes weak. That is, the magnitude of the frequency components above some frequency can be considered as insignificant and the signal can be considered as band-limited for practical purposes. Therefore, continuous signals can be adequately processed by their samples. Of course, we want to keep the number of samples to be the minimum to reduce the processing cost. The sampling theorem dictates that if more than 2 samples of the highest frequency content of a signal are taken, then the signal can be reconstructed from its samples exactly. In practice, due to the limitation of the response of physical devices, a somewhat higher number of samples are taken. If the sampling theorem is not satisfied, the representation of the signal by samples is distorted.

If a signal is slowly varying, the sampling interval can be longer. For rapidly varying signals, the sampling interval should be sufficiently short. If we sample a signal component with a longer sampling interval than necessary, that component loses its identity and impersonates as a lower frequency component, a phenomenon called aliasing. To avoid aliasing, either sufficient number of samples must be taken or, with a given number of samples, restrict the number of frequency components in the signal by prefiltering. In practice, aliasing effect cannot be eliminated but limited so that the accuracy of representation is adequate.

Figure 1.12 shows the aliasing effect with sinusoids

$$\cos\left(2\frac{2\pi}{8}n + \frac{\pi}{6}\right) \quad \text{and} \quad x_a(n) = \cos\left(6\frac{2\pi}{8}n - \frac{\pi}{6}\right)$$

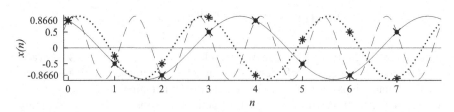

Fig. 1.12 $x(n) = \cos(2\frac{2\pi}{8}n + \frac{\pi}{6})$ (solid line), $x_a(n) = \cos(6\frac{2\pi}{8}n - \frac{\pi}{6})$ (dashed line) and $x_3(n) = \cos(3\frac{2\pi}{8}n - \frac{\pi}{6})$ (dotted line)

having the same set of samples. In their continuous form, they have distinct amplitude profiles and we can discriminate them easily. However, their discrete versions have the same set of sample points. With 8 samples, we can distinguish sinusoids with frequency up to 3/8 cycles per sample. With a higher frequency, the spectrum folds at the midpoint. That is, $x_a(n)$ alias as $\cos(2\frac{2\pi}{8}n+\frac{\pi}{6})$. The minimum number of samples to represent a sinusoid with cyclic frequency f is $2f+1$ samples. In this case, with the sinusoid completing 6 cycles, the minimum number of samples required is $2(6)+1=13$ samples. Due to inadequate number of samples, we get

$$x_a(n) = \cos\left(\frac{2\pi}{8}(8-2)n - \frac{\pi}{6}\right) = \cos\left(\frac{2\pi}{8}(-2)n - \frac{\pi}{3}\right)$$

$$= \cos\left(2\frac{2\pi}{8}n + \frac{\pi}{6}\right) = x(n)$$

Representing a signal with a finite number of samples limits the effective range of the frequencies also to a finite value. It is assumed that the magnitude of the frequency components outside this range is negligible. Figure 1.12 also shows the plot of $x_3(n) = \cos(3\frac{2\pi}{8}n - \frac{\pi}{6})$. It has got 8 samples in 3 cycles, which is adequate for proper representation. With further increase in the frequency, proper representation is not possible.

For complex signals, the aliasing effect is characterized by

$$e^{j(\frac{2\pi}{N}(k+pN)n+\theta)} = e^{j(2\pi pn)}e^{j(\frac{2\pi}{N}kn+\theta)} = e^{j(\frac{2\pi}{N}kn+\theta)}$$

where N is the period. There are only N distinct complex exponential signals with N number of complex-valued samples. The sampling theory constraint of two samples per cycle is satisfied, since the samples of a complex-valued signal are composed of two values.

For real signals, there are only about $N/2$ distinct real sinusoidal signals with N number of real-valued samples. The sampling theory constraint of two samples per cycle is satisfied, since the samples of a real-valued signal are composed of just one real value. The frequency spectrum is folded back and forth at half the frequency with index $N/2$, called the folding frequency. At this frequency, the cosine component of the waveform only can be represented since the sine component has zero samples. Therefore, the frequency spectrum has increasing values up to $N/2$ and, beyond that, has decreasing values up to N. From the frequency index starting at N, the pattern is repeated indefinitely. Therefore, in general, with an even N, the aliasing effect is characterized by three formulas for real-valued signals $\cos(\frac{2\pi}{N}kn + \theta)$.

$$x(n) = \cos\left(\frac{2\pi}{N}(k+pN)n + \theta\right) = \cos\left(\frac{2\pi}{N}kn + \theta\right), \quad k = 0, 1, \ldots, \frac{N}{2} - 1$$

$$x(n) = \cos\left(\frac{2\pi}{N}(k + pN)n + \theta\right) = \cos(\theta)\cos\left(\frac{2\pi}{N}kn\right), \quad k = \frac{N}{2}$$

$$x(n) = \cos\left(\frac{2\pi}{N}(pN - k)n + \theta\right) = \cos\left(\frac{2\pi}{N}kn - \theta\right), \quad k = 1, 2, \ldots, \frac{N}{2} - 1$$

where N and index p are positive integers. In the first case, there is no aliasing. In the second case, there is no aliasing but the sine component is lost. All the sinusoids in the third case get aliased.

With an odd N, the aliasing effect is characterized by two formulas,

$$x(n) = \cos\left(\frac{2\pi}{N}(k + pN)n + \theta\right) = \cos\left(\frac{2\pi}{N}kn + \theta\right), \quad k = 0, 1, \ldots, \frac{N-1}{2}$$

$$x(n) = \cos\left(\frac{2\pi}{N}(pN - k)n + \theta\right) = \cos\left(\frac{2\pi}{N}kn - \theta\right), \quad k = 1, 2, \ldots, \frac{N-1}{2}$$

where N and index p are positive integers.

1.3   Signal Operations

In addition to point-by-point addition, subtraction, multiplication and division of signal samples, operations on the independent variable are often used. While we present the operations assuming that the independent variable is time, the description is valid for other independent variables, such as frequency and distance.

1.3.1   Linear Time Shifting

If we replace the independent variable n in $x(n)$ by $n - k$, with k positive, then the origin of the signal is shifted to $n = k$ (delayed). That is, the locations of the samples become $n + k$. With k negative, the samples of the signals in $x(n)$ appear at $n - k$ (advanced). Consider the signal shown in Fig. 1.13a. Except for the 8 samples shown, the rest of $x(n)$ are zero. In the shifted signal $x(n - 2)$, shown in Fig. 1.13b, the appearance of samples of $x(n)$ are delayed by 2 sample intervals. For example, $x(-4) = -4$ is located at $n = (-4 + 2) = -2$ in the shifted signal. It amounts to shifting the plot forward by 2 sample intervals, while keeping the axes of the plot the same or shifting the axes backward by 2 sample intervals, while keeping the plot the same.

In the shifted signal $x(n + 1)$, shown in Fig. 1.13(c), the appearance of samples of $x(n)$ are advanced by 1 sample interval. For example, $x(3) = 3$ is located at $n = (3 - 1) = 2$ in the shifted signal. It amounts to shifting the plot backward by 1 sample interval, while keeping the axes of the plot the same or shifting the axes forward by 1 sample interval, while keeping the plot the same.

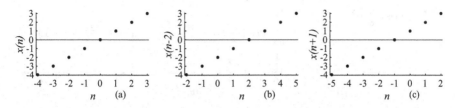

Fig. 1.13 Linear time shifting. (a) $x(n)$; (b) $x(n-2)$; (c) $x(n+1)$

1.3.2 Circular Time Shifting

Both the linear and the circular shifting are basically the change of the location of the samples of a signal. The difference is due to the nature of aperiodic and periodic signals. Therefore, the definitions are the same with the locations of the samples of a periodic signal computed using the *mod* function. The result is that the samples wrap around in the same range of the independent variable defining the signal in circular shifting. Operations on periodic signals can be easily visualized by placing its samples on a circle. The circular shift of a N-point signal $x(n)$ by k sample intervals results in

$$x((n-k) \bmod N)$$

The shift index k is a positive integer for right circular shift and it is negative for left circular shift. Consider the sinusoid $x(n) = \cos(\frac{2\pi}{8}n)$ shown in Fig. 1.14a. The left shift by 2 sample intervals results in

$$\cos\left(\frac{2\pi}{8}(n+2)\right) = \cos\left(\frac{2\pi}{8}n + \frac{\pi}{2}\right) = -\sin\left(\frac{2\pi}{8}n\right)$$

The locations of the signal shown in Fig. 1.14a with indices

$$(n+2) \bmod 8 = \{2, 3, 4, 5, 6, 7, 0, 1\}$$

are shifted to locations with indices

$$\{0, 1, 2, 3, 4, 5, 6, 7\}$$

in the shifted signal shown in Fig. 1.14b.
    The right shift by 1 sample interval results in

$$\cos\left(\frac{2\pi}{8}(n-1)\right) = \cos\left(\frac{2\pi}{8}n - \frac{\pi}{4}\right)$$

The locations of the signal shown in Fig. 1.14a with indices

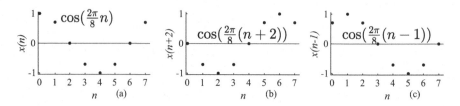

Fig. 1.14 Circular time shifting. (a) $x(n)$; (b) $x(n+2)$; (c) $x(n-1)$

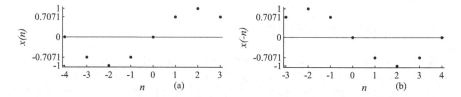

Fig. 1.15 (a) Signal $x(n)$; (b) the time inversion of $x(n)$ to get $x(-n)$

$$(n-1) \bmod 8 = \{7, 0, 1, 2, 3, 4, 5, 6\}$$

are shifted to locations with indices

$$\{0, 1, 2, 3, 4, 5, 6, 7\}$$

in the shifted signal shown in Fig. 1.14c. If the samples of a periodic signal with period N are not given in the range 0 to $N-1$, we can get the values in that range by periodic extension.

1.3.3 Time Reversal

The replacement of the independent variable n of a signal $x(n)$ by $-n$ to get $x(-n)$ is the time reversal operation. The values of $x(n)$ appear at $-n$ in its time-reversed version $x(-n)$, which is the mirror image of $x(n)$ about the vertical axis at the origin. The reversed signal $x(-n)$ of $x(n)$, shown in Fig. 1.15a, is shown in Fig. 1.15b. For example, $x(1) = 0.7071$ in $x(n)$ appears as $x(-1) = 0.7071$ in its time-reversed version $x(-n)$.

1.3.4 Circular Time Reversal

This operation results in plotting the samples of $x(n)$ in the other direction on the unit-circle. That is, the indexing is *mod N* and, hence the shift is circular. The circular time reversal of a N-point signal $x(n)$ is given by

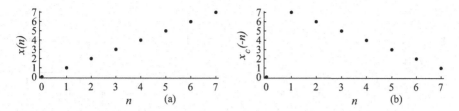

Fig. 1.16 (a) Periodic signal $x(n)$; (b) circular time reversal of $x(n)$ to get $x(-n \bmod 8)$

$$\begin{cases} x(0) & \text{for } n = 0 \\ x(N - n) & 1 \leq n \leq N - 1 \end{cases}$$

Sample $x(0)$ is left alone and the order of the other samples are reversed. The circularly time-reversed signal $x_c(n)$ of $x(n)$, shown in Fig. 1.16a, is shown in Fig. 1.16b. For example,

$$x(-1 \bmod 8) = x(7)$$

1.3.5 Time Scaling

The expansion and compression of a signal in time is known as time scaling. For multirate digital signal processing, time scaling operation is essential. Of course, the operation also applies in the frequency domain.

1.3.6 Downsampling

When the independent variable n in $x(n)$ is replaced by nK to get $x(Kn) = x_d(n)$, called downsampling the signal by an integer factor K, the samples at index n in $x(n)$ appear at index n/K in the downsampled signal $x_d(n)$. That is,

$$x_d(n) = x(Kn)$$

At $n = 0$, $x_d(0) = x(0)$. Both the signals have the same sample value at $n = 0$. On either side, take every Kth sample of $x(n)$ to find $x_d(n)$. The number of samples in $x_d(n)$ is $1/K$ of that in $x(n)$. Consider the signal $x(n)$ shown in Fig. 1.17a with 8 samples. In the downsampled signal by a factor of 2, shown in Fig. 1.17b, there are only 4 samples. The samples of $x_d(n)$ are

$$\{x_d(-2) = x(-4), \ x_d(-1) = x(-2), \ x_d(0) = x(0), \ x_d(1) = x(2)\}$$

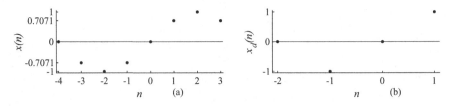

Fig. 1.17 (a) Signal $x(n)$; (b) $x_d(n)$, the downsampled version of $x(n)$ by a factor of 2

The odd-indexed samples of $x(n)$ are lost in the operation and only the even-indexed samples are retained. If the samples of $x(n)$ are obtained by sampling an analog signal $x(t)$, then it amounts to decreasing the sampling rate by a factor of 2.

1.3.7 Upsampling

Upsampling is required, when we want a signal to be represented by more number of samples (interpolation). For example, it is easier to reconstruct an analog signal from its sampled version with more number of samples. When the independent variable n in $x(n)$ is replaced by n/K to get $x(n/K) = x_u(n)$, called upsampling by an integer factor of K, the samples at index n in $x(n)$ appear at index nK in the upsampled signal $x_u(n)$. $K - 1$ zeros are inserted between every two adjacent samples of $x(n)$. That is,

$$x_u(n) = \begin{cases} x\left(\frac{n}{K}\right) & \text{for } n = 0, \pm K, \pm 2K, \ldots, \\ 0 & \text{otherwise} \end{cases}$$

At $n = 0$, $x_u(0) = x(0)$. Both the signals have the sample value at $n = 0$. On either side, each sample of $x(n)$ is located in $x_u(n)$ at index nK. The in-between samples are zero-valued. The number of samples in $x_u(n)$ is nK. Consider the signal $x(n)$ shown in Fig. 1.18a with 8 samples. In the upsampled signal by a factor of 2, shown in Fig. 1.18b, there are 16 samples. The samples of $x_u(n)$ are

$$\{, \ldots, 0, x(-2), 0, x(-1), 0, x(0), 0, x(1), 0, x(2), 0 \ldots, \}$$

All the samples of $x(n)$ are used to form the upsampled signal, in addition to zero-valued samples. If the samples of $x(n)$ are obtained by sampling an analog signal $x(t)$, then it amounts to increasing the sampling rate by a factor of 2.

Often, we need transformation involving all the three operations, shifting, reversal, and scaling. Let $x(n)$ be the ramp signal $nu(n)$ shown in Fig. 1.19a only for the range $n = 0$ to $n = 7$. We want to find the samples of the signal $y(n) = x(-2n + 2)u(-2n + 2)$. Of course, we can find the values of $y(n)$ for various values of n using the expression $(-2n + 2)$. With $n = -2, -1, 0, 1$, for example, we get

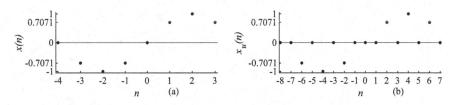

Fig. 1.18 (a) Signal $x(n)$; (b) its upsampled version by a factor of 2, $x_u(n)$

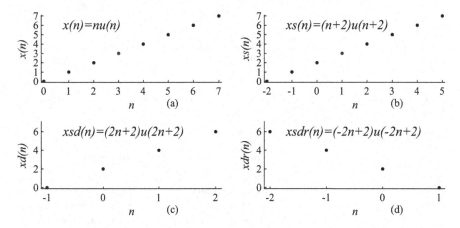

Fig. 1.19 (a) $x(n) = nu(n)$; (b) $xs(n) = (n + 2)u(n + 2)$; (c) $xsd(n) = (2n + 2)u(2n + 2)$; (d) $xsdr(n) = (-2n + 2)u(-2n + 2)$

$$y(-2) = x(-2(-2) + 2) = x(6) = 6, \quad y(-1) = x(-2(-1) + 2) = x(4) = 4,$$

$$y(0) = x(2) = 2, \quad y(1) = x(-2(1) + 2) = x(0) = 0\}$$

as shown in Fig. 1.19d. With a decimation factor of 2, we get only 4 samples from the original 8. Further, as time reversal is also involved, the samples are reversed compared with $x(n)$. It is instructive to find $y(n)$ using a sequence of operations. We can first shift the signal to get $x(n + 2)u(n + 2)$, shown in Fig. 1.19b. Decimating this signal by a factor of 2, we get $x(2n + 2)u(2n + 2)$, shown in Fig. 1.19c. Time reversing this signal, we get $x(-2n + 2)u(-2n + 2)$, shown in Fig. 1.19d. Other infinite values of $y(n)$ can be found similarly.

Let us find $y(n)$ using another sequence of operations. We can first time reverse the signal $xr(n) = (-n)u(-n)$, shown in Fig. 1.20b. Decimating this signal by a factor of 2, we get $xr(n) = (-2n)u(-2n)$, shown in Fig. 1.20c. Time shifting this signal to the right by one sample interval (replace n by $n - 1$) we get $x(-2n + 2)u(-2n + 2)$, shown in Fig. 1.20d.

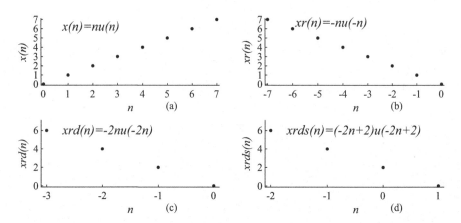

Fig. 1.20 (a) $x(n) = nu(n)$; (b) $xr(n) = (-n)u(-n)$; (c) $xrd(n) = (-2n)u(-2n)$; (d) $xrds(n) = (-2(n-1))u(-2(n-1))$

Table 1.2 Basic signals

Constant	$x(n) = c$
Unit-impulse	$x(n) = \delta(n)$
Unit-step	$x(n) = u(n)$
Real exponential	$x(n) = r^n$
Real sinusoid	$x(n) = A\cos(\omega n + \theta)$
Damped real sinusoid	$x(n) = Ar^n\cos(\omega n + \theta)$
Complex exponential	$x(n) = Ae^{j(\omega n + \theta)}$
Damped complex exponential	$x(n) = Ar^n e^{j(\omega n + \theta)}$

Zero Padding

Often it is required to append a sequence with some number of zero-valued samples. For example, regular and fast algorithms for the computation of the DFT are available only for sequence lengths those are powers of 2. Let a sequence have M samples and the zero-padded version $x_z(n)$ be of length N with $M \leq N$. Then,

$$x_z(n) = \begin{cases} x(n) \text{ for } n = 0, 1, \ldots, M-1 \\ 0 \quad \text{ for } n = M, M+1, \ldots, N-1 \end{cases}$$

For example, $x_z(n) = \{2, 1, 3, 0\}$ is the zero-padded version of $x(n) = \{2, 1, 3\}$ to make its length equal to the nearest power of 2.

Table 1.2 shows a list of basic signals.

1.4   Summary

- Signals are used for communication of information about some behavior or nature of some physical phenomenon, such as temperature and pressure.

- In the mathematical form, it is a function of one or more independent variables.
- Signals usually need some processing for their effective use.
- Most of the naturally occurring signals have arbitrary amplitude and are difficult to process as such. Basically, in signal processing, these signals are suitably approximated by a few basis signals so that signals are represented compactly and the processing becomes simpler and efficient.
- The approximation divides signal processing into two major types, processing in the time domain with the time as the independent variable and processing in the frequency domain with the frequency as the independent variable.
- Depending on the nature of sampling, quantization, and other characteristics, signals are classified.
- Continuous-time signal is defined at each and every instant of time over the period of its occurrence.
- The values of discrete signals are available only at discrete intervals.
- The values of quantized continuous-time signals are quantized to certain levels.
- Both the dependent and independent variables of digital signals are sampled values.
- A periodic signal repeats its values over any one period indefinitely. A signal that is not periodic is aperiodic.
- An even signal is symmetric with respect to the vertical axis at $n = 0$. An odd signal is antisymmetric with respect to the vertical axis at $n = 0$.
- Signals with finite energy are energy signals. Signals with finite average power are power signals.
- The unit-impulse signal has a value of 1, when its argument is zero and zero otherwise.
- The value of the unit-step signal is 1 for positive values of its argument and zero otherwise.
- Starting with value 0 at $n = 0$, the unit-ramp signal increases linearly with increment 1 for positive values of its argument and it is zero otherwise.
- The mathematical definitions of the sinusoidal waveforms are those of the trigonometric sine and cosine functions.
- The polar form of a discrete sinusoid is given by $x(n) = A \cos(\omega n + \theta)$, $-\infty < n < \infty$.
- An exponential signal is defined as $x(n) = a^n$, where a is a constant and n is the independent variable.
- In the complex exponential representation, a sinusoid is represented by a single complex-valued signal with a single complex coefficient.
- The sampling theorem dictates that if more than 2 samples of the highest frequency content of a signal are taken, then the signal can be reconstructed from its samples exactly.
- If we sample a signal component with a longer sampling interval than necessary, that component loses its identity and impersonates as a lower frequency component, a phenomenon called aliasing.
- If we replace the independent variable n in $x(n)$ by $n - k$, then the origin of the signal is shifted to $n = k$.

- The replacement of the independent variable n of a signal $x(n)$ by $-n$ to get $x(-n)$ is the time reversal operation.
- The expansion and compression of a signal in time is known as time scaling.
- Zero padding is appending a sequence with some number of zero-valued samples.

Exercises

1.1 Given a continuous-time signal $x(t)$, find the corresponding samples of the digital signal in decimal form. Assume rounding and represent the samples with 3-digit decimal precision for the fractional part. For rounding, add 0.0005 to the magnitude of the number and truncate the result to 3-digit decimal precision for the fractional part. Assume sampling interval $T_s = 0.25$ s.

1.1.1 $x(t) = e^{-t}, \quad 0 \le t < 2$.

* 1.1.2 $x(t) = \cos(\pi t + \frac{\pi}{5}), \quad 0 \le t < 2$.

1.1.3 $x(t) = \sin(\pi t + \frac{\pi}{7}), \quad 0 \le t < 2$.

1.1.4 $x(t) = t^2 u(t), \quad 0 \le t < 2$.

1.2 If the waveform $x(n)$ is periodic, what is its period N? Find the sample values of $x(n)$ in the range $n = 0$ to $n = N$. Verify that $x(n) = x(n+N)$. Find the sample value $x(78)$.

* 1.2.1 $x(n) = \sin(0.4\pi n)$.

1.2.2 $x(n) = \sin(\frac{2\sqrt{2}\pi}{8}n + \frac{\pi}{4})$.

1.2.3 $x(n) = -6 + \sin(\frac{2\pi}{5}n + \frac{\pi}{3})$.

1.3 Is $x(n)$ even-symmetric, odd-symmetric, or neither? List the values of $x(n)$ for $n = -3, -2, -1, 0, 1, 2, 3$.

1.3.1 $x(n) = \delta(n)$.

1.3.2 $x(n) = u(n)$, the unit-step signal.

1.3.3 $x(n) = \cos(\frac{2\pi}{8}n + \frac{\pi}{6})$.

1.3.4 $x(n) = \sin(\frac{2\pi}{7}n)$.

*1.3.5 $x(n) = \frac{\sin(\frac{2\pi}{6}n)}{\frac{2\pi}{6}n}$.

1.4 Find the even and odd components of the signal. Verify that, for $n = -3, -2, -1, 0, 1, 2, 3$, the two components add up to the values of the signal. Verify that the sum of the values of the odd component is zero and that those of the even component and the signal are equal.

1.4.1 $x(0) = 1, x(1) = 2, x(2) = -3, x(3) = 4$, and $x(n) = 0$ otherwise.

1.4.2 $x(n) = u(n - 1)$.

1.4.3 $x(n) = \sin(\frac{2\pi}{7}n - \frac{\pi}{3})$.

* 1.4.4 $x(n) = (-0.8)^{n+1} u(n + 1)$.

1.4.5 $x(n) = n^2 u(n)$.

1.5 If $x(n)$ is an energy signal, find its energy. If $x(n)$ is a power signal, find its average power.

1.5.1 $x(-1) = -1, x(0) = 2, x(1) = 3, x(2) = 4$, and $x(n) = 0$ otherwise.

1.5.2 $x(n) = 3(0.9)^n u(n)$.

1.5.3 $x(n) = 3^n$.

1.5.4 $x(n) = 2\sin(\frac{2\pi}{8}n - \frac{\pi}{3})$.

* 1.5.5 $x(n) = 2e^{-j(\frac{2\pi n}{4})}$.

1.6 Using scaled and shifted impulses, obtain an analytical description of the signal $x(n)$.

1.6.1 $x(0) = 1, x(-3) = 2, x(2) = -4, x(5) = 7$, and $x(n) = 0$ otherwise.

* 1.6.2 $x(0) = 3, x(-11) = 1, x(12) = -4, x(5) = 3$, and $x(n) = 0$ otherwise.

1.6.3 $x(2) = 3, x(15) = 3, x(9) = -2, x(6) = 1$, and $x(n) = 0$ otherwise.

1.7 Using scaled and shifted unit-step signals, obtain an analytical description of the signal $x(n)$.

1.7.1 $x(2) = 1, x(3) = 1, x(4) = 1$ and $x(n) = 0$ otherwise.

* 1.7.2 $x(-1) = 1, x(0) = 1, x(1) = 1, x(2) = 2, x(3) = 2, x(4) = -5, x(5) = 3$, and $x(n) = 0$ otherwise.

1.7.3 $x(-2) = 3, x(-1) = 3, x(0) = -2, x(1) = -2$ and $x(n) = 0$ otherwise.

1.8 Find the rectangular form of the sinusoid. List the sample values of one cycle, starting from $n = 0$, of the sinusoid. Convert the rectangular form back to polar form and verify that it is the same as the given sinusoid.

* 1.8.1 $x(n) = 2\sin(\frac{2\pi}{8}n + \frac{\pi}{3})$.

1.8.2 $x(n) = 2\cos(\frac{2\pi}{8}n + \frac{\pi}{6})$.

1.8.3 $x(n) = 3\cos(\frac{2\pi}{8}n - \frac{\pi}{6})$.

1.8.4 $x(n) = \cos(\frac{2\pi}{8}n + \frac{\pi}{4})$.

1.9 Given the sinusoids $x(n)$ and $y(n)$, find the polar form of the sinusoid $z(n) = x(n) + y(n)$. Find the sample values of one cycle, starting from $n = 0$, of all the three sinusoids and verify that the sample values of $x(n) + y(n)$ are the same as those of $z(n)$.

* 1.9.1 $x(n) = \sin(\frac{2\pi}{6}n - \frac{\pi}{6}), y(n) = 2\cos(\frac{2\pi}{6}n + \frac{\pi}{3})$.

1.9.2 $x(n) = 3\cos(\frac{2\pi}{6}n + \frac{\pi}{6}), y(n) = 2\cos(\frac{2\pi}{6}n - \frac{\pi}{3})$.

1.9.3 $x(n) = \cos(\frac{2\pi}{6}n + \frac{\pi}{4}), y(n) = -\cos(\frac{2\pi}{6}n + \frac{\pi}{3})$.

1.9.4 $x(n) = 2\sin(\frac{2\pi}{6}n - \frac{\pi}{2}), y(n) = 1\cos(\frac{2\pi}{6}n + \frac{\pi}{6})$.

1.10 Given a real sinusoid, express it in terms of complex exponentials $xc(n)$. Find the sample values of one cycle, starting from $n = 0$, of the two equivalent forms and verify that the sample values of of both of them are the same.

1.10.1 $x(n) = 3\cos(\frac{2\pi}{8}n - \frac{\pi}{6})$.

1.10.2 $x(n) = \sin(\frac{2\pi}{8}n + \frac{\pi}{4})$.

* **1.10.3** $x(n) = -2\sin(\frac{2\pi}{8}n + \frac{\pi}{2})$.

1.10.4 $x(n) = -4\cos(\frac{2\pi}{8}n + \frac{\pi}{3})$.

1.11 Given the sinusoid $x(n)$, find the sample values of one cycle, starting from $n = 0$, of $x(n)$ and $x(an + k)$. Assume interpolation using zero-valued samples, if necessary. First find the samples of $x(an + k)$ by replacing n by $an + k$ in $x(n)$. Then, subject $x(n)$ to the required shifting and then do the required scaling on the shifted signal. Verify that either procedure produces the same result.

1.11.1 $x(n) = \sin(\frac{2\pi}{8}n - \frac{\pi}{3})$, $a = -2, k = 1$.

1.11.2 $x(n) = \cos(\frac{2\pi}{8}n + \frac{\pi}{6})$, $a = -\frac{1}{2}, k = -1$.

* **1.11.3** $x(n) = \cos(\frac{2\pi}{3}n + \frac{\pi}{6})$, $a = \frac{1}{3}, k = -2$.

1.11.4 $x(n) = \cos(\frac{2\pi}{12}n - \frac{\pi}{3})$, $a = 3, k = 3$.

1.12 Given the signal $x(n)$, $x(an + k)$. Assume interpolation using zero-valued samples, if necessary. First find the samples of $x(an + k)$ by replacing n by $an + k$ in $x(n)$. Then, subject $x(n)$ to the required shifting and then do the required scaling on the shifted signal. Verify that either procedure produces the same result.

* **1.12.1** $x(0) = 1, x(1) = -2, x(2) = 3, x(3) = 4$ and $x(n) = 0$ otherwise. $a = 2, k = -2$.

1.12.2 $x(n) = (-0.6)^n u(n)$. $a = 2, k = 1$.

1.12.3 $x(n) = (-0.9)^n u(n)$. $a = \frac{1}{2}, k = -1$.

1.12.4 $x(n) = (1.1)^n u(-n)$. $a = -2, k = 1$.

Discrete-Time Systems

2

A system is an interconnection of components, hardware, and/or software, which performs an action or produces desired output signals in response to input signals. Processing of a signal is required to improve its quality, such as reducing its noise content or extracting some information from the input signal. Radio receivers, telephones, amplifiers, and oscillators are examples of systems. While practical input and output signals are mostly analog, DSP systems are mostly used for processing by converting analog signals into digital and converting them back after processing. Digital systems are preferred due to their overwhelming advantages. There are two basic components in signal processing: signals and systems. Both the amplitude profiles of signals and system responses are arbitrary. Signals are decomposed in terms of basic signals in order to facilitate their analysis. Similarly, the response of a system is also characterized in terms basic signals. The most often used basic signals are the impulse in the time domain, where time is the independent variable, and the sinusoid in the frequency domain, where frequency is the independent variable. It is similar to testing a new car with little bit of braking, acceleration, and turning. In this chapter, we find the response of the system using its impulse response. Systems are classified in terms of their characteristics.

2.1 Difference Equations and System Response

Each component of a system has its own input–output relationship. Using these relationships, we can derive the relation between the input and output of a system. This relationship is used to find the response of the system to basic signals, such as impulse or sinusoid. The response to these signals characterizes a system, from which the response to any arbitrary signal can be derived. This is the essence of linear system analysis. In practical systems, the components are physical, such as resistor, inductor, and capacitor. In linear system theory, it is assumed that the input–

Fig. 2.1 A resistor–capacitor filter circuit

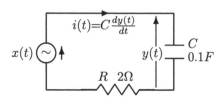

output relationship is linear. Of course, the response of practical systems is nonlinear to some extent, since the response of components are not strictly linear. Despite all these constraints, linear system analysis provides results with an acceptable tolerance. Nonlinear system analysis is extremely difficult and only possible in restricted conditions.

Consider the resistor–capacitor filter circuit shown in Fig. 2.1. The circuit consists of a $2\ \Omega$ resistor and a 0.1 F capacitor excited by the voltage source $x(t)$. The output voltage across the capacitor is $y(t)$. The input–output relationship for the capacitor is that the current through it is the derivative of the voltage across it multiplied by its value. That is,

$$i(t) = C\frac{dy(t)}{dt}$$

The input–output relationship for the resistor is that the voltage across it is the current through it multiplied by its value. As the sum of the voltage drops around a loop is zero according to Kirchhoff's voltage law, the circuit response is governed by the equation

$$RC\frac{dy(t)}{dt} + y(t) = x(t) \tag{2.1}$$

This equation is called a first-order differential equation, as it has only a first derivative term. The response of components, such as inductor and capacitor, are governed by differential equations. The more the number of such components in the circuit, the higher is the order of the differential equation governing it. As the order of differential equation governing practical circuits is usually very high, solving the equations becomes practically very difficult. Further, the amplitude profiles of practical signals are usually arbitrary. Therefore, we have to use the samples of the signals so that we can use the digital systems to find the solution of the equation to a required accuracy. The sampling interval must be sufficiently short so that the response is approximated satisfying the specified tolerance. The approximation of a differential equation using samples of the signals is called a difference equation. Difference between samples is used to approximate the derivative, which becomes more accurate as the sampling interval between samples is reduced and becomes exact as the interval tends to zero.

To make the problem more general, let us replace RC by $1/\tau$ in Eq. (2.1) to get

$$\frac{dy(t)}{dt} + \tau y(t) = \tau x(t), \quad \tau = \frac{1}{RC} \tag{2.2}$$

Although we can solve a first-order differential equation easily by analytical methods for well-defined inputs, the problem is that, in practice, the input variable usually varies arbitrarily with time.

In order to use digital systems to solve the problem, we have to sample the continuous variables $x(t)$ and $y(t)$ at intervals of a suitable sampling interval T_s. That is, we replace t by nT_s to get $x(nT_s)$ and $y(nT_s)$, where n is an integer. Then, Eq. (2.2) becomes a difference equation amenable to solving using digital systems.

$$\frac{dy(nT_s)}{dt} + \tau y(nT_s) = \tau x(nT_s), \quad n = 0, 1, 2, \dots \tag{2.3}$$

The derivative of a function is its instantaneous rate of change. One way of approximating the derivative at a point is

$$\frac{dy(nT_s)}{dt} \approx \frac{y(nT_s) - y((n-1)T_s)}{T_s}$$

Using this approximation, Eq. (2.3) becomes

$$\frac{y(nT_s) - y((n-1)T_s)}{T_s} + \tau y(nT_s) = \tau x(nT_s), \quad n = 0, 1, 2, \dots \tag{2.4}$$

Let

$$b = \frac{\tau T_s}{\tau T_s + 1} \text{ and } a = -\frac{1}{\tau T_s + 1}$$

Simplifying and dropping T_s in the variables, we get

$$y(n) = bx(n) - ay(n-1), \quad n = 0, 1, 2, \dots \tag{2.5}$$

While T_s is important in sampling, reconstruction, and coefficient determination, it does not involve in the computation. The sampling interval should not be too short, since the processing time and round-off errors become large. On the other hand, if it is too long the accuracy of the samples representing the corresponding continuous signal gets reduced. Selecting a suitable sampling interval is critical in approximating a continuous system by a digital system.

Iteratively, we can find the samples of the output, if the input and initial conditions are specified. The iterative nature of the solution makes it highly suitable for implementation using digital systems. From the pattern of the output, it is also possible to deduce a closed-form solution. The complete output of a system can be considered as the sum of two independent components. One component is the zero-input response. This response is due to initial conditions alone, assuming that the input is zero. The second component is the zero-state response. This response is due to the input alone, assuming that the initial conditions are zero. That is, no constraints on the state of the system, when the input is applied. To put it another way, the system is relaxed at the time the input is applied.

2.1.1   Zero-Input Response

The input is zero. Let the initial condition be $y(-1) = 2$. Higher-order systems require more number of initial conditions. The output is defined, for the example system, by

$$y_{zi}(n) = bx(n) - ay_{zi}(n-1), \quad n = 0, 1, 2, \ldots$$

Since there is no input, the first term on the right side becomes zero and can be dropped. Then, we get

$$y_{zi}(n) = -ay_{zi}(n-1), \quad n = 0, 1, 2 \ldots$$

Therefore,

$$y_{zi}(0) = 2(-a), \quad y_{zi}(1) = 2(-a)^2, \ldots, y_{zi}(n) = 2(-a)^{(n+1)}$$

2.1.2   Zero-State Response

With the unit-step signal, $u(n)$, as the input and zero initial condition, we get, by iteration,

$$y_{zs}(0) = bx(0) + (-a)y_{zs}(-1) = b$$
$$y_{zs}(1) = bx(1) + (-a)y_{zs}(0) = b(1 + (-a))$$

$$\vdots$$

$$y_{zs}(n) = b(1 + (-a) + (-a)^2 + \cdots + (-a)^n)$$
$$= b\left(\frac{1 - (-a)^{(n+1)}}{1 - (-a)}\right), \quad (-a) \neq 1, \quad n = 0, 1, 2, \ldots$$

2.1.3   The Complete Response

The complete response $y(n)$ is the sum of zero-input and zero-state responses

$$y(n) = y_{zi}(n) + y_{zs}(n)$$

2.1.4 Solution of the Differential Equation

Assuming the input $u(t)$ and initial condition $y(0) = 2$, the equation

$$\frac{dy(t)}{dt} + \tau y(t) = \tau u(t)$$

can be solved analytically. Let us find the transient response. Then, with the input zero, we get

$$\frac{dy(t)}{dt} + \tau y(t) = 0$$

The solution is expected to be of the form

$$y_{zi}(t) = Ce^{-\tau t}u(t)$$

Since at $t = 0$, $y(0) = 2$, $C = 2$ and the solution is

$$y_{zi}(t) = 2e^{-\tau t}u(t)$$

For the particular solution, let us use the integrating factor method. The integrating factor is

$$e^{\int \tau \, dt} = e^{\tau t}$$

The solution is

$$y_{zs}(t) = e^{-\tau t} \int_0^t \tau e^{\tau u} \, du = e^{-\tau t}(e^{\tau t} - 1) = (1 - e^{-\tau t})u(t)$$

The total solution is

$$y(t) = 2e^{-\tau t}u(t) + (1 - e^{-\tau t})u(t) = (1 + e^{-\tau t})u(t) = (1 + e^{-\frac{t}{RC}})u(t)$$

Figure 2.2a shows the resistor–capacitor circuit response for unit-step input signal with the initial condition $y(0) = 2$ and sampling interval $T_s = 0.005$ s using a longer wordlength. The continuous line shows the exact response. Using sampled data with $T_s = 0.005$ s, the response of the circuit is quite close to the exact response. The first few values of the exact response are

$$\{2, 1.9753, 1.9512, 1.9277, 1.9048, 1.8825, 1.8607, 1.8395, 1.8187, 1.7985\}$$

The first few values of the approximate response are

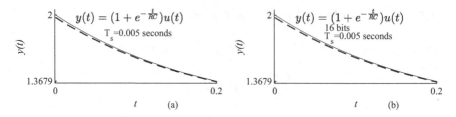

Fig. 2.2 (**a**) The complete exact response in solid line. Response with sampling interval $T_s =$ 0.005 s in dashed line; (**b**) response with 16-bit representation of the samples with the same sampling interval

{1.9756, 1.9518, 1.9285, 1.9059, 1.8837, 1.8622, 1.8411, 1.8206, 1.8005, 1.7810}

The transient response drops to $(1/e)$ (about 37% of its initial value) in one time constant at $t = 1/\tau = 1/5 = 0.2$ s. The value of the exponential is reduced to 37% of its initial value over any time interval of duration $(1/\tau)$. The difference between the responses gets reduced as the sampling interval is reduced.

Using 16-bits wordlength to represent the samples and with the same T_s, the response of the circuit is still the same, as shown in Fig. 2.2b. The quantization step and the sampling interval should be chosen so that the accuracy of the response is just sufficient. Too short a sampling interval results in increased processing time unnecessarily and also increases truncation errors. On the other hand, if the sample does not represent the signal adequately with a longer sampling interval, the response becomes corrupt. The sampling interval can be chosen using the sampling theorem and the frequency content of the signal. The choice of the quantization step depends on the precision required to represent the coefficients and the samples. Based on these considerations, an initial choice for the sampling interval and the quantization step should be made. The final values for these two parameters have to be fixed by some trial.

This example shows that, by sampling and quantization of continuous variables, a digital system can be used to approximate the response of continuous systems. Quantization is required to convert a continuous-amplitude signal into a digital signal, whose wordlength is finite and this type of signals alone can be processed by digital systems. With a suitable sampling interval and an appropriate wordlength, the response of a continuous system can be approximated adequately by digital systems. This is the basis of digital signal processing.

2.1.5 Transient Response

The complete response of a system can also be expressed in terms of transient and steady-state components. The component of the response of a system due to its natural modes only is its transient response. The transient response of a stable system always decays to insignificant levels in a finite time. The transient response is its natural response while the steady-state response is due to the natural modes of the excitation. The natural response is a sum of exponential signals, the exponents and the coefficients depend on the characteristics of the system. The system initially resists the instantaneous establishment of the steady-state behavior by the excitation. This resisting tendency exists at a significant level only for a short time, called the effective transient interval, for stable systems. The complete response $y(t)$ is the sum of the transient response $y_{tr}(t)$ and the steady-state response, $y_{ss}(t)$. That is,

$$y(t) = y_{tr}(t) + y_{ss}(t)$$

Figure 2.3 shows the various components of the response of the circuit shown in Fig. 2.1 for the unit-step input with the initial condition, $y(0) = 2$. With $T_s = 0.005$, $\tau = 5$, $a = -0.9756$, and $b = 0.0244$,

$$y(n) = 0.0244x(n) + 0.9756y(n-1), \quad n = 0, 1, 2 \ldots$$

The complete solution is

$$y(n) = 0.0244 \underbrace{\left(\frac{1 - (0.9756)^{(n+1)}}{1 - 0.9756} \right)}_{\text{zero-state}} + \underbrace{2(0.9756)^{(n+1)}}_{\text{zero-input}}, \quad n = 0, 1, 2, \ldots$$

$$= \underbrace{(1)}_{\text{steady-state}} \quad \underbrace{-(0.9756)^{(n+1)} + 2(0.9756)^{(n+1)}}_{\text{transient}}, \quad n = 0, 1, 2, \ldots$$

$$= \underbrace{1}_{\text{steady-state}} \quad + \underbrace{0.9756^{(n+1)}}_{\text{transient}}, \quad n = 0, 1, 2, \ldots$$

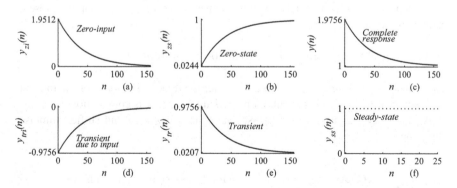

Fig. 2.3 The response of a first-order system to the unit-step input signal. (**a**) zero-input response; (**b**) zero-state response; (**c**) complete response; (**d**) transient response due to input; (**e**) transient response; (**f**) steady-state response

The responses of a first-order system to the unit-step input signal are shown in Fig. 2.3. The component of the response are: (a) zero-input response; (b) zero-state response; (c) complete response; (d) transient response due to input; (e) transient response; (f) steady-state response

2.2 Convolution

Physical systems are analyzed using discrete mathematical models in digital signal processing. Difference equation is a model. Transfer function is another model. Now, we present the convolution-summation model. Each model is advantageous in certain types of analysis. Therefore, all the commonly used models have to be studied. Practical signals have arbitrary amplitude profile. They are represented in terms of scaled and shifted impulse signals in the time domain. In the frequency domain, sinusoids are used. Similarly, systems are characterized by their responses to basic signals, such as the impulse and the sinusoid. Then, using the linearity property of systems, the response of systems to arbitrary input signals can be found systematically. In the convolution model, we find the response, called the impulse response, of a relaxed system to the unit-impulse input. Then, the response of the system for arbitrary input signals can be found by representing them in terms of a sum of scaled and shifted impulses. Since the impulse response is determined assuming that the system is relaxed (zero initial conditions), the response computed using the convolution is the zero-state response.

In the convolution model, the present output is computed using the present and all the past input samples only. It is basically a sum of products of two sequences, after one of them is time-reversed. The operation is repeated to find all the output samples of interest. Let us say, we have deposited 100$ last year and 200$ this year. The value of the money deposited now is product of the value of the deposit multiplied by 1. The value of the money deposited a year ago is product of the value of the deposit multiplied by 1.1, assuming yearly interest is 10%. Then, the present balance in our account is

$$100 \times 1.1 + 200 \times 1 = 310 \text{ or } 200 \times 1 + 100 \times 1.1 = 310$$

The balance is the sum of products, after either the deposit sequence or the interest rate sequence is time-reversed but not both. This operation is convolution.

An arbitrary signal can be expressed as a sum of scaled and shifted impulse components as

$$x(n) = \sum_{k=-\infty}^{\infty} x(k)\delta(n-k) = \cdots + x(-1)\delta(k+1) + x(0)\delta(k) + x(1)\delta(k-1) + \cdots$$

For example, $x(1)\delta(k-1)$ is a scaled and shifted impulse with value $x(1)$ located at $k = 1$ with the rest of the samples zero-valued. The impulse response $h(k)$ of a system is its response to the unit-impulse signal $\delta(k)$ with the initial conditions zero. With the knowledge of $h(k)$, the response of a system to an arbitrary input signal can be found by summing its responses to all the constituent impulse components of the input signal. It is assumed that the system is linear and time-invariant. The task is to find the zero-state response $y(n)$ of the system to the input signal $x(n)$, indicated as $x(n) \rightarrow y(n)$. The impulse response of the system is $\delta(n) \rightarrow h(n)$. Since the system is linear,

$$x(k)\delta(n) \rightarrow x(k)h(n)$$

Since the system is time-invariant,

$$x(k)\delta(n-k) \rightarrow x(k)h(n-k)$$

The sum of the responses to all the impulse components of the input signal is the total response, again due to linearity. Therefore, the convolution of sequences $x(n)$ and $h(n)$ resulting in the output sequence $y(n)$ is defined as

$$y(n) = \sum_{k=-\infty}^{\infty} x(k)h(n-k) = \sum_{k=-\infty}^{\infty} h(k)x(n-k) = x(n) * h(n)$$

The operation is commutative.

Let us find the convolution of the impulse response $\{h(k), k = 0, 1, 2, 3\} = \{3, 1, 2, 4\}$ and the input $\{x(k), k = 0, 1, 2, 3\} = \{2, 1, 3, 4\}$ shown in Fig. 2.4. The output $y(0)$, from the definition, is

$$y(0) = x(k)h(0-k) = (2)(3) = 6$$

where $h(0-k)$ is the time reversal of $h(k)$. Only one term of the two sequences overlaps. Shifting the $h(k)$ sequence to further left results in no overlap of the terms and the convolution output is zero for $y(-1)$, $y(-2)$, etc. Shifting $h(0-k)$ to the right one sample at a time, we get the remaining valid outputs. The output is zero for $y(7)$, $y(8)$, etc., since there are no overlap of the terms.

$$y(1) = x(k)h(1-k) = (2)(1) + (1)(3) = 5$$

$$y(2) = x(k)h(2-k) = (2)(2) + (1)(1) + (3)(3) = 14$$

$$y(3) = x(k)h(3-k) = (2)(4) + (1)(2) + (3)(1) + (4)(3) = 25$$

$$y(4) = x(k)h(3-k) = (1)(4) + (3)(2) + (4)(1) = 14$$

$$y(5) = x(k)h(4-k) = (3)(4) + (4)(2) = 20$$

$$y(6) = x(k)h(5-k) = (4)(4) = 16$$

Fig. 2.4 The linear
convolution operation

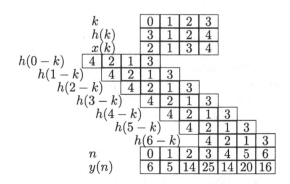

The convolution operation involves repeated operations of folding one of the sequences, shifting the folded signal, multiplying and adding of the overlapping terms of the two sequences for each output sample. The maximum number of nonzero elements in the convolution output of 2 sequences $x(n)$ of length N and $h(n)$ length M is length $N+M-1$. The length of both the sequences in the example presented is 4 and the length of the output over which nonzero samples can occur is $4+4-1=7$.

The coefficients of the product polynomial of the product of two polynomials are the convolution sum of the coefficients of the two polynomials getting multiplied. Using the sequences $x(n)$ and $h(n)$ as the coefficients of two polynomials and multiplying, the coefficients of the product polynomial are the same as the convolution output of the two sequences.

$$(2+q+3q^2+4q^3)(3+q+2q^2+4q^3) = 6+5q+14q^2+25q^3+14q^4+20q^5+16q^6$$

The convolution sum can be checked as follows. The product of the sum of the terms of $x(n)$ and $h(n)$ must equal to the sum of the terms of the sequence $y(n)$. For example,

$$(3+1+2+4)(2+1+3+4) = 10 \times 10 = 100 = (6+5+14+25+14+20+16)$$

The same test, with the sign of the odd-indexed terms of all the 3 sequences changed, also holds.

$$(3-1+2-4)(2-1+3-4) = 0 \times 0 = 0 = (6-5+14-25+14-20+16)$$

Example 2.1 Find the closed-form expression of the convolution of the sequences $x(n) = (0.8)^n u(n)$ and $h(n) = (0.7)^n u(n)$. Find the first four values of the convolution of $x(n)$ and $h(n)$ using their sample values and verify that they are the same as those obtained from the closed-form expression.

Solution The first few values of the sequences are

$$\{\hat{1}, 0.8000, 0.6400, 0.5120, 0.4096, 0.3277\}$$

$$\{\hat{1}, 0.7000, 0.4900, 0.3430, 0.2401, 0.1681\}$$

$$y(n) = \sum_{k=-\infty}^{\infty} x(k)h(n-k) = \sum_{k=0}^{n} (0.8)^k (0.7)^{n-k}, \ n \geq 0$$

$$= (0.7)^n \sum_{k=0}^{n} \left(\frac{0.8}{0.7}\right)^k = (0.7)^n \left(\frac{1 - \left(\frac{0.8}{0.7}\right)^{n+1}}{1 - \left(\frac{0.8}{0.7}\right)}\right)$$

$$= (8(0.8)^n - 7(0.7)^n)u(n)$$

The first four values of the convolution of $x(n)$ and $h(n)$ are

$$\{y(0) = 1, \quad y(1) = 1.5000, \quad y(2) = 1.6900, \quad y(3) = 1.6950\}$$

■

2.2.1 Zero-State Response by Convolution

Let us derive the closed-form formula of the zero-state response $y_{zs}(n)$ of the circuit in Fig. 2.1 using convolution. First, we need the impulse response $h(n)$ of the system characterized by

$$y(n) = 0.0244x(n) + 0.9756y(n-1), \quad n = 0, 1, 2 \ldots$$

Let us determine $h(n)$ by iteration. Let $b = 0.0244$ and $a = -0.9756$. The input is $x(n) = \delta(n)$.

$$h(0) = b$$
$$h(1) = b(-a)$$
$$h(2) = b(-a)^2$$
$$\vdots$$
$$h(n) = b(-a)^n$$

Since convolution of a signal $h(n)$ with the unit-step signal is its running sum, the zero-state response the system is

$$y_{zs}(n) = \sum_{k=0}^{n} h(k) = b\left(\frac{1 - (-a)^{(n+1)}}{1 - (-a)}\right), \quad (-a) \neq 1 \ , \ n = 0, 1, 2, \ldots$$

as found earlier by iteration. The first four values of $y(n)$ are

$$\{y(0) = 0.0244, \; y(1) = 0.0482, \; y(2) = 0.0714, \; y(3) = 0.0941\}$$

Now, let us find the impulse response by solving the difference equation. As the system is initially relaxed (initial conditions zero), we get from the difference equation $h(0) = a$ and $h(1) = ba$ by iteration. As the values of the impulse signal is one only for $n = 0$ and zero otherwise, the response for $n > 0$ can be considered as zero-input response. The solution is expected to be of the form

$$h(n) = K \, (-a)^n \, u(n - 1) = K \, (-a)^n \, u(n) - K\delta(n), \; n > 0$$

Since $h(1) = b(-a)$ by iteration, for $n = 1$, we get $K = b$. The impulse response is the sum of the response of the system at $n = 0$ and the zero-input response for $n > 0$. Therefore,

$$h(n) = b\delta(n) + b \, (-a)^n \, u(n) - b\delta(n) = b \, (-a)^n \, u(n)$$

as found by iteration.

Properties of Convolution

A binary operation is commutative, if we get the same result after changing the order of the operands. Polynomial addition and multiplication are commutative. As convolution of two sequences is like multiplication of two polynomials, the convolution-summation is commutative

$$x(n) * h(n) = h(n) * x(n)$$

We demonstrate the properties using polynomials, as the operations are easily understandable. For example, let $x(0) = 2$ and $x(1) = 4$ and $y(0) = 1$ and $y(1) = -2$. Then, the product of the corresponding polynomials is

$$(2 + 4p)(1 - 2p) = (1 - 2p)(2 + 4p) = 2 + 0p - 8p^2$$

and

$$\{2, 4\} * \{1, -2\} = \{1, -2\} * \{2, 4\} = \{2, 0, -8\}$$

The convolution-summation is distributive.

$$x(n) * (h_1(n) + h_2(n)) = x(n) * h_1(n) + x(n) * h_2(n)$$

Convolution with a sum is the same as convolution by each addend separately and then adding the convolutions. In the case of multiplication and addition,

$$a(b+c) = ab + ac$$

Value a is distributed to b and then to c. For example, with 3 polynomials,

$$(2+p)((1+2p)+(2-3p)) = ((2+p)(1+2p)+(2+p)(2-3p)) = (6+p-p^2)$$

The convolution-summation is associative.

$$x(n) * (h_1(n) * h_2(n)) = (x(n) * h_1(n)) * h_2(n)$$

It does not matter where we put the parentheses. In the case of addition,

$$a + (b+c) = (a+b) + c$$

For example, with 3 polynomials,

$$((2+p)(1+2p))(2-3p) = (2+p)((1+2p)(2-3p)) = (4+4p-11p^2-6p^3)$$

The shift property of convolution states that

$$\text{if} \quad x(n) * h(n) = y(n), \quad \text{then} \quad x(n-p) * h(n-q) = y(n-p-q)$$

We can convolve two sequences assuming that the starting index is zero and then shift the resulting sequence by the algebraic sum of their shift values. For example, with 2 polynomials,

$$(p^5 + 2p^6)(-p^{-4} + p^{-3}) = p^5(1+2p)p^{-4}(-1+p) = p(1+2p)(-1+p)$$
$$= (0 - p - p^2 + 2p^3)$$

Convolution of a sequence $x(n)$ with the unit-impulse just translates the origin of the sequence to the location of the impulse.

$$y(n) = x(n) * \delta(n) = x(n)$$

Due to shift property,

$$x(n) * \delta(n-m) = y(n-m) = x(n-m)$$

Convolution of $x(n)$ with the unit-step is the cumulative sum of $x(n)$.

$$x(n) * u(n) = \sum_{k=-\infty}^{n} x(k)$$

For example, with the two polynomials,

$$(2+p+3p^2+4p^3)(1+p+p^2+\cdots,) = 2+3p+6p^2+10p^3+10p^4+10p^5+\cdots$$

That is, the convolution output of $\{2, 1, 3, 4\}$ with the unit-step is

$$\{2, 3, 6, 10, 10, \ldots\}$$

2.3 Classification of Systems

Systems can be classified according to their general properties. Depending on the class of a system, the most appropriate procedure for its analysis can be selected.

2.3.1 Discrete Systems

The difference equation of a Nth order discrete system, relating the output $y(n)$ to the input $x(n)$, is given by

$$y(n) + a_1 y(n-1) + a_2 y(n-2) + \cdots + a_N y(n-N)$$
$$= b_0 x(n) + b_1 x(n-1) + \cdots + b_M x(n-M) \tag{2.6}$$

where the coefficients a's and b's are real constants, and N and M are fixed integers. The classification depends on the nature of the terms. This equation represents a Nth order discrete system, since the present output $y(n)$ depends on N previous values of the output. A first-order system depends on one previous output, $y(n-1)$. The difference equation can be concisely written as

$$y(n) = \sum_{k=0}^{M} b_k x(n-k) - \sum_{k=1}^{N} a_k y(n-k)$$

The difference equation can be implemented by hardware and/or software digital systems. Therefore, discrete-time system transforms a discrete input signal into another discrete signal with desirable characteristics or extract some information about the signal or takes some action specified by the input signal. That is, the output $y(n)$ is a function the input $x(n)$. Since most naturally occurring signals are of continuous nature, the input signal has to be converted to a discrete signal first by an analog-to-digital converter, processed by a discrete system and reconstructed to the analog form by a digital-to-analog converter.

2.3.2 Linear Systems

Let $y_1(n)$ be the response of a system to an input signal $x_1(n)$. Let $y_2(n)$ be the response of the system to an input signal $x_2(n)$. Then, the system is linear if the response to the input $ax_1(n) + bx_2(n)$ is $ay_1(n) + by_2(n)$. That is, with

$$x_1(n) \rightarrow y_1(n) \quad \text{and} \quad x_2(n) \rightarrow y_2(n),$$

$$ax_1(n) + bx_2(n) \rightarrow ay_1(n) + by_2(n)$$

where a and b are arbitrary constants. A system that does not satisfy the linearity condition is a nonlinear system.

This definition is designed for the purpose of system analysis. The input and output signals have arbitrary amplitude. As such, it is difficult to analyze systems. Therefore, we are forced to specify certain characteristics for the systems. The basic way is to decompose arbitrary signals into well-defined signals, such as the impulse and the sinusoid, and find the response of the system to the well-defined signals only once. Then, assuming linearity of systems, we are able to analyze systems efficiently. For example, a signal can be decomposed into scaled and shifted impulses. While we have the response of the system for just the unit-impulse, the constituent impulses of a signal will have arbitrary amplitudes. Therefore, we need the scaling property. As there is a system response for each impulse, we have to sum all the responses to find the total response. Therefore, we need the summation property. The combination of these two properties is the linearity property. Equation (2.6) characterizes a linear system. Fortunately, most practical systems can be characterized by such an equation with adequate accuracy.

Now, let us find indications of nonlinearity in the difference equation of a system. Let the difference equation be $y(n) = mx(n) + C$, where m and C are arbitrary constants. Due to scaling property of a linear system, the response of a system must be zero for a zero input. Replacing $x(n)$ by zero results in $y(n) = C$. Therefore, a constant term is ruled out in the difference equation of a linear system. Let the difference equation be $y(n) = x^2(n)$. The response of the system for $x(0) = 1$ is 1 and for $x(0) = 2$ it is 4. For a linear system, the output should be 2. Therefore, higher powers or product of input and output terms are ruled out in the difference equation of a linear system. $y(n) = mx(n)$ is the difference equation of a linear system, since

$$Px_1(n) + Qx_2(n) = Py_1(n) + Qy_2(n)$$

Then,

$$m(Px_1(n) + Qx_2(n)) = m(Py_1(n) + Qy_2(n))$$

where m, P, and Q are arbitrary constants. For example, an amplifier is a linear system for a certain frequency range of input signals. A full-wave rectifier characterized by $y(n) = |x(n)|$ is a nonlinear system.

2.3.3 Time-Invariant Systems

A system is time-invariant, if

$$x(n) \rightarrow y(n) \quad \text{and} \quad x(n + N) \rightarrow y(n + N)$$

where N is any arbitrary integer. That is, the operation of the system is not time-dependent. The temperature in an air-conditioned room is a constant at any time. The same output appears, but shifted in time. We can delay the input signal and apply to the system or delay the output signal for getting the same response. For example, $y(n) = mx(n)$ is the difference equation of a time-invariant system, since

$$x(n) \rightarrow y(n) \quad \text{and} \quad x(n - N) \rightarrow y(n - N) = y(n)|_{n=n-N}$$

Morning or evening, a digital computer will give the same sum for adding the same two numbers, which is an indispensable time-invariant discrete system. If any coefficient or the time-index n is a function of time, n, then the system is time-variant. For example, $y(n) = nx(n)$ is the difference equation of a time-variant system, since

$$x(n) \rightarrow y(n) = nx(n) \quad \text{and}$$
$$x(n - N) \rightarrow nx(n - N) \neq y(n)|_{n=n-N} = (n - N)y(n - N)$$

$y(n) = 3x(-2n)$ is the difference equation of a time-variant system, since

$$x(n) \rightarrow y(n) = 3x(-2n) \quad \text{and}$$
$$x(n - N) \rightarrow 3x(-2n - N) \neq y(n)|_{n=n-N} = 3y(-2n + 2N)$$

The linearity and time-invariance properties of systems enable the use of operations such as convolution in the time domain and linear transforms in the frequency domain facilitating system analysis. Such systems are called linear and time-invariant (LTI) systems.

2.3.4 Causal Systems

A system is causal if its output does not depend on future values of the input. Otherwise it is noncausal. A system is switched on at a finite time, usually chosen as $n = 0$. If its response depends on an input sample occurring at $n < 0$, then it is noncausal. To implement such systems in practice, we can shift the input to the right by sufficient number of samples to make it causal. Consequently, the output will be delayed. The unit-impulse signal is nonzero only at $n = 0$. Therefore, if the impulse response of a system is causal, then the system is a causal system. If the index of the present output $y(n)$, n is greater than or equal to the indices of all the input terms and greater than the other output terms in the difference equation, then the system is causal. For example, the difference equation of a system

$$y(n) = x(n + 1) + 2x(n) + y(n - 1)$$

is noncausal, as the future input sample $x(n + 1)$ is required to find the output. The system characterized by

$$y(n) = x(n) + x(n - 1) - 2y(n - 1)$$

is causal.

2.3.5 Instantaneous and Dynamic Systems

A system is instantaneous, if its response depends only on the present input values. Such type of systems has no memory. For example, the discrete model of a circuit with resistors only is an instantaneous systems since the signals at any part of the circuit depends on present inputs only. Such type of systems is characterized by the difference equation

$$y(n) = Kx(n)$$

where K is a constant.

On the other hand, circuits composed of inductors and capacitors are dynamic systems and their outputs depend on past input values also. Such systems require memory. Obviously, instantaneous systems are a specific case of dynamic systems. The difference equation that characterizes a system with finite memory is

$$y(n) = \sum_{k=0}^{N} x(n - k)$$

The difference equation that characterizes a system with infinite memory is

$$y(n) = \sum_{k=0}^{\infty} x(n - k)$$

2.3.6 Inverse Systems

Inverse systems are often used to remove known distortions of a signal. Systems with unique output for each unique input are invertible. For example, $y(n) = x^2(n)$ is not invertible. Consider the difference equation $y(n) = 0.5x(n)$. Its impulse response is 0.5. Its inverse system is $x(n) = 2y(n)$ with impulse response 2. When a system and its inverse are connected in cascade, we get back the input. That is, the impulse response of the cascade system must be $\delta(n)$. The convolution output of the impulse responses of a system and its inverse is $\delta(n)$.

Consider a system and its inverse with impulse responses

$$h(n) = 0.5^n u(n) \quad \text{and} \quad h_i(n) = \delta(n) - 0.5\delta(n-1)$$

Now, $h(n) * h_i(n) = \delta(n)$.

2.4   Response to Complex Exponential Input

The convolution operation is a sum of products. The longer the width of the impulse response, the more is the number products evaluated and summed to find the output at each point. Except for very short impulse responses, the execution time becomes prohibitively high. Therefore, it becomes a necessity to evaluate the operation in an alternate way. That is, the input signal and the impulse response are decomposed in terms of sinusoids or, equivalently, complex exponentials. Then, we just need one multiplication to find the output corresponding to each frequency component. The reason is that the frequency of an input sinusoid remains the same at any point in a linear system. Only its amplitude and phase are changed. It turns out that the overall computational complexity of implementing the convolution in this indirect way reduces significantly in practical applications.

Let the impulse response of a stable system be $h(n)$. The system response to a complex exponential $e^{j\omega_0 n}$, $-\infty < n < \infty$, with frequency $j\omega_0$, is the convolution of the impulse response and the exponential. That is,

$$y(n) = h(n) * e^{j\omega_0 n}$$

$$= \sum_{k=-\infty}^{\infty} h(k)e^{j\omega_0(n-k)}$$

$$= e^{j\omega_0 n} \sum_{k=-\infty}^{\infty} h(k)e^{-j\omega_0 k}$$

$$= H(e^{j\omega_0})e^{j\omega_0 n}, \qquad H(e^{j\omega_0}) = \sum_{k=-\infty}^{\infty} h(k)e^{-j\omega_0 k}$$

$H(e^{j\omega_0})$ is a complex constant. If the complex amplitude of the exponential is $X(e^{j\omega_0})$, then the coefficient of the exponential at the output is $H(e^{j\omega_0})X(e^{j\omega_0})$. The steady-state output for an input complex exponential or a real sinusoid is the magnitude of the input is multiplied by the magnitude of $H(e^{j\omega_0})$ and the phase of $H(e^{j\omega_0})$ added. That is,

$$e^{j\omega_0 n} \rightarrow |H(e^{j\omega_0})|e^{j(\omega_0 n + \angle H(e^{j\omega_0}))}$$

and

$$\cos(\omega_0 n) \rightarrow |H(e^{j\omega_0})| \cos(\omega_0 n + \angle H(e^{j\omega_0}))$$

Therefore, only one complex multiplication is required for each complex exponential. Of course, we have to take the transform of the signal and the impulse response, as shown in later chapters, for following this procedure. For a causal sinusoidal input, the output is the zero-state response. That is, the output is the sum of a transient component and a steady-state component. As the transient component of a stable system decays with time, what persists is the steady-state component.

2.5 System Stability

A stable system only is useful in practical applications. The response of such systems is bounded for bounded input. This stability criterion is called the bounded input bounded output stability criterion (BIBO). For example, the unit-step signal is a bounded input, since it is bounded by the finite value 1. An unbounded input is the unit-ramp signal, since it is not bounded by any finite value. An unbounded response of an unstable system could burn out the system or lead to other dangerous effects. Therefore, system stability must be ensured. One of the stability criteria is the BIBO stability. The output of a system is given by the convolution of the input and the impulse response. The basic operation in convolution is sum of products. Assuming that the input, $|x(n)| < \infty$, is bounded to some finite value, for the sum of the products to be bounded, the impulse response $h(n)$ must be absolutely summable. That is,

$$\sum_{n=-\infty}^{\infty} |h(n)| < \infty$$

This stability criterion ensures stable zero-state response.

2.6   Realization and Interconnection of Systems

The three basic operations, addition, multiplication, and delaying, repeatedly appear in a difference equation characterizing a discrete system. The multiplier unit, shown in Fig. 2.5a, is characterized by $y(n) = Kx(n)$, where K is a constant. The output $y(n)$ is the input multiplied by K. Instead of a constant, the second operand K could be a signal. For example, a signal can be modified by multiplying with another signal, called a window signal.

The adder unit, shown in Fig. 2.5b, is characterized by $y(n) = x_1(n) + x_2(n)$. The output $y(n)$ is the sum of the two sequences $x_1(n)$ and $x_2(n)$. By changing the sign of subtrahend sequence, the subtraction operation can be implemented by the adder unit.

The delay unit, shown in Fig. 2.5c, is characterized by $y(n) = x(n - 1)$. The output $y(n)$ is the one sample delayed version of the input $x(n)$. By cascading several units, a longer delay can be achieved.

By using the necessary number of the basic units, any arbitrary difference equation can be realized. Consider the implementation of the difference equation

$$y(n) = 3x(n) - 2x(n - 1) - 0.8y(n - 1)$$

shown in Fig. 2.6. Let the input be

$$x(0) = 1, \ x(2) = 3, \ x(3) = -2, \ x(4) = 4, \ x(5) = 3$$

By iteration, the first few outputs, with $x(-1) = y(-1) = 0$, are

$$y(0) = 3, \ y(1) = 4.6, \ y(2) = -15.68, \ y(3) = 28.5440, \ y(4) = -21.8352$$

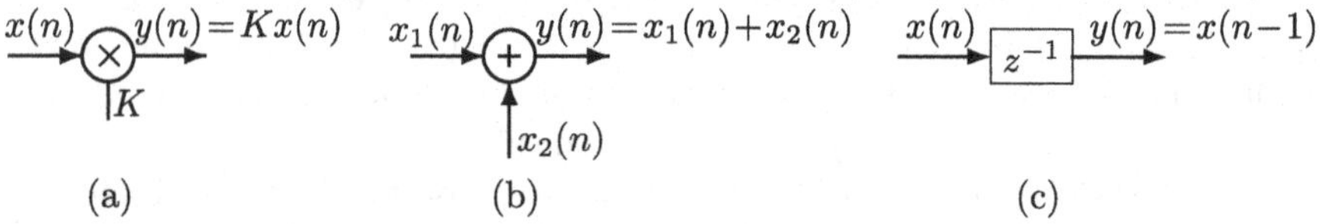

$x(n) \xrightarrow{} \times \xrightarrow{y(n)=Kx(n)}$ $x_1(n) \xrightarrow{} + \xrightarrow{y(n)=x_1(n)+x_2(n)}$ $x(n) \xrightarrow{} \boxed{z^{-1}} \xrightarrow{y(n)=x(n-1)}$
$\quad \uparrow K$ $\quad \uparrow x_2(n)$

$\quad$ (a)   (b)   (c)

Fig. 2.5 (a) Multiplier; (b) adder; (c) delay unit

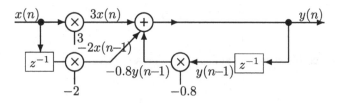

Fig. 2.6 Block diagram of the realization of a discrete system

The order of the systems used in practice is usually high. The realization of systems directly has several disadvantages. Therefore, higher-order systems are decomposed into several lower-order systems, usually as first- or second-order systems, and realized as connected in parallel and series. Figure 2.7a and b show, respectively, two systems connected in parallel and its equivalent single system. In a parallel system, a higher-order system is expressed as a sum of lower-order systems. The inputs to all the lower-order systems are the same and the outputs are added up to get the total output. Due to the distributive property of convolution, the impulse response of the equivalent single system is the sum of the individual responses.

Figure 2.8a and b show, respectively, two systems connected in cascade or series and its equivalent single system. In a series system, a higher-order system is expressed as a product of lower-order systems. The given input is applied to the first system. Afterwards, the input to a system is the output of the preceding one. The total output is the output of the last system in cascade. Due to the associative property of convolution, the impulse response of the equivalent single system is the convolution of the individual responses. In the cascade form, obviously, there are various ways of grouping the numerator and denominator factors of the original transfer function. One significant advantage of cascade and parallel forms is the lower coefficient sensitivity due to the interaction of a coefficient with lower number of other coefficients. That is, any change in a coefficient affects the coefficients in that section only.

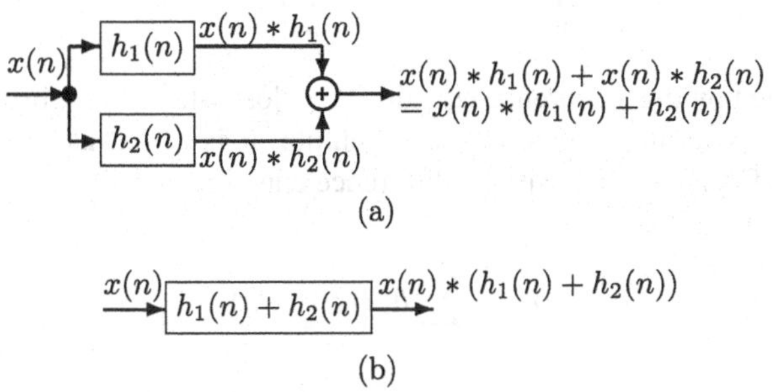

(a)

(b)

Fig. 2.7 (a) Two systems connected in parallel; (b) an equivalent single system

(a)

(b)

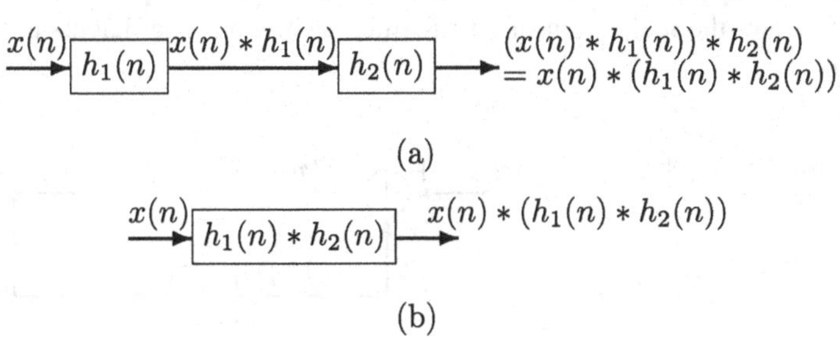

Fig. 2.8 (a) Two systems connected in series; (b) an equivalent single system

2.7   Feedback Systems

Linear filters come under two major types, one with feedback and another without feedback. Filters with feedback require lesser number of coefficients for the same specification, but introduce the stability problem that has to be taken care of. A general block diagram of the feedback system is shown in Fig. 2.9. In a feedback system, a part of the output signal $y(n)$, $f(n)$ is fed back and subtracted from the input signal $x(n)$ to form the error signal $e(n)$. The equation governing the feedback system are

$$f(n) = \sum_{k=1}^{\infty} a(k)y(n-k)$$

$$e(n) = x(n) - f(n)$$

$$y(n) = \sum_{k=0}^{\infty} b(k)e(n-k)$$

The difference governing the linear filters in the discrete time domain is given by

$$y(n) = \sum_{k=0}^{M} b(k)x(n-k) - \sum_{k=1}^{N} a(k)y(n-k)$$

A filter governed by this difference equation is a closed-loop filter, usually referred as IIR filters. As its impulse response is infinite, it is called as infinite impulse response (IIR) filters. A filter with its difference equation

$$y(n) = \sum_{k=0}^{M} b(k)x(n-k)$$

is an open-loop filter, usually referred as FIR filters, with all its a coefficients zero. As its impulse response is finite, it is called as finite impulse response (FIR) filters. Figure 2.6 shows the realization of an IIR filter with two b coefficients and one a coefficient.

Fig. 2.9 Block diagram of a feedback system

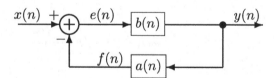

2.8   Summary

- A system is an interconnection of components, hardware, and/or software, which performs an action or produces desired output signals in response to input signals.
- Each component of a system has its own input–output relationship. Using these relationships, we can derive the relation between the input and output of a system. This relationship is used to find the response of the system to basic signals, such as impulse or sinusoid. The response to these signals characterizes a system, from which the response to any arbitrary signal can be derived.
- The approximation of a differential equation using samples of the signals is called a difference equation. Difference between samples is used to approximate the derivative, which becomes more accurate as the sampling interval between samples is reduced and becomes exact as the interval tends to zero.
- Iteratively, we can find the samples of the output from the difference equation of a system, if the input and initial conditions are specified. The iterative nature of the solution makes it highly suitable for implementation using digital systems.
- Linear system response can be expressed as the sum of zero-input and zero-state components. The zero-input response is due to the initial conditions alone with the input to the system zero. The zero-state response is due to the input alone with the initial conditions of the system zero.
- Linear system response can also be expressed as the sum of transient and steady-state components. The steady-state component is that part of the response caused by the natural modes of the input signal. The transient component is that part of the response caused by the natural modes of the system difference equation.
- Physical systems are analyzed using discrete mathematical models in DSP and other studies. Convolution-summation is one of the often used model.
- In the convolution model, we find the response, called the impulse response, of a relaxed system to the unit-impulse input. Then, the response of the system for arbitrary input signals can be found by representing them in terms of a sum of scaled and shifted impulses. Since the impulse response is determined assuming that the system is relaxed (zero initial conditions), the response computed using the convolution is the zero-state response.
- In the convolution model, the present output is computed using the present and all the past input samples only. It is basically a sum of products of two sequences, after one of them is time-reversed. The operation is repeated to find all the output samples of interest.
- Convolution of discrete signals is like polynomial multiplication.
- A LTI system satisfies both the linear and time-invariance properties. They are easy to characterize, analyze, and design.
- The output of a causal system depends only on the present and past input values.
- The sum of two complex exponentials (or real sinusoids) is an exponential of the same frequency, with different magnitude and phase. Differentiating a complex exponential any number of times does not change its frequency. For

these reasons, the output of a LTI system, for a complex exponential, is also of the same form multiplied by a constant.

- The BIBO stability criterion states that if the impulse response $h(n)$ of a system is absolutely summable, then the system is stable.
- The three basic operations, adder, multiplier, and delay operator, are used in the realization of systems.
- It is advantageous to decompose higher-order systems into smaller systems and realize in cascade or parallel.
- In a feedback system, a part of the output signal is fed back and used in producing the output.

Exercises

2.1 A system is characterized by the differential equation

$$\frac{dy(t)}{dt} + \tau y(t) = \tau u(t)$$

where the input $u(t)$ is the unit-step signal. Given the value of τ, find the complete response of the system by solving the differential equation. The initial condition is $y(0) = 2$.

The difference equation characterizing the system is

$$\frac{dy(nT_s)}{dnT_s} + \tau y(nT_s) = \tau u(nT_s), \quad n = 0, 1, 2, \ldots$$

with the input $u(nT_s)$, the sampled unit-step signal. Find the zero-input, zero-state, and complete responses of the discrete system for the specified sampling intervals. Compare these responses with the samples of the output of the continuous-time response. Verify your results by finding the first 10 total output samples using the difference equation by iteration.

2.1.1 $\tau = 4$. The sampling intervals are: (1) $T_s = 0.004$ s and (2) $T_s = 0.04$ s.

2.1.2 $\tau = 3$. The sampling intervals are: (1) $T_s = 0.003$ s and (2) $T_s = 0.03$ s.

* 2.1.3 $\tau = 2$. The sampling intervals are: (1) $T_s = 0.002$ s and (2) $T_s = 0.02$ s.

2.2 Find the linear convolution $y(n)$ of the two finite sequences $x(n)$ and $h(n)$. The symbol ^ on the element indicates that its index is zero. Verify the convolution sum by using the sum and alternating sum methods.

2.2.1

$$x = \{\hat{2}, 1, 3, 4\} \quad \text{and} \quad \{h = \hat{1}, -1, 4, 3\}.$$

2.2.2

$$x = \{1, \hat{2}, 1, 3, 4\} \quad \text{and} \quad \{h = 2, 3, \hat{1}, -1, 4, 3\}.$$

2.2.3

$$x = \{\hat{0}, 0, 2, 1, 3, -4\} \quad \text{and} \quad \{h = 2, 3, \hat{1}, -1, 4, -3\}.$$

* 2.2.4

$$x = \{2, 1, 3, \hat{-4}\} \quad \text{and} \quad \{h = 1, -2, \hat{4}, 3\}.$$

2.2.5

$$x = \{\hat{2}, 0, 3, 0\} \quad \text{and} \quad \{h = \hat{0}, -2, 0, 3\}.$$

2.3 Find the closed-form expression of the convolution of the sequences $x(n)u(n)$ and $h(n)u(n)$. Find the first few values of the convolution of $x(n)$ and $h(n)$ using their sample values and verify that they are the same as those obtained from the closed-form expression.

2.3.1

$$x(n) = (0.9)^n u(n) \text{ and } h(n) = (0.6)^n u(n).$$

* 2.3.2

$$x(n) = (0.6)^{(n-1)} u(n-1) \text{ and } h(n) = (0.5)^{(n-2)} u(n-2).$$

2.3.3

$$x(n) = (0.7)^n u(n-1) \text{ and } h(n) = (0.6)^n u(n+1).$$

2.3.4

$$x(n) = (0.9)^n u(n) \text{ and } h(n) = \delta(n-2).$$

2.3.5

$$x(n) = (0.8)^n u(n) \text{ and } h(n) = u(n).$$

2.4 A system is characterized by the difference equation with the given input signal $x(n)$ and the initial condition $y(-1)$. First, find the impulse response of the system and using which find the zero-input, zero-state, and total responses of the system. Verify your results by finding the first 10 total output samples using the difference equation by iteration. Deduce expressions for the transient and steady-state response of the system.

2.4.1 The difference equation is

$$y(n) = 3x(n) - 2x(n-1) - 0.8y(n-1) \quad n = 0, 1, 2, \ldots$$

The input is $x(n) = u(n)$. The initial condition is $y(-1) = 3$.

2.4.2 The difference equation is

$$y(n) = x(n) + x(n-1) + 0.9y(n-1) \quad n = 0, 1, 2, \ldots$$

The input is $x(n) = (0.5)^n u(n)$. The initial condition is $y(-1) = 2$.

* 2.4.3 The difference equation is

$$y(n) = 2x(n) + x(n-1) - 0.7y(n-1)$$

The initial condition $y(-1) = -2$ and the input $x(n) = \cos(\frac{2\pi}{8}n - \frac{\pi}{6})u(n)$.

2.5 The difference equation of the system is given. Apply the given inputs

$$xa(n) = \{\hat{2}, -1, 3, 4\}, \quad xb(n) = \{\hat{3}, 2, 0, -4\}, \quad \text{and}$$

$$xc(n) = 2xa(n) + 3xb(n)$$

and find the outputs

$$ya(n), \quad yb(n) \text{ and } y(n)$$

Determine whether the system is linear. Assume that the system is initially relaxed.

2.5.1. $y(n) = 3 + 2x(n) + 0.8y(n-1)$.

2.5.2. $y(n) = -x(n) - 0.2y^2(n-1)$.

2.5.3. $y(n) = 2x(n) - 0.4(-n)y(n-1) - 2\sin(\pi)$.

* 2.5.4. $y(n) = x(n) - 0.7|y(n-1)|$.

2.6 The difference equation of the system is given. Apply the given input

$$x(n) = \{\hat{2}, -1, 3, 4\}$$

starting at $n = 0$. Apply the same input values starting at $n = 2$. Find the outputs

$$y(n) \text{ and } yd(n)$$

Determine whether the system is time-invariant. Assume that the system is initially relaxed at the instant of applying the input.

2.6.1. $y(n) = 3 + 2x(n) + 0.8y(n - 1)$.

2.6.2. $y(n) = 2x(n) - 0.4(-n)y(n - 1)$.

2.6.3. $y(n) = x(n) + \cos(\frac{2\pi}{8}n)$.

* 2.6.4. $y(n) = x(2n)$.

2.7 The difference equation of the system is given. Determine whether the system is causal.

2.7.1. $y(n) = 3 + 2x(n) + 0.8x(n - 1)$.

2.7.2. $y(n) = 2x(n) + 0.8x(n - 1) - 0.6x(n - 2)$.

2.7.3. $y(n) = 2x(n) + 0.8x(n - 1) - 0.6x(n + 1)$.

* 2.7.4. $y(n) = 2x(-2n) + 0.8x(n - 1) - 0.6x(n - 2)$.

2.7.5. $y(n) = 2x(n^2) + 0.8x(n - 1) - 0.6x(n - 2)$.

2.8 Find the closed-form expression for the zero-state response of the system, with the impulse response

$$h(n) = (a)^n u(n), \ n = 0, 1, 2, \ldots$$

to the input $x(n) = e^{j\omega_0 n} u(n)$ by convolution. Find the first few samples of the signal and the impulse response and convolve in the time domain to find the first 4 values of the output. Verify that the outputs by both methods are the same. Deduce the steady-state response.

* 2.8.1. $\omega_0 = 1$ and $h(n) = (a)^n u(n), a = 0.8$.

2.8.2. $\omega_0 = -2$ and $h(n) = (a)^n u(n), a = -0.6$.

2.9 Given the impulse response of a LTI discrete system, is the system BIBO stable?

2.9.1 $h(n) = (-0.8)^n u(n)$.

2.9.2 $h(n) = (0.6)^n u(n)$.

2.9.3 $h(n) = u(n)$.

2.9.4 $h(n) = h(0) = 0, \ h(n) = \frac{1}{n}, \ n = 1, 2, \ldots$

* 2.9.5 $h(0) = 0, \ h(n) = \frac{1}{n^2}, \ n = 1, 2, \ldots$

2.10 The impulse response of a second-order system has been decomposed into those of the two first-order systems and are given. The impulse responses of two systems connected in parallel are

$$hp1(n) = 4(0.8^n)u(n) \quad \text{and} \quad hp2(n) = (-3)(0.6^n)u(n)$$

The impulse responses of two systems connected in cascade are

$$hc1(n) = (0.8^n)u(n) \quad \text{and} \quad hc2(n) = (0.6^n)u(n)$$

Find the impulse response of the single equivalent system from both the cascade and parallel configuration impulse responses and verify that they are the same. List the first four values of the impulse response of the single equivalent system.

* 2.11 The impulse response of a second-order system has been decomposed into those of the two first-order systems and are given. The impulse responses of two systems connected in parallel are

$$hp1(n) = -5(-0.7)^{(n-1)}u(n-1) \quad \text{and} \quad hp2(n) = 5(-0.5^n)u(n)$$

The impulse responses of two systems connected in cascade are

$$hc1(n) = (-0.7)^{(n-1)}u(n-1) \quad \text{and} \quad hc2(n) = (-0.5)^n u(n)$$

Find the impulse response of the single equivalent system from both the cascade and parallel configuration impulse responses and verify that they are the same. List the first four values of the impulse response of the single equivalent system.

Discrete Fourier Transform

3

The exponential function is of the form b^n, where b is a constant, called the base, and the exponent n is the independent variable. The exponent indicates the number of times the base is multiplied by itself. It inherently appears in problems such as computing the compound amount from a principal amount of money invested with a compounded interest rate, population growth, etc. In general, it occurs whenever a quantity is increasing or decreasing at a constant percentage rate. For example, the voltage across a charging or discharging capacitor is given by an exponential function. One important property is that its derivative is also of the form. Further, it is compact and easy to manipulate. In addition, the representation of quantities in exponential form often reduces the complexity of manipulating them. For example, when two real numbers are expressed in exponential form, the problem of finding their product is reduced to much simpler addition. In a similar way, when two functions are expressed in complex exponential form, the problem of finding their convolution is reduced to much simpler multiplication. When combined appropriately, the complex exponential function represents a real sinusoid, getting physical significance.

Most of the naturally occurring signals have arbitrary amplitude profile. As such, it is difficult to process. Therefore, the first major problem in signal processing is appropriate signal representation. Fourier analysis and the related transforms represent arbitrary signals in terms of complex exponentials. The difference between the transforms is that each one is more suitable for certain class of signals, such as periodic or aperiodic and continuous or sampled. Further, the exponent in the basis complex exponential could be complex or purely imaginary. The transform representation of signals is more efficient in general and, therefore, more often used than the time-domain methods. Although we are going to use different transforms depending on the signal characteristics, it should be remembered that the basic principle behind all the transforms is the same. That is, signal representation is obtained by the integral or sum of the product of the signal and the basis complex exponentials.

3.1   The DFT

In Fourier analysis, the basis functions are exponentials with purely imaginary exponents. DFT is the only version, in which the signal representation is discrete and finite in both the time and frequency domains. Since it is inherently suitable for implementing on a digital computer or using digital hardware, it is the most often used version of the Fourier analysis in practice. The complex exponential with the exponent $j\omega$, using Euler's formula, is given by

$$e^{\pm j\omega} = \cos(\omega) \pm j\sin(\omega)$$

With $\cos(2\pi) = 1$ and $\sin(2\pi) = 0$, $e^{j2\pi} = 1$. Since sine and cosine functions are periodic with period 2π,

$$e^{j\omega} = e^{j(\omega + 2k\pi)}, \quad k = 0, 1, \ldots$$

Therefore, in the DFT, the given finite data and its finite spectrum are periodically extended. The periodically extended data can be expressed in terms of complex exponentials. Cosine and sine functions can be expressed in terms of exponentials as

$$\cos(\omega) = \frac{e^{j\omega} + e^{-j\omega}}{2} \quad \text{and} \quad \sin(\omega) = \frac{e^{j\omega} - e^{-j\omega}}{j2}$$

An arbitrary signal is composed of an arbitrary number of complex exponentials. The time-domain representation of a signal is the superposition sum of all its constituent complex exponentials. In order to find the complex amplitude of each exponential, the orthogonality property of the complex exponentials is used. That is, the sum of products of an exponential with its conjugate, over an integral number of periods, is a nonzero constant. On the other hand, the sum of products of two exponentials, with different harmonically related frequencies and one of them conjugated, is zero. Harmonically related frequencies are frequencies that are integer multiples of the frequency of a periodic signal, called the fundamental. Let the two discrete complex exponential signals be $e^{j\frac{2\pi}{N}kn}$ and $e^{j\frac{2\pi}{N}ln}$ with the fundamental frequency $j\frac{2\pi}{N}$. Then, the orthogonality condition is given by

$$\sum_{n=0}^{N-1} e^{j\frac{2\pi}{N}kn} \left(e^{j\frac{2\pi}{N}ln} \right)^* = \sum_{n=0}^{N-1} e^{j\frac{2\pi}{N}(k-l)n} = \begin{cases} N & \text{for } k = l \\ 0 & \text{for } k \neq l \end{cases}$$

where $k, l = 0, 1, \ldots, N - 1$. For $k = l$, the summation evaluates to N. Using the closed-form identity for the geometric summation, we get

$$\sum_{n=0}^{N-1} e^{j\frac{2\pi}{N}(k-l)n} = \frac{1 - e^{j2\pi(k-l)}}{1 - e^{j\frac{2\pi(k-l)}{N}}} = 0, \text{ for } k \neq l$$

The sum of the samples, over an integral number of cycles, of $\cos(\frac{2\pi}{N}(k)n)$ and $\sin(\frac{2\pi}{N}(k)n)$, with $k \neq 0$, is zero. Therefore, the sum of the samples, over an integral number of cycles, of

$$e^{j\frac{2\pi}{N}(k)n} = \cos\left(\frac{2\pi}{N}(k)n\right) + j\sin\left(\frac{2\pi}{N}(k)n\right)$$

is zero, with $k \neq 0$.

In the DFT, an arbitrary aperiodic signal, defined as $x(n)$, $n = 0, 1, \ldots, N - 1$ and periodically extended, can be expressed by a complex exponential polynomial of order $N - 1$ as

$$x(n) = X(0)e^{j\frac{2\pi}{N}(0)n} + X(1)e^{j\frac{2\pi}{N}(1)n} +, \cdots, +X(n-1)e^{j\frac{2\pi}{N}(N-1)n} \tag{3.1}$$

This summation of the complex exponentials $e^{jk\frac{2\pi}{N}n}$ multiplied by their respective DFT coefficients gets back the time-domain samples. Therefore, the inverse DFT (IDFT) equation, the Fourier reconstruction or synthesis of the input signal, is defined as

$$x(n) = \frac{1}{N}\sum_{k=0}^{N-1} X(k)e^{jk\frac{2\pi}{N}n}, \quad n = 0, 1, 2, \ldots N - 1 \tag{3.2}$$

Due to the constant N in the orthogonality property, the DFT coefficients are scaled by N and, therefore, the factor $\frac{1}{N}$ appears in the IDFT definition. The constant N can be split between the DFT and the IDFT definitions and in other ways also.

Due to the orthogonality condition, we can multiply Equation (3.1) by $e^{-jk\frac{2\pi}{N}n}$, for each k, from 0 to $N - 1$, by the input signal $x(n)$ and sum the product to get all the DFT coefficients $X(k)$ of the signal $x(n)$. Therefore, the DFT of a sequence $x(n)$ of length N is defined as

$$X(k) = \sum_{n=0}^{N-1} x(n)e^{-jk\frac{2\pi}{N}n}, \quad k = 0, 1, 2, \ldots N - 1 \tag{3.3}$$

The complex exponential is often written in its abbreviated form as

$$e^{-j\frac{2\pi}{N}kn} = W_N^{kn}, \quad \text{where} \quad W = e^{-j\frac{2\pi}{N}}$$

Let us do a 4-point example. Figure 3.1a shows one period of a discrete periodic waveform,

$$x(n) = 1 + \cos\left(\frac{2\pi}{4}n + \frac{\pi}{3}\right) - \cos\left(2\frac{2\pi}{4}n\right),$$

with period 4 samples. The sample values are

$$x(0) = 0.5000, \quad x(1) = 1.1340, \quad x(2) = -0.5000, \quad x(3) = 2.8660$$

This waveform can be represented by the polynomial

$$x(n) = X(0)e^{j\frac{2\pi}{4}(0)n} + X(1)e^{j\frac{2\pi}{4}(1)n} + X(2)e^{j\frac{2\pi}{N}(2)n} + X(3)e^{j\frac{2\pi}{N}(3)n} \qquad (3.4)$$

In order to find the value of $X(0)$, we multiply $x(n)$ by

$$e^{-j\frac{2\pi}{4}(0)n}, n = 0, 1, 2, 3 = \{1, 1, 1, 1\}$$

point by point and sum to get

$$X(0) = (0.5)(1) + (1.1340)(1) + (-0.5)(1) + (2.866)(1) = 4,$$

as shown in Fig. 3.1b. In order to find the value of $X(1)$, we multiply $x(n)$ by

$$e^{-j\frac{2\pi}{4}(1)n}, n = 0, 1, 2, 3 = \{1, -j, -1, j\}$$

point by point and sum to get

$$X(1) = (0.5)(1) + (1.1340)(-j) + (-0.5)(-1) + (2.866)(j) = 1.0000 + j1.7321$$

In order to find the value of $X(2)$, we multiply $x(n)$ by

$$e^{-j\frac{2\pi}{4}(2)n}, n = 0, 1, 2, 3 = \{1, -1, 1, -1\}$$

point by point and sum to get

$$X(2) = (0.5)(1) + (1.1340)(-1) + (-0.5)(1) + (2.866)(-1) = -4$$

In order to find the value of $X(3)$, we multiply $x(n)$ by

$$e^{-j\frac{2\pi}{4}(3)n}, n = 0, 1, 2, 3 = \{1, j, -1, -j\}$$

point by point and sum to get

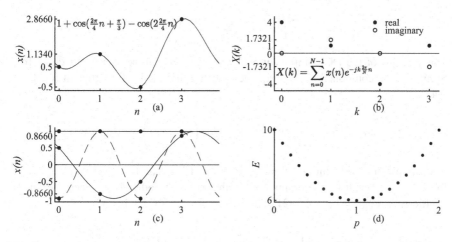

Fig. 3.1 (**a**) One period of a discrete periodic waveform, $x(n) = 1 + \cos(\frac{2\pi}{4}n + \frac{\pi}{3}) - \cos(2\frac{2\pi}{4}n)$, with period 4 samples and (**b**) its DFT coefficients representing it in the frequency domain; (**c**) the time-domain components of the waveform in (**a**); (**d**) the square error in approximating the waveform in (**a**) using only the DC component with different amplitudes

$$X(3) = (0.5)(1) + (1.1340)(j) + (-0.5)(-1) + (2.866)(-j) = 1.0000 - j1.7321$$

Note that $X(3) = X^*(1)$. About half the coefficients are redundant for real $x(n)$. Therefore, DFT coefficients of $x(n)$, representing it in the frequency domain, are

$$X(0) = 4, \ \ X(1) = 1.0000 + j1.7321, \ \ X(2) = -4, \ \ X(3) = 1.0000 - j1.7321$$

A plot of these coefficients versus frequency index k is called the spectrum of $x(n)$, as shown in Fig. 3.1b. Alternatively, the spectrum can be expressed as an amplitude spectrum and a phase spectrum. The DFT gets the spectrum, which is the frequency-domain representation of $x(n)$. The IDFT gets back $x(n)$ from the spectrum. What one does, the other undoes. Both the time- and frequency-domain representations are equivalent representation of $x(n)$. In the processing of $x(n)$, we use the more advantageous representation. Essentially, the signal is represented using time as the independent variable in the time domain, and it is represented using frequency as the independent variable in the frequency domain. As it is the change of the independent variable, the operations are called forward (DFT) and inverse (IDFT) transforms.

Figure 3.1c shows the three frequency components of the input waveform in Fig. 3.1a. We used the DFT to separate these components from the given set of samples. The DFT coefficient $X(0)$ is 4. As the coefficients are scaled by 4, the number of input samples, we divide 4 by 4 to get 1 as the coefficient of the DC component. That is, the DC component is

$$\frac{X(0)}{4}e^{j\frac{2\pi}{4}(0)n} = 1e^{j\frac{2\pi}{4}(0)n}$$

The samples of the DC component are $\{1, 1, 1, 1\}$, as can be seen from Fig. 3.1c. Similarly, the component, with frequency index 2 and the DFT coefficient $X(2) = -4$, is

$$\frac{X(2)}{4}e^{j\frac{2\pi}{4}(2)n} = -1e^{j\frac{2\pi}{4}(2)n} = -\cos\left(\frac{2\pi}{4}(2)n\right) = -\cos(\pi n)$$

The samples of the component are $\{-1, 1, -1, 1\}$, as can be seen from Fig. 3.1c. In order to reconstruct the component with frequency index 1, we need the DFT coefficients $X(1) = 1 + j1.7321$ and $X(3) = 1 - j1.7321$. That is,

$$0.25(1 + j1.7321)e^{j\frac{2\pi}{4}n} + 0.25(1 - j1.7321)e^{j3\frac{2\pi}{4}n}$$

$$= 0.25(1 + j1.7321)e^{j\frac{2\pi}{4}n} + 0.25(1 - j1.7321)e^{-j\frac{2\pi}{4}n}$$

$$= \cos\left(\frac{2\pi}{4}n + \frac{\pi}{3}\right)$$

Remember,

$$e^{j3\frac{2\pi}{4}n} = e^{-j\frac{2\pi}{4}n}$$

due to periodicity. The samples of the component are $\{0.5, -0.866, -0.5, 0.866\}$, as can be seen from Fig. 3.1c.

The Fourier reconstruction of a signal is with respect to least squares error criterion. That is, the sum of the squared magnitude of the difference between the input and reconstructed signals must be minimum. If a N-point waveform $x(n)$ is reconstructed using fewer than N DFT coefficients, $x_a(n)$, then the error E is given by

$$E = \sum_{n=0}^{N-1} |x(n) - x_a(n)|^2$$

Let us reconstruct the waveform shown in Fig. 3.1a by the DC waveform alone. Then, the sample values of $x_a(n)$ are

$$x(0) = 1, \ x(1) = 1, \ x(2) = 1, \ x(3) = 1$$

The exact sample values of $x(n)$ are

$$x(0) = 0.5000, \ x(1) = 1.1340, \ x(2) = -0.5000, \ x(3) = 2.8660$$

Now, the error is

$$E = (0.5 - 1)^2 + (1.134 - 1)^2 + (-0.5 - 1)^2 + (2.866 - 1)^2 = 6$$

as shown in Fig. 3.1d. For values other than for $X(0)$, the error increases as shown in the figure. Let $X(0)$ be p. Then,

$$E = (0.5 - p)^2 + (1.134 - p)^2 + (-0.5 - p)^2 + (2.866 - p)^2$$

must be minimum. Differentiating this expression with respect to p, equating it to zero, and solving for p, we get $p = 1$ as given by Fourier analysis.

3.1.1 Center-Zero Format of the DFT and IDFT

The DFT coefficients $X(k)$ are periodic with period N, the number of samples of $x(n)$. Therefore, N coefficients, over any consecutive part of the spectrum, define the spectrum. However, the coefficients are usually specified in two ranges. In the standard format, the N coefficients are specified over one period starting from index $k = 0$. That is,

$$X(0), X(1), \ldots, X(N - 1)$$

For example, with $N = 4$,

$$X(0), X(1), X(2), X(3)$$

In the other format, called center-zero format, the coefficient $X(0)$ is placed approximately in the middle. That is, with N even,

$$X\left(-\frac{N}{2}\right), X\left(-\frac{N}{2} + 1\right), \ldots, X(-1),$$

$$X(0), X(1), \ldots, X\left(\frac{N}{2} - 2\right), X\left(\frac{N}{2} - 1\right)$$

For example, with $N = 4$,

$$X(-2), X(-1), X(0), X(1)$$

One format from the other can be obtained by circularly shifting the spectrum by $N/2$ number of sample intervals. The use of the center-zero format is convenient in some derivations. Further, the display of the spectrum in this format is better for viewing. The definitions of the DFT and IDFT in this format, with N even, are

$$X(k) = \sum_{n=\left(-\frac{N}{2}\right)}^{\frac{N}{2}-1} x(n)e^{-jk\frac{2\pi}{N}n}, \quad k = -\left(\frac{N}{2}\right), -\left(\frac{N}{2}-1\right), \ldots, \frac{N}{2}-1 \qquad (3.5)$$

$$x(n) = \frac{1}{N} \sum_{k=-\left(\frac{N}{2}\right)}^{\frac{N}{2}-1} X(k)e^{jk\frac{2\pi}{N}n}, \quad n = -\left(\frac{N}{2}\right), -\left(\frac{N}{2}-1\right), \ldots, \frac{N}{2}-1 \qquad (3.6)$$

3.2   Matrix Formulation of the DFT

We computed the DFT coefficients one by one in the earlier example. The matrix formulation is a symbolic representation of a set of simultaneous linear equations. All the N coefficients of the N-point DFT can be computed simultaneously using the matrix formulation of the DFT. The values of the input data and those of the DFT coefficients are stored in $N \times 1$ matrices (column vectors). To find the DFT, we multiplied the input data by the samples of the complex exponential of the form $e^{-j\frac{2\pi}{N}kn}$ for each k with $n = 0, 1, \ldots, N-1$. These samples are located on the unit circle equidistantly. As the samples of the complex exponentials are located on a unit circle, the exponentials and, hence, the samples are periodic. They are the Nth roots of unity. Let us construct the $N \times N$ transform matrix of the samples, called twiddle factors. Twiddle factors with magnitude 1, when multiplied with a number, change the angle of the number but not its magnitude. The vector representing a number on the unit circle is rotated with the same magnitude.

3.2.1   The Roots of Unity

There are N complex numbers such that each raised to the power N is equal to 1. The Nth roots of unity are given by

$$1^{\frac{1}{N}} = \cos\left(\frac{2k\pi}{N}\right) + j\sin\left(\frac{2k\pi}{N}\right), \quad k = 0, 1, 2, \ldots, N-1$$

They form the samples of the DFT basis functions. For example, with $N = 4$, we get the roots as

$$k = 0: \ \cos(0) + j\sin(0) = 1$$

$$k = 1: \ \cos\left(\frac{2\pi}{4}\right) + j\sin\left(\frac{2\pi}{4}\right) = j$$

$$k = 2: \quad \cos\left(\frac{2(2)\pi}{4}\right) + j\sin\left(\frac{2(2)\pi}{4}\right) = -1$$

$$k = 3: \quad \cos\left(\frac{2(3)\pi}{4}\right) + j\sin\left(\frac{2(3)\pi}{4}\right) = -j$$

The roots are $\{1, j, -1, -j\}$. Each root raised to the power of 4 will yield 1. The magnitude of the roots is 1. Their angles add to $\{0, 2\pi, 4\pi, -2\pi\}$. The complex number with these arguments is 1.

3.2.2 The Periodicity of the Twiddle Factors and the DFT Basis Functions

The basis functions are of the form

$$e^{j\frac{2\pi}{N}kn} = W_N^{-kn}, \quad n = 0, 1, \ldots, N-1$$

for each k, the frequency index. The fundamental radian frequency is $2\pi/N$ radians per sample. The cyclic frequency is $1/N$ cycles per sample. The DFT computes the frequency content of the input samples at cyclic frequencies, for example with $N = 8$,

$$\left\{0, \frac{1}{N}, \frac{2}{N}, \frac{3}{N}, \frac{4}{N}, \frac{5}{N}, \frac{6}{N}, \frac{7}{N}\right\}$$

The frequencies are shown in Fig. 3.2. The twiddle factors are the conjugate of the basis functions

$$e^{-j\frac{2\pi}{N}kn} = W_N^{kn}, \quad n = 0, 1, \ldots, N-1$$

for each k, the frequency index. As k increases, the unit circle is traversed at a faster rate. Therefore, the values of the roots of N are used repeatedly. Due to the periodicity of the twiddle factors, shown in Fig. 3.2, the values are mapped to the values of the roots of N for any value of nk in $e^{\frac{2\pi}{N}kn} = W_N^{-kn}$ and $e^{-\frac{2\pi}{N}kn} = W_N^{kn}$.

3.2.3 The Transform Matrix

With $N = 8$, the transform matrix W_8^{nk} is

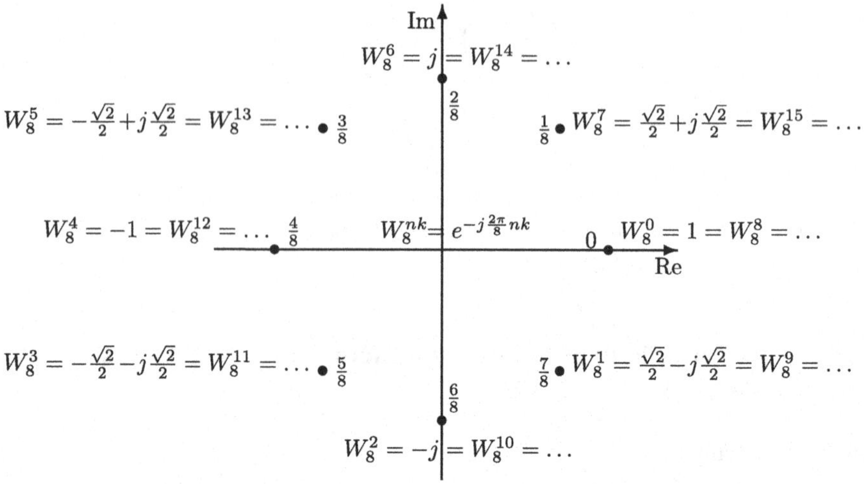

$$W_8^6 = j = W_8^{14} = \cdots$$

$$W_8^5 = -\frac{\sqrt{2}}{2} + j\frac{\sqrt{2}}{2} = W_8^{13} = \cdots \bullet \tfrac{3}{8} \qquad \tfrac{2}{8} \qquad \tfrac{1}{8} \bullet W_8^7 = \frac{\sqrt{2}}{2} + j\frac{\sqrt{2}}{2} = W_8^{15} = \cdots$$

$$W_8^4 = -1 = W_8^{12} = \cdots \bullet \tfrac{4}{8} \qquad W_8^{nk} = e^{-j\frac{2\pi}{8}nk} \qquad {}_0 \bullet W_8^0 = 1 = W_8^8 = \cdots$$
$$\text{Re}$$

$$W_8^3 = -\frac{\sqrt{2}}{2} - j\frac{\sqrt{2}}{2} = W_8^{11} = \cdots \bullet \tfrac{5}{8} \qquad \tfrac{7}{8} \bullet W_8^1 = \frac{\sqrt{2}}{2} - j\frac{\sqrt{2}}{2} = W_8^9 = \cdots$$

$$\tfrac{6}{8}$$

$$W_8^2 = -j = W_8^{10} = \cdots$$

Fig. 3.2 Periodicity of the twiddle factors with $N = 8$ and the discrete frequencies at which the DFT coefficients are computed

$$W_8^{nk} = \begin{bmatrix}
1 & 1 & 1 & 1 & 1 & 1 & 1 & 1 \\
1 & \frac{\sqrt{2}}{2}-j\frac{\sqrt{2}}{2} & -j & -\frac{\sqrt{2}}{2}-j\frac{\sqrt{2}}{2} & -1 & -\frac{\sqrt{2}}{2}+j\frac{\sqrt{2}}{2} & j & \frac{\sqrt{2}}{2}+j\frac{\sqrt{2}}{2} \\
1 & -j & -1 & j & 1 & -j & -1 & j \\
1 & -\frac{\sqrt{2}}{2}-j\frac{\sqrt{2}}{2} & j & \frac{\sqrt{2}}{2}-j\frac{\sqrt{2}}{2} & -1 & \frac{\sqrt{2}}{2}+j\frac{\sqrt{2}}{2} & -j & -\frac{\sqrt{2}}{2}+j\frac{\sqrt{2}}{2} \\
1 & -1 & 1 & -1 & 1 & -1 & 1 & -1 \\
1 & -\frac{\sqrt{2}}{2}+j\frac{\sqrt{2}}{2} & -j & \frac{\sqrt{2}}{2}+j\frac{\sqrt{2}}{2} & -1 & \frac{\sqrt{2}}{2}-j\frac{\sqrt{2}}{2} & j & -\frac{\sqrt{2}}{2}-j\frac{\sqrt{2}}{2} \\
1 & j & -1 & -j & 1 & j & -1 & -j \\
1 & \frac{\sqrt{2}}{2}+j\frac{\sqrt{2}}{2} & j & -\frac{\sqrt{2}}{2}+j\frac{\sqrt{2}}{2} & -1 & -\frac{\sqrt{2}}{2}-j\frac{\sqrt{2}}{2} & -j & \frac{\sqrt{2}}{2}-j\frac{\sqrt{2}}{2}
\end{bmatrix}$$

The entries in the first row with $k = 0$ are all 1s. With $k = 1$, the entries in the second row are just the roots of unity, shown in Fig. 3.2, starting from the value 1. With $k = 2$, the entries in the third row are the every second value of the roots of unity in traversing the unit circle two times in the clockwise direction. Similarly, the other row entries are obtained by selecting every kth value of the roots of unity in traversing the unit circle as many times as necessary to get 8 values for each row.

With $N = 4$, the transform matrix W_4^{nk} is

$$W_4^{nk} = \begin{bmatrix}
1 & 1 & 1 & 1 \\
1 & -j & -1 & j \\
1 & -1 & 1 & -1 \\
1 & j & -1 & -j
\end{bmatrix}$$

This matrix is obtained by deleting every 2nd row the matrix for $N = 8$ and taking the first 4 columns of the remaining rows.

With $N = 2$, the transform matrix W_2^{nk} is

$$W_2^{nk} = \begin{bmatrix} 1 & 1 \\ 1 & -1 \end{bmatrix}$$

This matrix is obtained by deleting every 2nd row of the matrix for $N = 4$ and taking the first 2 columns of the remaining rows.

For computing the IDFT, the transform matrix is the conjugate of that for the DFT, in addition to the constant N in the denominator. Therefore, the inverse and forward transform matrices are orthogonal. That is, for example, with $N = 4$,

$$\frac{1}{4} \begin{bmatrix} 1 & 1 & 1 & 1 \\ 1 & j & -1 & -j \\ 1 & -1 & 1 & -1 \\ 1 & -j & -1 & j \end{bmatrix} \begin{bmatrix} 1 & 1 & 1 & 1 \\ 1 & -j & -1 & j \\ 1 & -1 & 1 & -1 \\ 1 & j & -1 & -j \end{bmatrix} = \begin{bmatrix} 1 & 0 & 0 & 0 \\ 0 & 1 & 0 & 0 \\ 0 & 0 & 1 & 0 \\ 0 & 0 & 0 & 1 \end{bmatrix}$$

That is if we compute the DFT of the data and compute the IDFT of the result, we get back the original data, which proves that the DFT and IDFT operations form a transform pair.

With the transform matrices and input data known, the DFT and IDFT operations for various transform lengths can be defined using the matrix formulation. The DFT is defined for any finite sequence of an integer length, odd or even. However, due to the availability of fast practical algorithms only for lengths that are an integral power of 2, we give emphasis to DFT of those lengths. With $N = 2$, the DFT and IDFT definitions for the input $\{x(0), x(1)\}$ are

$$\begin{bmatrix} X(0) \\ X(1) \end{bmatrix} = \begin{bmatrix} 1 & 1 \\ 1 & -1 \end{bmatrix} \begin{bmatrix} x(0) \\ x(1) \end{bmatrix}$$

where $X(0)$ and $X(1)$ are the corresponding DFT coefficients. The IDFT gets the input back from the coefficients.

$$\begin{bmatrix} x(0) \\ x(1) \end{bmatrix} = \frac{1}{2} \begin{bmatrix} 1 & 1 \\ 1 & -1 \end{bmatrix} \begin{bmatrix} X(0) \\ X(1) \end{bmatrix}$$

For example, the DFT of $\{x(0) = -2, x(1) = 5\}$ is $\{X(0) = 3, X(1) = -7\}$. Using the IDFT, we can get back the input samples. From the DFT coefficients, we get the time-domain samples of the DC component by multiplying the coefficient 3 with the sample values of the basis function $\{1, 1\}$ and dividing by 2 as $\{3, 3\}/2$. The samples of the frequency component with frequency index 1 are $\{-7, 7\}/2$. The sum of the samples of the components yields the input samples.

With $N = 4$, the DFT definition is

$$
\begin{bmatrix} X(0) \\ X(1) \\ X(2) \\ X(3) \end{bmatrix} = \begin{bmatrix} 1 & 1 & 1 & 1 \\ 1 & -j & -1 & j \\ 1 & -1 & 1 & -1 \\ 1 & j & -1 & -j \end{bmatrix} \begin{bmatrix} x(0) \\ x(1) \\ x(2) \\ x(3) \end{bmatrix}
$$

The IDFT definition is

$$
\begin{bmatrix} x(0) \\ x(1) \\ x(2) \\ x(3) \end{bmatrix} = \frac{1}{4} \begin{bmatrix} 1 & 1 & 1 & 1 \\ 1 & j & -1 & -j \\ 1 & -1 & 1 & -1 \\ 1 & -j & -1 & j \end{bmatrix} \begin{bmatrix} X(0) \\ X(1) \\ X(2) \\ X(3) \end{bmatrix}
$$

The response to the impulse signal characterizes systems in the time domain. The transform of it characterizes systems in the frequency domain. As the signal is decomposed into frequency components, the transform must be uniform. The response of systems to all complex exponentials with unit complex amplitude is the frequency response. Let us find the DFT of the 4-point impulse signal. Using the 4-point DFT, we get

$$
\begin{bmatrix} X(0) \\ X(1) \\ X(2) \\ X(3) \end{bmatrix} = \begin{bmatrix} 1 & 1 & 1 & 1 \\ 1 & -j & -1 & j \\ 1 & -1 & 1 & -1 \\ 1 & j & -1 & -j \end{bmatrix} \begin{bmatrix} 1 \\ 0 \\ 0 \\ 0 \end{bmatrix} = \begin{bmatrix} 1 \\ 1 \\ 1 \\ 1 \end{bmatrix}
$$

As expected, the DFT spectrum of the impulse $\{X(0) = 1, X(1) = 1, X(2) = 1, X(3) = 1\}$ is unity at all frequencies. The magnitude of all the frequency components is 1 and the phase is zero. Therefore, by summing the constituent complex exponentials, we reconstruct the impulse signal. That is,

$$
x(n) = \frac{1}{4}\left(1 + e^{j\frac{2\pi}{4}n} + e^{j2\frac{2\pi}{4}n} + e^{j3\frac{2\pi}{4}n}\right) = \frac{1}{4}\left(1 + 2\cos\left(\frac{\pi}{2}n\right) + \cos(\pi n)\right)
$$

$$
= \begin{cases} 1 & \text{for } n = 0 \\ 0 & \text{for } n = 1, 2, 3 \end{cases}
$$

Formally, we can get back the input signal by computing the 4-point IDFT of the spectrum

$$
\begin{bmatrix} x(0) \\ x(1) \\ x(2) \\ x(3) \end{bmatrix} = \frac{1}{4} \begin{bmatrix} 1 & 1 & 1 & 1 \\ 1 & j & -1 & -j \\ 1 & -1 & 1 & -1 \\ 1 & -j & -1 & j \end{bmatrix} \begin{bmatrix} 1 \\ 1 \\ 1 \\ 1 \end{bmatrix} = \begin{bmatrix} 1 \\ 0 \\ 0 \\ 0 \end{bmatrix}
$$

By shifting a signal circularly by n_0 number of sample intervals clockwise or anticlockwise, the magnitude of the spectrum of the signal remains the same with phase increments $e^{\pm kn_0 \frac{2\pi}{4}}$. With $n_0 = 1, 2, 3$, we get the DFT of the delayed impulses as

$$\delta(n-1) = \{0, 1, 0, 0\} \leftrightarrow X(k) = \{1, -j, -1, j\}$$
$$\delta(n-2) = \{0, 0, 1, 0\} \leftrightarrow X(k) = \{1, -1, 1, -1\}$$
$$\delta(n-3) = \{0, 0, 0, 1\} \leftrightarrow X(k) = \{1, j, -1, -j\}$$

If we use these 4 transforms as rows or columns of a 4×4 matrix, we get the 4×4 DFT transform matrix, Therefore, the DFT of the rows or columns of the identity matrix and the DFT transform matrix form a transform pair

$$\begin{bmatrix} 1 & 0 & 0 & 0 \\ 0 & 1 & 0 & 0 \\ 0 & 0 & 1 & 0 \\ 0 & 0 & 0 & 1 \end{bmatrix} \quad \begin{matrix} \text{row-by-row or} \\ \text{column-by-column} \\ \text{DFT} \\ \rightarrow \end{matrix} \quad \begin{bmatrix} 1 & 1 & 1 & 1 \\ 1 & -j & -1 & j \\ 1 & -1 & 1 & -1 \\ 1 & j & -1 & -j \end{bmatrix}$$

In programming, the DFT transform matrix can be generated this way for any N. First, we construct the identity matrix of the required order by programming or using the command available in the software, such as MATLAB. Then, row by row or column by column of the identity matrix results in the DFT transform matrix.

With frequency 0, the DC signal $x(n) = e^{j0\frac{2\pi}{4}n}$, has all 1s, as its samples. That is, the samples of the 4-point DC signal are

$$\{x(0) = 1, x(1) = 1, x(2) = 1, x(3) = 1\}$$

Let us compute its DFT.

$$\begin{bmatrix} X(0) \\ X(1) \\ X(2) \\ X(3) \end{bmatrix} = \begin{bmatrix} 1 & 1 & 1 & 1 \\ 1 & -j & -1 & j \\ 1 & -1 & 1 & -1 \\ 1 & j & -1 & -j \end{bmatrix} \begin{bmatrix} 1 \\ 1 \\ 1 \\ 1 \end{bmatrix} = \begin{bmatrix} 4 \\ 0 \\ 0 \\ 0 \end{bmatrix}$$

The DFT spectrum is

$$\{X(0) = 4, X(1) = 0, X(2) = 0, X(3) = 0\}$$

with the only nonzero coefficient 4 at frequency index 0. The spectrum of the DC signal is nonzero only at $k = 0$, since its frequency index is zero. The IDFT of the spectrum gets back the input $x(n)$.

$$
\begin{bmatrix} x(0) \\ x(1) \\ x(2) \\ x(3) \end{bmatrix} = \frac{1}{4} \begin{bmatrix} 1 & 1 & 1 & 1 \\ 1 & j & -1 & -j \\ 1 & -1 & 1 & -1 \\ 1 & -j & -1 & j \end{bmatrix} \begin{bmatrix} 4 \\ 0 \\ 0 \\ 0 \end{bmatrix} = \begin{bmatrix} 1 \\ 1 \\ 1 \\ 1 \end{bmatrix}
$$

The 4-point alternating signal, $x(n) = e^{j2\frac{2\pi}{4}n} = (-1)^n$, has the pattern $\{1, -1\}$ repeated twice. Its samples are

$$
\{x(0) = 1, x(1) = -1, x(2) = 1, x(3) = -1\}
$$

Let us compute its DFT.

$$
\begin{bmatrix} X(0) \\ X(1) \\ X(2) \\ X(3) \end{bmatrix} = \begin{bmatrix} 1 & 1 & 1 & 1 \\ 1 & -j & -1 & j \\ 1 & -1 & 1 & -1 \\ 1 & j & -1 & -j \end{bmatrix} \begin{bmatrix} 1 \\ -1 \\ 1 \\ -1 \end{bmatrix} = \begin{bmatrix} 0 \\ 0 \\ 4 \\ 0 \end{bmatrix}
$$

The DFT spectrum is

$$
X(k) = 4\delta(k - 2) = \{X(0) = 0, X(1) = 0, X(2) = 4, X(3) = 0\}
$$

As its frequency index is 2, its spectrum is 4 at $k = 2$ and zero otherwise. Computing the IDFT of the spectrum, we get back the input $x(n)$.

$$
\begin{bmatrix} x(0) \\ x(1) \\ x(2) \\ x(3) \end{bmatrix} = \frac{1}{4} \begin{bmatrix} 1 & 1 & 1 & 1 \\ 1 & j & -1 & -j \\ 1 & -1 & 1 & -1 \\ 1 & -j & -1 & j \end{bmatrix} \begin{bmatrix} 0 \\ 0 \\ 4 \\ 0 \end{bmatrix} = \begin{bmatrix} 1 \\ -1 \\ 1 \\ -1 \end{bmatrix}
$$

The DFT transform pair for the complex exponential $e^{j\frac{2\pi}{N}k_0 n}$, which is the standard unit in Fourier analysis, is

$$
e^{j\frac{2\pi}{N}k_0 n} \leftrightarrow N\delta(k - k_0)
$$

and that of $e^{j(\frac{2\pi}{N}k_0 n + \psi)}$ is

$$
e^{j\left(\frac{2\pi}{N}k_0 n + \psi\right)} \leftrightarrow Ne^{j\psi}\delta(k - k_0)
$$

A real sinusoid, in terms of complex exponentials, can be expressed as

$$
\cos\left(\frac{2\pi}{N}k_0 n + \phi\right) = \frac{1}{2}\left(e^{j\phi}e^{j\frac{2\pi}{N}k_0 n} + e^{-j\phi}e^{-j\frac{2\pi}{N}k_0 n}\right)
$$

Therefore, its DFT transform pair is given by

$$\cos\left(\frac{2\pi}{N}k_0n + \phi\right) \leftrightarrow \frac{N}{2}\left(e^{j\phi}\delta(k - k_0) + e^{-j\phi}\delta(k + k_0)\right)$$

$$= \frac{N}{2}\left(e^{j\phi}\delta(k - k_0) + e^{-j\phi}\delta(k - (N - k_0))\right)$$

As special cases, the DFT transform pairs for cosine and signals, with $\phi = 0$ and $\phi = -\frac{\pi}{2}$, are

$$\cos\left(\frac{2\pi}{N}k_0n\right) \leftrightarrow \frac{N}{2}\left(\delta(k - k_0) + \delta(k - (N - k_0))\right)$$

$$\sin\left(\frac{2\pi}{N}k_0n\right) \leftrightarrow \frac{N}{2}\left(-j\delta(k - k_0) + j\delta(k - (N - k_0))\right)$$

The samples of one period of the complex exponential $x(n) = e^{j1\frac{2\pi}{4}n}$ are

$$\{x(0) = 1, x(1) = j, x(2) = -1, x(3) = -j\}$$

Let us compute its DFT.

$$\begin{bmatrix} X(0) \\ X(1) \\ X(2) \\ X(3) \end{bmatrix} = \begin{bmatrix} 1 & 1 & 1 & 1 \\ 1 & -j & -1 & j \\ 1 & -1 & 1 & -1 \\ 1 & j & -1 & -j \end{bmatrix} \begin{bmatrix} 1 \\ j \\ -1 \\ -j \end{bmatrix} = \begin{bmatrix} 0 \\ 4 \\ 0 \\ 0 \end{bmatrix}$$

Since the frequency is 1, we get $X(1) = 4$ and the rest zero. Computing the IDFT, we get back the exponential

$$\begin{bmatrix} x(0) \\ x(1) \\ x(2) \\ x(3) \end{bmatrix} = \frac{1}{4}\begin{bmatrix} 1 & 1 & 1 & 1 \\ 1 & j & -1 & -j \\ 1 & -1 & 1 & -1 \\ 1 & -j & -1 & j \end{bmatrix} \begin{bmatrix} 0 \\ 4 \\ 0 \\ 0 \end{bmatrix} = \begin{bmatrix} 1 \\ j \\ -1 \\ -j \end{bmatrix}$$

The samples of one period of the complex exponential $x(n) = 2e^{j(\frac{2\pi}{4}n - \frac{\pi}{3})} = 2e^{-j\frac{\pi}{3}}e^{j\frac{2\pi}{4}n}$ are

$$\{x(0) = 1 - j\sqrt{3}, x(1) = \sqrt{3} + j1, x(2) = -1 + j\sqrt{3}, x(3) = -\sqrt{3} - j1\}$$

Let us compute its DFT.

$$
\begin{bmatrix} X(0) \\ X(1) \\ X(2) \\ X(3) \end{bmatrix} = \begin{bmatrix} 1 & 1 & 1 & 1 \\ 1 & -j & -1 & j \\ 1 & -1 & 1 & -1 \\ 1 & j & -1 & -j \end{bmatrix} \begin{bmatrix} 1-j\sqrt{3} \\ \sqrt{3}+j1 \\ -1+j\sqrt{3} \\ -\sqrt{3}-j1 \end{bmatrix} = \begin{bmatrix} 0 \\ 4(1-j\sqrt{3}) \\ 0 \\ 0 \end{bmatrix}
$$

$$
\begin{bmatrix} x(0) \\ x(1) \\ x(2) \\ x(3) \end{bmatrix} = \frac{1}{4} \begin{bmatrix} 1 & 1 & 1 & 1 \\ 1 & j & -1 & -j \\ 1 & -1 & 1 & -1 \\ 1 & -j & -1 & j \end{bmatrix} \begin{bmatrix} 0 \\ 4(1-j\sqrt{3}) \\ 0 \\ 0 \end{bmatrix} = \begin{bmatrix} 1-j\sqrt{3} \\ \sqrt{3}+j1 \\ -1+j\sqrt{3} \\ -\sqrt{3}-j1 \end{bmatrix}
$$

The samples of one period of the cosine waveform $x(n) = 2\cos(\frac{2\pi}{4}n)$ are $\{x(0) = 2, x(1) = 0, x(2) = -2, x(3) = 0\}$. Let us compute its DFT.

$$
\begin{bmatrix} X(0) \\ X(1) \\ X(2) \\ X(3) \end{bmatrix} = \begin{bmatrix} 1 & 1 & 1 & 1 \\ 1 & -j & -1 & j \\ 1 & -1 & 1 & -1 \\ 1 & j & -1 & -j \end{bmatrix} \begin{bmatrix} 2 \\ 0 \\ -2 \\ 0 \end{bmatrix} = \begin{bmatrix} 0 \\ 4 \\ 0 \\ 4 \end{bmatrix}
$$

Since

$$
2\cos\left(\frac{2\pi}{4}n\right) = e^{j\frac{2\pi}{4}n} + e^{-j\frac{2\pi}{4}n},
$$

$X(1) = 4$ and $X(3) = 4$. Computing the IDFT of the spectrum, we get back the cosine waveform

$$
\begin{bmatrix} x(0) \\ x(1) \\ x(2) \\ x(3) \end{bmatrix} = \frac{1}{4} \begin{bmatrix} 1 & 1 & 1 & 1 \\ 1 & j & -1 & -j \\ 1 & -1 & 1 & -1 \\ 1 & -j & -1 & j \end{bmatrix} \begin{bmatrix} 0 \\ 4 \\ 0 \\ 4 \end{bmatrix} = \begin{bmatrix} 2 \\ 0 \\ -2 \\ 0 \end{bmatrix}
$$

The samples of one period of the sine waveform $x(n) = 4\sin(\frac{2\pi}{4}n)$ are $\{x(0) = 0, x(1) = 4, x(2) = 0, x(3) = -4\}$. Let us compute its DFT.

$$
\begin{bmatrix} X(0) \\ X(1) \\ X(2) \\ X(3) \end{bmatrix} = \begin{bmatrix} 1 & 1 & 1 & 1 \\ 1 & -j & -1 & j \\ 1 & -1 & 1 & -1 \\ 1 & j & -1 & -j \end{bmatrix} \begin{bmatrix} 0 \\ 4 \\ 0 \\ -4 \end{bmatrix} = \begin{bmatrix} 0 \\ -j8 \\ 0 \\ j8 \end{bmatrix}
$$

Since

$$
4\sin\left(\frac{2\pi}{4}n\right) = -j\left(2e^{j\frac{2\pi}{4}n} - 2e^{-j\frac{2\pi}{4}n}\right)
$$

$X(1) = -j8$ and $X(3) = j8$. Computing the IDFT of the spectrum, we get back the sine waveform

$$
\begin{bmatrix} x(0) \\ x(1) \\ x(2) \\ x(3) \end{bmatrix} = \frac{1}{4} \begin{bmatrix} 1 & 1 & 1 & 1 \\ 1 & j & -1 & -j \\ 1 & -1 & 1 & -1 \\ 1 & -j & -1 & j \end{bmatrix} \begin{bmatrix} 0 \\ -j8 \\ 0 \\ j8 \end{bmatrix} = \begin{bmatrix} 0 \\ 4 \\ 0 \\ -4 \end{bmatrix}
$$

The samples of one period of the sinusoid $x(n) = 2\cos(\frac{2\pi}{4}n + \frac{\pi}{6})$ are $\{x(0) = \sqrt{3}, x(1) = -1, x(2) = -\sqrt{3}, x(3) = 1\}$. Let us compute its DFT

$$
\begin{bmatrix} X(0) \\ X(1) \\ X(2) \\ X(3) \end{bmatrix} = \begin{bmatrix} 1 & 1 & 1 & 1 \\ 1 & -j & -1 & j \\ 1 & -1 & 1 & -1 \\ 1 & j & -1 & -j \end{bmatrix} \begin{bmatrix} \sqrt{3} \\ -1 \\ -\sqrt{3} \\ 1 \end{bmatrix} = \begin{bmatrix} 0 \\ 2\sqrt{3} + j2 \\ 0 \\ 2\sqrt{3} - j2 \end{bmatrix}
$$

Let us compute the IDFT of the spectrum

$$
\begin{bmatrix} x(0) \\ x(1) \\ x(2) \\ x(3) \end{bmatrix} = \frac{1}{4} \begin{bmatrix} 1 & 1 & 1 & 1 \\ 1 & j & -1 & -j \\ 1 & -1 & 1 & -1 \\ 1 & -j & -1 & j \end{bmatrix} \begin{bmatrix} 0 \\ 2\sqrt{3} + j2 \\ 0 \\ 2\sqrt{3} - j2 \end{bmatrix} = \begin{bmatrix} \sqrt{3} \\ -1 \\ -\sqrt{3} \\ 1 \end{bmatrix}
$$

The DFT of the samples shown in Fig. 3.1a

$$\{x(0) = 0.5000, \ x(1) = 1.1340, \ x(2) = -0.5000, \ x(3) = 2.8660\}$$

is computed as

$$
\begin{bmatrix} X(0) \\ X(1) \\ X(2) \\ X(3) \end{bmatrix} = \begin{bmatrix} 1 & 1 & 1 & 1 \\ 1 & -j & -1 & j \\ 1 & -1 & 1 & -1 \\ 1 & j & -1 & -j \end{bmatrix} \begin{bmatrix} 0.5 \\ 1.1340 \\ -0.5 \\ 2.8660 \end{bmatrix} = \begin{bmatrix} 4 \\ 1.0000 + j1.7321 \\ -4 \\ 1.0000 - j1.7321 \end{bmatrix}
$$

The DFT spectrum, as shown in Fig. 3.1b, is

$$\{X(0) = 4, X(1) = 1.0000 + j1.7321, X(2) = -4, X(3) = 1.0000 - j1.7321\}$$

Using the IDFT, we get back the input $x(n)$.

$$
\begin{bmatrix} x(0) \\ x(1) \\ x(2) \\ x(3) \end{bmatrix} = \frac{1}{4} \begin{bmatrix} 1 & 1 & 1 & 1 \\ 1 & j & -1 & -j \\ 1 & -1 & 1 & -1 \\ 1 & -j & -1 & j \end{bmatrix} \begin{bmatrix} 4 \\ 1.0000 + j1.7321 \\ -4 \\ 1.0000 - j1.7321 \end{bmatrix} = \begin{bmatrix} 0.5 \\ 1.1340 \\ -0.5 \\ 2.8660 \end{bmatrix}
$$

From the closed-form identity for the geometric summation, we get, with $M \leq N$,

$$
\sum_{n=0}^{M-1} e^{-j\frac{2\pi}{N}kn} = \frac{1 - e^{-j\frac{2\pi}{N}Mk}}{1 - e^{-j\frac{2\pi}{N}k}}
$$

The rectangular waveform is defined as

$$
x(n) = \begin{cases} 1 & \text{for } 0 \leq n \leq M - 1 \\ 0 & \text{for } M \leq n \leq N - 1 \end{cases}
$$

Let us compute its DFT

$$
X(k) = \sum_{n=0}^{M-1} e^{-j\frac{2\pi}{N}nk} = \frac{1 - e^{-j\frac{2\pi}{N}Mk}}{1 - e^{-j\frac{2\pi}{N}k}} = \frac{e^{j\frac{2\pi}{N}\frac{M}{2}k} - e^{-j\frac{2\pi}{N}\frac{M}{2}k}}{e^{j\frac{2\pi}{N}\frac{k}{2}} - e^{-j\frac{2\pi}{N}\frac{k}{2}}}
$$

$$
= e^{-j\frac{\pi}{N}(M-1)k} \frac{\sin\left(\frac{\pi}{N}Mk\right)}{\sin\left(\frac{\pi}{N}k\right)}
$$

Let us use this formula to compute the DFT of the DC signal. With $M = N$, $X(0) = N$ and zero otherwise, as obtained earlier. Let us use this formula to compute the DFT of the impulse signal $\delta(n)$. With $M = 1$, $X(k) = 1$ for all k, as obtained earlier.

Let us compute the DFT of the signal $\{x(0) = 1, x(1) = 1, x(2) = 1, x(3) = 0\}$

$$
\begin{bmatrix} X(0) \\ X(1) \\ X(2) \\ X(3) \end{bmatrix} = \begin{bmatrix} 1 & 1 & 1 & 1 \\ 1 & -j & -1 & j \\ 1 & -1 & 1 & -1 \\ 1 & j & -1 & -j \end{bmatrix} \begin{bmatrix} 1 \\ 1 \\ 1 \\ 0 \end{bmatrix} = \begin{bmatrix} 3 \\ -j \\ 1 \\ j \end{bmatrix}
$$

Let us compute the IDFT of the spectrum

$$
\begin{bmatrix} x(0) \\ x(1) \\ x(2) \\ x(3) \end{bmatrix} = \frac{1}{4} \begin{bmatrix} 1 & 1 & 1 & 1 \\ 1 & j & -1 & -j \\ 1 & -1 & 1 & -1 \\ 1 & -j & -1 & j \end{bmatrix} \begin{bmatrix} 3 \\ -j \\ 1 \\ j \end{bmatrix} = \begin{bmatrix} 1 \\ 1 \\ 1 \\ 0 \end{bmatrix}
$$

While practical signals are often real-valued, the complex signals are also often used in signal and system analysis. The DFT is formulated using the complex exponentials as the basis function. The samples of a complex signal are $\{x(0) = 1 - j1, x(1) = 2 + j1, x(2) = 3 + j1, x(3) = 1 - j2\}$ and

$$
\begin{bmatrix} X(0) \\ X(1) \\ X(2) \\ X(3) \end{bmatrix} = \begin{bmatrix} 1 & 1 & 1 & 1 \\ 1 & -j & -1 & j \\ 1 & -1 & 1 & -1 \\ 1 & j & -1 & -j \end{bmatrix} \begin{bmatrix} 1 - j1 \\ 2 + j1 \\ 3 + j1 \\ 1 - j2 \end{bmatrix} = \begin{bmatrix} 7 - j1 \\ 1 - j3 \\ 1 + j1 \\ -5 - j1 \end{bmatrix}
$$

Let us compute the IDFT of the spectrum

$$
\begin{bmatrix} x(0) \\ x(1) \\ x(2) \\ x(3) \end{bmatrix} = \frac{1}{4} \begin{bmatrix} 1 & 1 & 1 & 1 \\ 1 & j & -1 & -j \\ 1 & -1 & 1 & -1 \\ 1 & -j & -1 & j \end{bmatrix} \begin{bmatrix} 7 - j1 \\ 1 - j3 \\ 1 + j1 \\ -5 - j1 \end{bmatrix} = \begin{bmatrix} 1 - j1 \\ 2 + j1 \\ 3 + j1 \\ 1 - j2 \end{bmatrix}
$$

3.3 Properties of the DFT

Properties make the analysis easier and also make the derivation of transform of signals easier by decomposing them in terms of simpler signals.

Linearity
Signal and system analysis using the DFT is based on the linearity property. The fast algorithms, due to which the DFT has become the workhorse in DSP, are also based on the linearity property

$$
x(n) \leftrightarrow X(k), \; y(n) \leftrightarrow Y(k) \rightarrow ax(n) + by(n) \leftrightarrow aX(k) + bY(k)
$$

where a and b are the arbitrary constants. The DFT of a linear combination of two signals is the same linear combination of the DFT of the two signals. For example,

$$
\{2, 3\} \leftrightarrow \{5, -1\}, \; \{1, -2\} \leftrightarrow \{-1, 3\} \rightarrow 3\{2, 3\} + 2\{1, -2\}
$$
$$
\leftrightarrow 3\{5, -1\} + 2\{-1, 3\} = \{13, 3\}
$$

That is,

$$
\{8, 5\} \leftrightarrow \{13, 3\}
$$

Periodicity
With $x(n) \leftrightarrow X(k)$, both $x(n)$ and $X(k)$ are periodic with period N. That is,

$$x(n + cN) = x(n), \text{ for all } n \quad \text{and} \quad X(k + cN) = X(k), \text{ for all } k$$

where c is an arbitrary integer. Let us replace k by $k + N$ in the DFT definition. Then, we get

$$X(k + N) = \sum_{n=0}^{N-1} x(n) e^{-j(k+N)\frac{2\pi}{N}n} = \sum_{n=0}^{N-1} x(n) e^{-jk\frac{2\pi}{N}n} = X(k),$$

due to the fact that

$$e^{-j(N)\frac{2\pi}{N}n} = 1, \quad \text{for all } n$$

Similarly, if we replace n by $n + N$ in the IDFT definition, we get the same $x(n)$.

Circular Time Shifting

In shifting a signal, the magnitude spectrum remains the same, while the phase spectrum gets changed. Let $x(n)$ be the signal with period N and $X(k)$ is its spectrum. If we shift a frequency component with frequency index 1 by one sample interval, the phase change is $2\pi/N$ radians. Therefore, if we shift a frequency component with frequency index k by n_0 sample intervals, the phase change is $2\pi k n_0/N$ radians. This property is given by

$$x(n \pm n_0) \leftrightarrow e^{\pm j\frac{2\pi}{N}kn_0} X(k)$$

The shift of a signal in the time domain by n_0 sample intervals shifts the origin of the signal by the same amount. In the frequency domain, the phase of each frequency component is changed proportionately depending on the frequency index. Such a proportional change does not affect the form of the signal, but the whole signal is relocated. There are only $N - 1$ unique shifts for a N-point periodic signal.

For example,

$$x(n) = \{\check{2}, 3, 1, 4\} \leftrightarrow X(k) = \{\check{10}, 1 + j1, -4, 1 - j1\}$$
$$x(n-1) = \{\check{4}, 2, 3, 1\} \leftrightarrow X(k) = \{\check{10}, (1 + j1)(-j), -4(-1), (1 - j1)(j)\}$$
$$= \{\check{10}, 1 - j1, 4, 1 + j1\}$$

The index of the signal sample with a check symbol $\check{}$ is zero. Note that $X(k)$ is the same for $x(n - 1) = x(n - 5) = x(n - 9) = \cdots$, due to periodicity.

Circular Frequency Shifting

Let $x(n)$ be the signal with period N and $X(k)$ is its spectrum. In shifting a spectrum by replacing the frequency index k by $k + k_0$, the spectrum just gets shifted, where k_0 is an arbitrary number of sampling intervals. To shift a spectrum by k_0, the time-domain signal has to be multiplied by a complex exponential with frequency index by k_0. Therefore,

$$e^{\pm j\frac{2\pi}{N}k_0 n} x(n) \leftrightarrow X(k \mp k_0)$$

There are only $N-1$ unique shifts for a N-point periodic signal.
    For example,

$$x(n) = \{\breve{1}, 2, 3, 4\} \leftrightarrow X(k) = \{\breve{10}, -2+j2, -2, -2-j2\}$$

$$e^{j\frac{2\pi}{4}n} x(n) = \{\breve{1}, j2, -3, -j4\} \leftrightarrow X(k) = \{-2-j2, \breve{10}, -2+j2, -2\}$$

$$e^{j\frac{2\pi}{4}2n} x(n) = (-1)^n x(n) = \{\breve{1}, -2, 3, -4\} \leftrightarrow X(k) = \{-2, -2-j2, \breve{10}, -2+j2\}$$

This frequency shift by $N/2$ sample intervals is often used to find the center-zero spectrum.

Circular Time Reversal
Let the N-point DFT of $x(n)$ be $X(k)$. The DFT of the time reversal $x(N-n)$ of $x(n)$ is

$$\begin{bmatrix} X(0) \\ X(3) \\ X(2) \\ X(1) \end{bmatrix} = \begin{bmatrix} 1 & 1 & 1 & 1 \\ 1 & -j & -1 & j \\ 1 & -1 & 1 & -1 \\ 1 & j & -1 & -j \end{bmatrix} \begin{bmatrix} x(0) \\ x(3) \\ x(2) \\ x(1) \end{bmatrix} = X(N-k)$$

Therefore,

$$x(N-n) \leftrightarrow X(N-k)$$

For example,

$$x(n) = \{\breve{4}, 0, 3, 1\} \leftrightarrow X(k) = \{\breve{8}, 1+j1, 6, 1-j1\}$$

$$x(4-n) = \{\breve{4}, 1, 3, 0\} \leftrightarrow X(4-k) = \{\breve{8}, 1-j1, 6, 1+j1\}$$

Duality
The DFT and IDFT operations are almost similar. The differences are that the sign of the exponents in the definitions differs and there is a constant in the IDFT definition. Let the DFT of $x(n)$ be $X(k)$ with period N. Then, due to the dual nature of the definitions, we can interchange the independent variables and make the time-domain function as the DFT of the frequency-domain function with some minor changes. That is, if we compute the DFT of $X(n)$, then we get $Nx(N-k)$. That is,

$$X(n) \leftrightarrow Nx(N-k)$$

For example, the DFT of $\{\check{-}1, 1, 3, 2\}$ is $\{\check{5}, -4 + j1, -1, -4 - j1\}$. The DFT of this is $4\{-1, 2, 3, 1\}$.

Transform of Complex Conjugates

Let the DFT of $x(n)$ be $X(k)$ with period N. Conjugating both sides of the DFT definition, we get

$$X^*(k) = \sum_{n=0}^{N-1} x^*(n) e^{j\frac{2\pi}{N}nk} = \sum_{n=0}^{N-1} x^*(N-n) e^{-j\frac{2\pi}{N}nk} \rightarrow x^*(N-n) \leftrightarrow X^*(k)$$

Conjugating both sides of the IDFT definition, we get

$$x^*(n) = \frac{1}{N} \sum_{k=0}^{N-1} X^*(k) e^{-jk\frac{2\pi}{N}n} = \frac{1}{N} \sum_{k=0}^{N-1} X^*(N-k) e^{jk\frac{2\pi}{N}n} \rightarrow x^*(n) \leftrightarrow X^*(N-k)$$

For example,

$$x(n) = \{1 \check{+} j2, 2 + j3, 0, -j\} \leftrightarrow X(k) = \{3 \check{+} j4, 5, -1, -3 + j4\}$$

$$x^*(n) = \{1 \check{-} j2, 2 - j3, 0, j\} \leftrightarrow X^*(4-k) = \{3 \check{-} j4, -3 - j4, -1, 5\}$$

$$x^*(4-n) = \{1 \check{-} j2, j, 0, 2 - j3\} \leftrightarrow X^*(k) = \{3 \check{-} j4, 5, -1, -3 - j4\}$$

Circular Convolution in the Time Domain

As the sequences are considered periodic in DFT, the convolution of two signals using the DFT is circular or periodic convolution. The circular convolution of two time-domain sequences $x(n)$ and $h(n)$, resulting in the output sequence $y(n)$ is defined as

$$y(n) = \sum_{m=0}^{N-1} x(m)h(n-m) = \sum_{m=0}^{N-1} h(m)x(n-m), \quad n = 0, 1, \ldots, N-1$$

Note that all the sequences are of the same length N. In practice, linear convolution is more often required. To speed up this computation, we use the DFT with zero-padded sequences.

In the last chapter, we found that the convolution of two complex exponential sequences of the same frequency is of the same form with the complex coefficient being the product of the individual coefficients of the exponentials convolved. The convolution of two complex exponentials of different harmonic frequencies is zero. Therefore, the IDFT of the product of the respective coefficients at each frequency component yields the convolution output. Let $x(n) \leftrightarrow X(k)$ and $h(n) \leftrightarrow H(k)$, both with period N. Then,

$$y(n) = \frac{1}{N} \sum_{k=0}^{N-1} X(k)H(k)e^{j\frac{2\pi}{N}nk}$$

That is, the IDFT of $X(k)H(k)$ is the convolution of $x(n)$ and $h(n)$

$$x(n) \circledast h(n) \leftrightarrow X(k)H(k)$$

Therefore, the convolution is reduced to much simpler multiplication in the frequency domain. This is one of the major advantages of frequency-domain representation of signals, justifying the dominance of frequency-domain analysis. For example,

$$x(n) = \left\{ \check{1}, 2, 4, -3 \right\} \leftrightarrow X(k) = \left\{ \check{4}, -3 - j5, 6, -3 + j5 \right\}$$

$$h(n) = \left\{ \check{3}, 4, 1, 2 \right\} \leftrightarrow X(k) = \left\{ \check{10}, 2 - j2, -2, 2 + j2 \right\}$$

$$x(n) \circledast h(n) = \left\{ \check{-1}, 15, 15, 11 \right\} \leftrightarrow X(k)H(k) = \left\{ \check{40}, -16 - j4, -12, -16 + j4 \right\}$$

Linear Convolution Using the DFT
Use of the DFT to implement the convolution is advantageous, except for very short sequences. The computational complexities of implementing convolution in the frequency and time domains are, respectively, $O(N \log_2 N)$ and $O(N^2)$. The linear convolution of the sequences

$$x(n) = \left\{ \check{2}, 1, 4, 3 \right\} \quad \text{and} \quad h(n) = \left\{ 2, \check{1}, -3 \right\}$$

yields the output sequence

$$y(n) = \left\{ 4, \check{4}, 3, 7, -9, -9 \right\}$$

There are some constraints to carry out the same convolution using the DFT. The length of the linear convolution output of convolving two sequences is the sum of the length of the sequences minus 1. Therefore, in this case with lengths 4 and 3, the length of the output $y(n)$ is $4 + 3 - 1 = 6$. As DFT transforms a N-point signal to a N-point spectrum, the length of the input sequences must be increased to 6 by appending 2 zeros, called zero padding. However, practically efficient algorithms for the fast computation of the DFT are available only for lengths that are a power of 2. Consequently, we have to zero pad the input sequences to make it equal to the nearest power of 2. For the example, the input sequences must be zero padded to make their length to 8. Due to zero padding, the output will have two zeros at the end of the effective output. The zero-padded sequences are

$$\{\check{2}, 1, 4, 3, 0, 0, 0, 0\} \quad \text{and} \quad \{2, \check{1}, -3, 0, 0, 0, 0, 0\}$$

As the origins of the two sequences must be aligned, the second sequence is circularly shifted by one position left, to get

$$\{\check{1}, -3, 0, 0, 0, 0, 0, 2\}$$

Taking the 8-point DFT of the zero-padded and aligned sequences, with a precision of 4 digits, we get

$$X(k) = \{\check{10}, 0.5858 - j6.8284, -2 + j2, 3.4142 + 1.1716,$$
$$2, 3.4142 - 1.1716, -2 - j2, 0.5858 + j6.8284\}$$

$$H(k) = \{\check{0}, 0.2929 + j3.5355, 1 + j5, 1.7071 + j3.5355,$$
$$2, 1.7071 - j3.5355, 1 - j5, 0.2929 - j3.5355\}$$

Multiplying $X(k)$ and $H(k)$ term by term, we get

$$X(k)H(k) = \{\check{0}, 24.3137 + j0.0711, -12 - j8, 1.6863 + j14.0711,$$
$$4, 1.6863 - j14.0711, -12 + j8, 24.3137 - j0.0711\}$$

Taking the IDFT of $X(k)H(k)$, we get

$$\{\check{4}, 3, 7, -9, -9, 0, 0, 4\}$$

In this example, we shifted the second sequence to the left by one position. Therefore, circularly shifting the convolution output to right by one position, we get the linear convolution output appended by 2 zeros.

$$y(n) = \{4, \check{4}, 3, 7, -9, -9, 0, 0\}$$

Convolution of Long Sequences

In practice, the impulse response $h(n)$ is relatively much shorter than the input sequence $x(n)$. Zero padding and implementing the convolution has two drawbacks. One problem is that a large amount of memory is required. Another problem is that the time difference between the occurrence of an input sample and the corresponding output sample gets increased. In order to alleviate these problems, the implementation of convolution is modified in two ways. In this section, one of the two methods, called the overlap-save method, is presented. The computational complexity and memory requirements of both the methods are about the same.

The basic idea is to decompose the long sequence into short segments with lengths much longer than the shorter sequence. Then, convolution of a segment with the shorter sequence is mostly the same as the required output. At the beginning and the end of the segment, the output is distorted. However, these distortions can be eliminated by patching the convolution outputs of segments appropriately. The patching is carried out by overlapping the input or the output segments. In the overlap-save method, the input segments are overlapped. Then, the circular convolution of the segments with the shorter sequence is carried out. Part of the output is saved, and all the saved outputs are concatenated to form the required linear convolution output.

Let the length of the input and impulse response sequences be, respectively, N and Q. Then, for computing the convolution efficiently, the block length B should satisfy the constraint

$$N \gg B \gg Q$$

For illustrative purposes, we use short sequences. Let the input and impulse response sequences, respectively, be $x(n)$ and $h(n)$ with values

$$x(n) = \left\{ \overset{\smallsmile}{1}, 2, 3, 4 \right\} \quad \text{and} \quad h(n) = \left\{ \overset{\smallsmile}{2}, -1, 4, 3 \right\}$$

The output of linearly convolving the sequences $x(n)$ and $h(n)$ is

$$y(n) = \left\{ \overset{\smallsmile}{2}, 3, 8, 16, 14, 25, 12 \right\}$$

The overlap-save method of convolution of these sequences is shown in Fig. 3.3. With $N = Q = 4$, the convolution output sequence length is $N + Q - 1 = 7$. Let the block length B be 8. As the first three values of circular convolution are corrupt, for each input block, we get 5 valid outputs. As there are 7 values in the output, we have to divide the input into 2 overlapping segments. These segments are circularly convolved with the zero-padded impulse response sequence $h(n)$. The input is prepended with $Q - 1$ zeros at the beginning and appended with zeros to make a complete block at the end.

The DFT of the zero-padded $h(n)$ is $\{2, -1, 4, 3, 0, 0, 0, 0\}$. The DFT of this sequence, which is computed only once and stored, is

$$\{8, -0.8284 - j5.4142, -2 + j4, 4.8284 + j2.5858, 4,$$

$$4.8284 - j2.5858, -2 - j4, -0.8284 + j5.4142\}$$

The first block of extended $x(n)$ is $\{0, 0, 0, 1, 2, 3, 4, 0\}$. The DFT of this block, with a precision of 4 digits, is

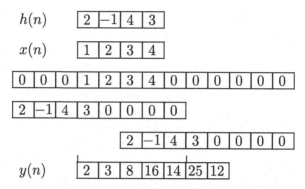

Fig. 3.3 Illustration of implementing convolution using the overlap-save method

$$\{10, -4.8284 + j5.4142, -2 - j2, 0.8284 - j2.5858,$$

$$2, 0.8284 + j2.5858, -2 + j2, -4.8284 - j5.4142\}$$

The pointwise product of the two DFTs is

$$\{80, 33.3137 + j21.6569, 12 - j4, 10.6863 - j10.3431,$$

$$8, 10.6863 + j10.3431, 12 + j4, 33.3137 - j21.6569\}$$

The IDFT of the product is

$$\{25, 12, 0, 2, 3, 8, 16, 14\}$$

The last five values $\{2, 3, 8, 16, 14\}$ are the first five values of linearly convolving $x(n)$ and $h(n)$. The second block has to overlap the first block by three samples. Therefore, the second block values are $\{3, 4, 0, 0, 0, 0, 0, 0\}$. The DFT of this block, with a precision of 4 digits, is

$$\{7, 5.8284 - j2.8284, 3 - j4, 0.1716 - j2.8284,$$

$$-1, 0.1716 + j2.8284, 3 + j4, 5.8284 + j2.8284\}$$

The pointwise product of this DFT with that of zero-padded $h(n)$ is

$$\{56, -20.1421 - j29.2132, 10 + j20, 8.1421 - j13.2132,$$

$$-4, 8.1421 + j13.2132, 10 - j20, -20.1421 + j29.2132\}$$

The IDFT of the product is

$$\{6, 5, 8, 25, 12, 0, 0, 0\}$$

The first two of the last five values $\{25, 12, 0, 0, 0\}$ are the last two·values of the linear convolution of $x(n)$ and $h(n)$.

Circular Convolution in the Frequency Domain
Let $x(n) \leftrightarrow X(k)$ and $h(n) \leftrightarrow H(k)$ with period N. Then, using the duality property of the DFT and the time-domain convolution property, we get

$$x(n)h(n) \leftrightarrow \frac{1}{N}(X(k) \circledast H(k))$$

For example,

$$x(n) = \{\check{2}, 3, -1, 4\} \leftrightarrow X(k) = \{\check{8}, 3 + j1, -6, 3 - j1\}$$

$$h(n) = \{\check{1}, 4, -2, 3\} \leftrightarrow H(k) = \{\check{6}, 3 - j1, -8, 3 + j1\}$$

$$x(n)h(n) = \{\check{2}, 12, 2, 12\} \leftrightarrow \frac{1}{4}X(k) \circledast H(k) = \frac{1}{4}\{\check{2}8, 0, -20, 0\}$$

Circular Correlation in the Time Domain
The correlation operation, while similar to convolution, is a similarity measure between two signals. As it is similar, it is usually implemented using the same way as convolution with some minor changes. In convolution, we repeatedly use time reverse, shift, multiply, and add operations. Except for the time reversal, the other three operations are involved in the computation of correlation. Therefore, if we take the time reversal of a sequence and convolve, we get the correlation output. One important difference is correlation operation, in general, is not commutative. Signal restoration and detection are two major applications of correlation. If both the signals are the same, correlation is called autocorrelation. Otherwise, it is a cross-correlation. Autocorrelation is related to power spectral density.

Let $x(n) \leftrightarrow X(k)$ and $h(n) \leftrightarrow H(k)$, both with period N. The circular cross-correlation of $x(n)$ and $h(n)$ in the time and frequency domains is given by

$$r_{xh}(n) = \sum_{p=0}^{N-1} x(p)h^*(p - n), n = 0, 1, \ldots, N - 1 \leftrightarrow X(k)H^*(k)$$

For complex signals, similarity is defined with one of them conjugated. Due to conjugate property, $h^*(N - n) \leftrightarrow H^*(k)$. Therefore, correlation operation is the same as convolution of $x(n)$ and $h^*(N - n)$.

$$r_{hx}(n) = r_{xh}(N - n) = \text{IDFT}(X^*(k)H(k))$$

For example,

$$x(n) = \{\overset{\vee}{4}, 3, 2, 1\} \leftrightarrow X(k) = \{\overset{\vee}{10}, 2 - j2, 2, 2 + j2\}$$

$$h(n) = \{\overset{\vee}{3}, 2, -1, 4\} \leftrightarrow H(k) = \{\overset{\vee}{8}, 4 + j2, -4, 4 - j2\}$$

The cross-correlation output of $x(n)$ and $h(n)$ and its DFT are

$$r_{xh}(n) = \{\overset{\vee}{20}, 28, 16, 16\} \leftrightarrow X(k)H^*(k) = \{\overset{\vee}{80}, 4 - j12, -8, 4 + j12\}$$

The cross-correlation output of $h(n)$ and $x(n)$ and its DFT are

$$r_{hx}(n) = \{\overset{\vee}{20}, 16, 16, 28\} \leftrightarrow H(k)X^*(k) = \{\overset{\vee}{80}, 4 + j12, -8, 4 - j12\}$$

$$r_{xx}(n) = \text{IDFT}(|X(k)|^2)$$

The autocorrelation of $x(n)$ is

$$\{\overset{\vee}{30}, 24, 22, 24\} \leftrightarrow |X(k)|^2 = \{\overset{\vee}{100}, 8, 4, 8\}$$

Sum and Difference of Sequences
As the twiddle factors are all 1s with the frequency index $k = 0$, the coefficient of the DC frequency component $X(0)$ is just the sum of the time-domain samples $x(n)$. Let the transform length N be even. As the twiddle factors are all alternating 1s and -1s with the frequency index $k = N/2$, the coefficient $X(N/2)$ is just the difference of the sum of the even and odd time-domain samples $x(n)$. These values computed this way are a check on the DFT computation

$$X(0) = \sum_{n=0}^{N-1} x(n) \quad \text{and} \quad X\left(\frac{N}{2}\right) = \sum_{n=0,2}^{N-2} x(n) - \sum_{n=1,3}^{N-1} x(n)$$

Similarly, in the frequency domain,

$$x(0) = \frac{1}{N}\sum_{k=0}^{N-1} X(k) \quad \text{and} \quad x\left(\frac{N}{2}\right) = \frac{1}{N}\left(\sum_{k=0,2}^{N-2} X(k) - \sum_{k=1,3}^{N-1} X(k)\right)$$

For example, let $\{\overset{\vee}{4}, 2, -1, 3\} \leftrightarrow \{\overset{\vee}{8}, 5 + j1, -2, 5 - j1\}$. We can verify the formulas using this transform pair.

Upsampling of a Sequence
Consider the sequence and its DFT

$$x(n) = \left\{2 \overset{\smile}{+} j3, 1 - j2\right\} \leftrightarrow X(k) = \left\{3 \overset{\smile}{+} j1, 1 + j5\right\}$$

Let us upsample $x(n)$ by a factor of 2 to get

$$x_u(n) = \left\{2 \overset{\smile}{+} j3, 0, 1 - j2, 0\right\} \leftrightarrow X_u(k) = \left\{3 \overset{\smile}{+} j1, 1 + j5, 3 + j1, 1 + j5\right\}$$

The spectrum is repeated.
 In general, with

$$x(n) \leftrightarrow X(k), \; n, k = 0, 1, \ldots, N - 1$$

and a positive integer upsampling factor L,

$$x_u(n) = \begin{cases} x(\frac{n}{L}) & \text{for } n = 0, L, 2L, \ldots, L(N-1) \\ 0 & \text{otherwise} \end{cases} \leftrightarrow X(k) = X(k \bmod N),$$

$$k = 0, 1, \ldots, LN - 1$$

The spectrum $X(k)$ is repeated L times.
 The same thing happens in the upsampling of a spectrum, except for a constant factor in the amplitude of the time-domain signal. Consider the sequence and its DFT

$$x(n) = \left\{2 \overset{\smile}{+} j3, 1 - j2\right\} \leftrightarrow X(k) = \left\{3 \overset{\smile}{+} j1, 1 + j5\right\}$$

Let us upsample $X(k)$ by a factor of 2 to get

$$X_u(k) = \left\{3 \overset{\smile}{+} j1, 0, 1 + j5, 0\right\} \leftrightarrow \left\{2 \overset{\smile}{+} j3, 1 - j2, 2 + j3, 1 - j2\right\}/2$$

The time-domain sequence is repeated.

Zero Padding the Data
Appending zeros to a sequence $x(n)$ is carried out mostly for two purposes. As the practically fast DFT algorithms are of length 2, $x(n)$ is zero padded to meet this constraint. Another purpose is to make the spectrum denser, as the frequency increment in the spectrum is inversely proportional to the length of the time-domain sequence. Sufficient number of zeros should be appended so that all the essential features, such as a peak, are adequately represented. While any number of zeros can be appended, we present the case of making the signal longer by an integer number of times, L.
 Let $L = 2$ and $x(n) = \{\overset{\smile}{3}, 2, -1, 4\}$. $X(k) = \{\overset{\smile}{8}, 4 + j2, -4, 4 - j2\}$. Then

$$x_z(n) = \left\{\overset{\smile}{3}, 2, -1, 4, 0, 0, 0, 0\right\} \leftrightarrow X_z(k) = \left\{\overset{\smile}{8}, *, 4 + j2, *, -4, *, 4 - j2, *\right\}$$

With lengths of the sequences being 4 and 8, the DFT computes the coefficients at frequencies

$$\{0, 1, 2, 3\}/4 \quad \text{and} \quad \{0, 1, 2, 3, 4, 5, 6, 7\}/8$$

Therefore, the even-indexed DFT coefficients of $x_z(n)$ are the same as that of $X(k)$. With 8 frequency components, the spectrum is denser.

In general, with $x(n) \leftrightarrow X(k)$, $n, k = 0, 1, \ldots, N - 1$,

$$x_z(n) = \begin{cases} x(n) \text{ for } n = 0, 1, \ldots, N - 1 \\ 0 \quad \text{ for } n = N, N + 1, \ldots, LN - 1 \end{cases} \leftrightarrow X_z(Lk)$$

$$= X(k), \ k = 0, 1, \ldots, N - 1$$

Similarly, the zero padding of a spectrum results in a denser and scaled time-domain signal. For example, let $x(n) = \{\check{2}, 3\} \leftrightarrow \{\check{5}, -1\}$. Then,

$$x_z(n) = \left\{\check{1}, 1.25 - j0.25, 1.5, 1.25 + j0.25\right\} \leftrightarrow \left\{\check{5}, -1, 0, 0\right\}$$

The even-indexed samples of $x_z(n)$ are the same as that of $x(n)$ divided by 2.

Symmetry Properties

Real-Valued Signals
Since a real signal is a linear combination of complex exponentials and their conjugates, we need two complex exponentials to represent a frequency component and, therefore, two coefficients with each of them being the conjugate of the other. The spectrum of real signal is also complex-valued. However, as there are only N independent values in a N-point real-valued signal, and its spectrum composed of N complex values, the spectrum has to be redundant by a factor of 2. For example,

$$x(n) = \{1, 4, 3, 2\} \leftrightarrow X(k) = \{10, -2 - j2, -2, -2 + j2\}$$

This symmetry is called conjugate or Hermitian symmetry. In general,

$$X(k) = X^*(N - k)$$

Remember that the DFT spectrum is N-periodic. In terms of real and imaginary parts, we get

$$X_r(k) = X_r(N - k) \quad \text{and} \quad X_i(k) = -X_i(N - k) \text{ or}$$

$$X_r\left(\frac{N}{2} - k\right) = X_r\left(\frac{N}{2} + k\right) \quad \text{and} \quad X_i\left(\frac{N}{2} - k\right) = -X_i\left(\frac{N}{2} + k\right)$$

$$x(n) \text{ real} \leftrightarrow X(k) \text{ hermitian}$$

This redundancy is not a problem, since the complex spectrum of a real signal requires the same amount of storage due to redundancy of the spectrum and the computational complexity can also be made about one-half of that of the complex signal.

Even and Odd Symmetry
The sine function is odd, while the cosine function is even. Therefore, the sum of the product of an even function with the sine function over an integral number of cycles is zero, and the DFT of an even function is even with cosine components only. For example,

$$x(n) = \{1, 4, 3, 2, 3, 2, 3, 4\} \leftrightarrow X(k)$$
$$= \{22, 0.8284, -2, -4.8284, -2, -4.8284, -2, 0.8284\}$$

That is,

$$x(n) \text{ real and even} \leftrightarrow X(k) \text{ real and even}$$

The sum of the product of an odd function with the cosine function over an integral number of cycles is zero, Therefore, the DFT of an odd function is odd with sine components only. For example,

$$x(n) = \{0, 4, 3, 2, 0 - 2 - 3 - 4\} \leftrightarrow X(k)$$
$$= j\{0, -14.4853, -4, -2.4853, 0, 2.4853, 4, 14.4853\}$$

That is,

$$x(n) \text{ real and odd} \leftrightarrow X(k) \text{ imaginary and odd}$$

Due to the symmetry, the range of summation can be reduced to about one-half. For complex signals,

$$x(n) \text{ complex and even} \leftrightarrow X(k) \text{ complex and even}$$

For example,

$$x(n) = \{2 + j3, 1 - j2, 1 + j1, 1 - j2\} \leftrightarrow X(k) = \{5, 1 + j2, 1 + j8, 1 + j2\}$$

$$x(n) \text{ complex and odd} \leftrightarrow X(k) \text{ complex and odd}$$

For example,

$$x(n) = \{0, 1 - j2, 0, -1 + j2\} \leftrightarrow X(k) = \{0, -4 - j2, 0, 4 + j2\}$$

Half-Wave Symmetry

If the values in a period of a periodic function $x(n)$ with period N are such that the sequence of values in the first half of any period are negatives of the values in the preceding or succeeding half-period, then it is said to be odd half-wave symmetric. That is,

$$x\left(n \pm \frac{N}{2}\right) = -x(n),$$

Then, its even-indexed DFT coefficients are zero, as the sum of the product of it with even frequency components is zero. For example,

$$x(n) = \{2, 1, -2, -1, -2, -1, 2, 1\} \leftrightarrow$$

$$X(k) = \{0, 6.8284 + j4, 0, 1.1716 - j4, 0, 1.1716 + j4, 0, 6.8284 - j4\}$$

If the values in a period of a periodic function $x(n)$ with period N are such that the sequences of values in the first half of any period are the same as the values in the preceding or succeeding half-period, then it is said to be even half-wave symmetric. That is,

$$x\left(n \pm \frac{N}{2}\right) = x(n),$$

Then, its odd-indexed DFT coefficients are zero, as the sum of the product of it with odd frequency components is zero. For example,

$$x(n) = \{3, 1, -2, -1, 3, 1, -2, -1\} \leftrightarrow$$

$$X(k) = \{2, 0, 10 - j4, 0, 2, 010 + j4, 0\}$$

Parseval's Theorem

Transforms based on orthogonality, such as the Fourier analysis, have the power preservation property. That is, the power can be computed either using the time-domain or frequency-domain representation. Using the definition for power for periodic signals given in an earlier chapter, the power of the signal $x(n)$ with its DFT $X(k)$ of length N is given by the identity, called Parseval's theorem,

$$\sum_{n=0}^{N-1} |x(n)|^2 = \frac{1}{N} \sum_{k=0}^{N-1} |X(k)|^2$$

In the DFT, the signal is represented by a set of complex exponentials with purely imaginary exponents. Therefore, the samples of a complex exponential lie on the unit circle. The power of an exponential with period N is N. The DFT coefficients $X(k)$ are the scaled (by N) amplitudes of the exponentials. As the power of a complex exponential is

$$\frac{|X(k)|^2}{N^2}N = \frac{|X(k)|^2}{N},$$

the sum of the power of all the constituent complex exponentials is the total power of the signal. For example, let

$$x(n) = \{1 - j2, 2 + j2, 3 - j1, 2 + j3\} \leftrightarrow X(k) = \{8 + j2, -3 - j1, -j8, -1 - j1\}$$

The power computed in the time and frequency domains are the same, 36.

For two different signals, the generalized form of this theorem is given by

$$\sum_{n=0}^{N-1} x(n)y^*(n) = \frac{1}{N}\sum_{k=0}^{N-1} X(k)Y^*(k),$$

which is the same as the circular correlation evaluated with the output index zero.

3.3.1 The Criterion of Approximation

The Fourier reconstruction of a waveform is with respect to the least squares error criterion. Let $X(k)$ be the DFT of $x(n)$ with period N (N even). Let us reconstruct the signal, called the partial sum $\bar{x}(n)$, with $M < N$ (M odd) with the DFT coefficients $\bar{X}(k)$. For convenience, we are making some assumptions. However, the criterion is valid for real and complex signals with any N and M. Then,

$$\bar{x}(n) = \frac{1}{N}\sum_{k=-\left(\frac{M-1}{2}\right)}^{\frac{M-1}{2}} \bar{X}(k)e^{jk\frac{2\pi}{N}n}, \quad n = -\left(\frac{N}{2}\right), -\left(\frac{N}{2}-1\right), \ldots, \frac{N}{2}-1$$

Now, the least squares error is defined as

$$E(\bar{X}(k)) = \sum_{n=-\left(\frac{N}{2}\right)}^{\frac{N}{2}-1} \left(x(n) - \frac{1}{N}\sum_{k=-\left(\frac{M-1}{2}\right)}^{\frac{M-1}{2}} \bar{X}(k)e^{jk\frac{2\pi}{N}n}\right)^2$$

Adding and subtracting a term, we get

$$E(\bar{X}(k)) = \sum_{n=-\left(\frac{N}{2}\right)}^{\frac{N}{2}-1} \left(x(n) - \frac{1}{N} \sum_{k=-\left(\frac{M-1}{2}\right)}^{\frac{M-1}{2}} \bar{X}(k)e^{jk\frac{2\pi}{N}n} \right.$$

$$\left. + \frac{1}{N} \sum_{k=-\left(\frac{M-1}{2}\right)}^{\frac{M-1}{2}} X(k)e^{jk\frac{2\pi}{N}n} - \frac{1}{N} \sum_{k=-\left(\frac{M-1}{2}\right)}^{\frac{M-1}{2}} X(k)e^{jk\frac{2\pi}{N}n} \right)^2$$

Simplifying, we get

$$E(\bar{X}(k)) = \sum_{n=-\left(\frac{N}{2}\right)}^{\frac{N}{2}-1} \left(x(n) - \frac{1}{N} \sum_{k=-\left(\frac{M-1}{2}\right)}^{\frac{M-1}{2}} (\bar{X}(k) - X(k))e^{jk\frac{2\pi}{N}n} \right.$$

$$\left. - \frac{1}{N} \sum_{k=-\left(\frac{M-1}{2}\right)}^{\frac{M-1}{2}} X(k)e^{jk\frac{2\pi}{N}n} \right)^2$$

With $x(n)$ and $X(k)$ fixed, the error is minimum only when $\bar{X}(k) = X(k)$, implying that, with respect to this criterion, there is no other better approximation possible. Further, when constrained to use a smaller number of coefficients to represent a signal, the minimum error is provided by selecting the largest coefficients. Figure 3.4a and b show, respectively, the least squares error in varying the amplitude A and angle θ in the sinusoid $\cos(\frac{2\pi}{4}n + \frac{\pi}{3})$, which is a component of $x(n)$ shown in Fig. 3.1a

$$x(n) = 1 + \cos\left(\frac{2\pi}{4}n + \frac{\pi}{3}\right) - \cos\left(2\frac{2\pi}{4}n\right).$$

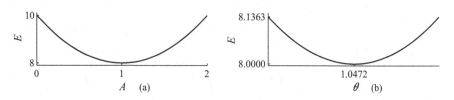

Fig. 3.4 The least squares error approximation of varying the magnitude (**a**) and phase (**b**) of a DFT coefficient

In Fig. 3.1c, we showed the error in approximating the waveform by the DC component. In Fig. 3.4, we showed the error in approximating the waveform by the sinusoidal component with frequency index 1. As the sinusoid has two parameters A and θ to characterize, two plots are required. As can be seen, with parameters obtained by Fourier analysis, we get the minimum least squares error.

3.4   Advantages of Frequency Domain in Signal and System Analysis

Signals mostly occur in time-domain form. But they are usually converted into frequency domain for efficient analysis and easier interpretation of the operations, such as filtering. Further, frequency-domain representation gives a better insight of the signal characteristics. Fourier analysis and other related transforms make the design of systems, finding the system response and stability analysis easier and efficient in most cases.

The features of Fourier analysis are as follows:

- Operations, such as convolution, reduce to much simpler multiplication operation in the frequency domain. The classification of the frequency content of signals into frequency bands, such as low and high frequencies, is important in designing signal processing systems. Signals can be compressed more efficiently in the frequency-domain representation, as the frequency content of practical signals becomes insignificant at high frequencies.
- The availability of fast computation of the DFT, which, in turn, approximates adequately and quickly infinite extent and continuous practical signals.
- All practical signals satisfy the constraints of the Fourier analysis.
- By increasing the number of frequency components to represent the signals, required accuracy of representation can be obtained.
- The criterion of least squares error holds for signal reconstruction by any number of coefficients.

3.5   Summary

- The exponential function is of the form b^n, where b is a constant, called the base, and the exponent n is the independent variable.
- Signals mostly occur in time-domain form. But they are usually converted into frequency domain for efficient analysis and easier interpretation of the operations, such as filtering. Further, frequency-domain representation gives a better insight of the signal characteristics.
- When combined appropriately, the complex exponential function represents a real sinusoid.
- Most of the naturally occurring signals have arbitrary amplitude profile. As such, it is difficult to process.

- Fourier analysis and the related transforms represent arbitrary signals in terms of complex exponentials.
- In most often used transforms, signal representation is obtained by the integral or sum of the product of the signal and the basis complex exponentials.
- DFT is the only version of Fourier analysis, in which the signal representation is discrete and finite in both the time and frequency domains.
- In the DFT, the given finite data and its finite spectrum are periodically extended. The periodically extended data can be expressed in terms of complex exponentials.
- In order to find the complex amplitude of each exponential constituting a signal, the orthogonality property of the complex exponentials is used.
- The DFT decomposes a signal in terms of complex exponentials, and IDFT reconstructs the signal back from its DFT coefficients.
- A plot of these coefficients $X(k)$ versus frequency index k is called the spectrum of $x(n)$. Alternatively, the spectrum can be expressed as an amplitude spectrum and a phase spectrum.
- The Fourier reconstruction of a signal is with respect to least squares error criterion. That is, the sum of the squared magnitude of the difference between the input and reconstructed signals must be minimum.
- All the N coefficients of the N-point DFT can be computed simultaneously using the matrix formulation of the DFT.
- Properties of transforms make the analysis easier and also make the derivation of transform of signals easier by decomposing it in terms of simpler signals.
- The convolution operation is reduced to much simpler multiplication in the frequency domain.
- Long sequences are decomposed into short segments for efficient convolution.
- Transforms based on orthogonality, such as the Fourier analysis, have the power preservation property. That is, the power can be computed either using the time-domain or frequency-domain representation.
- The availability of fast computation of the DFT, which, in turn, approximates adequately and quickly infinite extent and continuous practical signals.
- All practical signals satisfy the constraints of the Fourier analysis.

Exercises

3.1 Compute the DFT, $X(k)$, of $x(n)$ using the matrix form of the DFT. Find an expression for $x(n)$ in terms of its sinusoidal components. Compute the IDFT of $X(k)$ to get back $x(n)$. Compute the least squares error if $x(n)$ is represented by the DC component only with the values $X(0), 0.8X(0)$, and $1.2X(0)$. Verify the values $x(0)$ and $x(2)$ from the DFT values and $X(0)$ and $X(2)$ from the time-domain values using the sum and difference properties.

3.1.1 $x(n) = \{\overset{\wedge}{4}, 3, 2, 4\}$.

3.1.2 $x(n) = \{\overset{\wedge}{3}, -1, 3, 2\}$.

3.1.3 $x(n) = \{\overset{\wedge}{1}, -1, 2, 3\}$.

* 3.1.4 $x(n) = \{1 + j1, 2 + j3, 3 + j1, 3 - j4\}$.

3.1.5 $x(n) = \{2 + j3, 2 - j2, -1 + j2, 2 - j3\}$.

3.2 Find the samples of $x(n)$ over one period and use the matrix form of the DFT to compute the spectrum $X(k)$ of the set of samples. Using $X(k)$, find the exponential form of $x(n)$ and reduce the expression to real form. Verify that the input $x(n)$ is obtained.

3.2.1 $x(n) = 1 + 3\cos(\frac{2\pi}{4}n + \frac{\pi}{6}) + 2\cos(\pi n)$.

* 3.2.2 $x(n) = 1 - 2\sin(\frac{2\pi}{4}n - \frac{\pi}{3}) - 3\cos(\pi n)$.

3.2.3 $x(n) = 1 - \cos(\frac{2\pi}{4}n - \frac{\pi}{4}) + \cos(\pi n)$.

3.2.4 $x(n) = 3 + \sin(\frac{2\pi}{4}n - \frac{\pi}{4}) - \cos(\pi n)$.

3.2.5 $x(n) = -1 - 2\cos(\frac{2\pi}{4}n - \frac{\pi}{3}) - 3\cos(\pi n)$.

3.3 Find the circular convolution $y_c(n)$ of the sequences $x(n)$ and $h(n)$ using the DFT. Find also the linear convolution $y_l(n)$ of the sequences $x(n)$ and $h(n)$ using the DFT by zero padding the sequences. Verify $y_l(n)$ using the sum and difference tests.

3.3.1 $x(n) = \{\hat{3}, 1, 2, 4\}$ and $h(n) = \{\hat{1}, -1, 2, 3\}$.

3.3.2 $x(n) = \{\hat{3}, 1, -2, 4\}$ and $h(n) = \{1, \hat{-1}, 2, -3\}$.

* 3.3.3 $x(n) = \{2, 1, \hat{2}, 1\}$ and $h(n) = \{1, -1, -2, \hat{3}\}$.

3.3.4 $x(n) = \{\hat{-3}, 1, 2, 4\}$ and $h(n) = \{-1, -1, 2, \hat{3}\}$.

3.3.5 $x(n) = \{4, -1, 2, \hat{4}\}$ and $h(n) = \{2, -1, 2, \hat{3}\}$.

3.4 Find the circular correlation $y_{xh_c}(n)$ of the sequences $x(n)$ and $h(n)$ using the DFT. Find also the linear correlation $y_{xh_l}(n)$ of the sequences $x(n)$ and $h(n)$ using the DFT by zero padding the sequences. Verify $y_{xh_l}(n)$ using the sum and difference tests. Find the autocorrelation of $x(n)$. Compute the signal power and verify that the autocorrelation with lag zero is the same.

3.4.1 $x(n) = \{\hat{3}, 1, 2, 4\}$ and $h(n) = \{\hat{1}, -1, 2, 3\}$.

* 3.4.2 $x(n) = \{\hat{3}, 1, -2, 4\}$ and $h(n) = \{1, \hat{-1}, 2, -3\}$.

3.4.3 $x(n) = \{2, 1, \hat{2}, 1\}$ and $h(n) = \{1, -1, -2, \hat{3}\}$.

3.4.4 $x(n) = \{\hat{-3}, 1, 2, 4\}$ and $h(n) = \{-1, -1, 2, \hat{3}\}$.

3.4.5 $x(n) = \{4, -1, 2, \hat{4}\}$ and $h(n) = \{2, -1, 2, \hat{3}\}$.

3.5 Two sequences $x(n)$ and $h(n)$ are given. Using the frequency-domain convolution property, find the DFT $P_{xy}(k)$ of their product $y(n) = x(n)h(n)$. Verify that the IDFT of $P_{xy}(k)$ yields $y(n)$.

* 3.5.1

$$x(n) = \{\hat{1}, -1, 2, 4\}, \quad h(n) = \{\hat{4}, 2, -1, 3\}.$$

3.5.2

$$x(n) = \{\hat{1}, -3, 2, 2\}, \quad h(n) = \{\hat{4}, -1, 3, 2\}.$$

3.5.3

$$x(n) = \{\hat{2}, 1, 2, 4\}, \quad h(n) = \{\hat{3}, 1, 2, 4\}.$$

3.6 Find the even and odd components of the signal using the DFT and IDFT. Verify that, for $n = -3, -2, -1, 0, 1, 2, 3$, the two components add up to the values of the signal. Verify that the sum of the values of the odd component is zero and that those of the even component and the signal are equal.

3.6.1 $x(0) = 1, x(1) = 2, x(2) = -3, x(3) = 4$ and $x(n) = 0$ otherwise.

3.6.2 $x(n) = u(n - 1), n = -3 : 3$.

* **3.6.3** $x(n) = (-0.8)^{n+1} u(n + 1), n = -3 : 3$.

3.6.4 $x(n) = n^2 u(n)$.

3.7 Given $x(n)$, find its DFT $X(k)$. Now, find the DFT of $X(k)$ to get $Nx(N - n)$, and hence, verify the duality theorem of the DFT.

3.7.1 $x(n) = \{1, 2, -3, 4\}$.

3.7.2 $x(n) = \{2, 1, 4, -3\}$.

* **3.7.3** $x(n) = \{2, 1, 4 + j2, -3\}$.

3.8 Given the polar form of a sinusoid $x(n)$, list the sample values $x(n)$ of one cycle, starting from $n = 0$, of the sinusoid. Take the DFT of the samples and find the rectangular form $x_r(n)$ of the sinusoid from the DFT coefficients. List the sample values $x_r(n)$ of one cycle, starting from $n = 0$, of the sinusoid, and verify that they are the same as those of $x(n)$.

* **3.8.1** $x(n) = 2 \sin(\frac{2\pi}{8} n + \frac{\pi}{3})$.

3.8.2 $x(n) = 2 \cos(2 \frac{2\pi}{8} n + \frac{\pi}{6})$.

3.8.3 1.8.4 $x(n) = \cos(\frac{2\pi}{8} n + \frac{\pi}{4})$.

3.9 Given the sinusoids $x(n)$ and $y(n)$, find the polar form of the sinusoid $z(n) = x(n) + y(n)$ using the DFT. Find the sample values of one cycle, starting from $n = 0$, of all the three sinusoids, and verify that the sample values of $x(n) + y(n)$ are the same as those of $z(n)$.

3.9.1 $x(n) = \sin(\frac{2\pi}{6} n - \frac{\pi}{6}), y(n) = 2 \cos(\frac{2\pi}{6} n + \frac{\pi}{3})$.

3.9.2 $x(n) = 3 \cos(\frac{2\pi}{6} n + \frac{\pi}{6}), y(n) = 2 \cos(\frac{2\pi}{6} n - \frac{\pi}{3})$.

* **3.9.3** $x(n) = \cos(\frac{2\pi}{6} n + \frac{\pi}{4}), y(n) = - \cos(\frac{2\pi}{6} n + \frac{\pi}{3})$.

3.10 Given the sequence $x(n)$, find its DFT $X(k)$. Make the odd-indexed coefficients of $X(k)$ zero to get the DFT $XE(k)$ of the even half-wave symmetric component of $x(n)$. Make the even-indexed coefficients of $X(k)$ zero to get the DFT $XO(k)$ of the odd half-wave symmetric component of $x(n)$. Take the IDFT of the last two DFTs to get the symmetric components of $x(n)$. Verify that the sum of the sample values of the symmetric components is the same as that of $x(n)$.

* **3.10.1** $x(n) = \{\hat{2}, 1, -3, 4, \ 2, -1, 3, -1\}$.

  3.10.2 $x(n) = \{\hat{2}, 1, -3, -2, \ 1, -1, 3, -1\}$.

  3.10.3 $x(n) = \{\hat{4}, 1, 3, -2, \ 1, -1, 2, -1\}$.

Discrete-Time Fourier Transform

4

There are four versions of Fourier analysis, each one more suitable for a certain type of signals. In digital signal processing, we are primarily interested in two versions, the DFT and the DTFT. The input signal is discrete in these versions. No matter which of the versions is required, the other three versions are eventually approximated by the DFT that is inherently discrete and finite in both the domains. Therefore, digital hardware and/or software can be used for adequately approximating Fourier analysis. It is the fast algorithms for the computation of the DFT that makes the DSP practically efficient.

The DTFT finds the frequency content of aperiodic discrete signals. As the period of the time-domain signal is infinite, the frequency increment of the spectrum tends to zero and the spectrum is continuous. A sampled signal can only have a periodic spectrum, as the effective frequency range is finite. Therefore, the DTFT spectrum is continuous and periodic. The definition of the DTFT and its inverse can be obtained starting with that of the DFT and letting the period tend to infinity. The result is that the DTFT is a summation with range from plus infinity to minus infinity. The inverse DTFT is an integral, with limits over any range of 2π, since it is periodic.

4.1 The DTFT and Its Inverse

Consider the DFT transform pair, in the center-zero format,

$$X(k) = \sum_{m=-N}^{N} x(m) e^{-j\frac{2\pi}{(2N+1)}mk} \leftrightarrow x(n) = \frac{1}{2N+1} \sum_{k=-N}^{N} X(k) e^{j\frac{2\pi}{(2N+1)}nk},$$

$$n = 0, \pm 1, \pm 2, \ldots, \pm N$$

Replacing $X(k)$ by its definition in the IDFT expression, we get

© The Author(s), under exclusive license to Springer Nature Switzerland AG 2021
D. Sundararajan, *Digital Signal Processing*,
https://doi.org/10.1007/978-3-030-62368-5_4

$$x(n) = \frac{1}{2N+1} \sum_{k=-N}^{N} \left(\sum_{m=-N}^{N} x(m)e^{-j\frac{2\pi}{(2N+1)}mk} \right) e^{j\frac{2\pi}{(2N+1)}nk}$$

With a sampling interval of $T_s = 1$ s, the frequency increment of the spectrum $X(k)$ is

$$\Delta\omega = \frac{2\pi}{(2N+1)}$$

radians per second. As the time-domain range $2N + 1$ is increased by zero padding either side, the frequency increment is decreased, making the period $2N + 1$ of $x(n)$ longer and the spectrum denser. In the limit, the period becomes infinity and the frequency increment tends to zero. The range of the spectrum remains the same, as the sampling interval in the time domain is fixed. Then, $\Delta\omega$ is formally replaced by $d\omega$. The continuous frequency variable ω is substituted for the discrete frequency variable $\frac{2\pi}{(2N+1)}k$. The inverse DTFT becomes an integral with the range over any limits of 2π. The DTFT transform pair becomes

$$X(e^{j\omega}) = \sum_{n=-\infty}^{\infty} x(n)e^{-j\omega n} \leftrightarrow x(n)$$

$$= \frac{1}{2\pi} \int_{-\pi}^{\pi} X(e^{j\omega})e^{j\omega n}d\omega, \ n = 0, \pm1, \pm2, \ldots \qquad (4.1)$$

Example 4.1 The periodic DTFT spectrum $X(e^{j\omega})$ of a signal $x(n)$ is specified in terms of unit-step signals as

$$X(e^{j\omega}) = u\left(\omega + \frac{\pi}{5}\right) - u\left(\omega - \frac{\pi}{5}\right), \ -\pi < \omega \le \pi$$

over one period. Using the inverse DTFT, find $x(n)$.

Solution As the spectrum is even-symmetric,

$$x(n) = \frac{1}{\pi} \int_{0}^{\frac{\pi}{5}} \cos(\omega n)\, d\omega = \frac{\sin(\frac{\pi n}{5})}{n\pi}, \ -\infty < n < \infty \text{ and } n \ne 0$$

$$x(0) = \lim_{n\to 0} \frac{\sin(\frac{\pi n}{5})}{n\pi} = \frac{1}{5}$$

In practice, the amplitude profiles of the signals and their spectra are arbitrary and
of varying lengths, and, therefore, we cannot get a closed-form solution. Therefore,
the given spectrum is approximated by appropriate sampling to get the time-
domain signal $x(n)$. In the time domain, the infinite extent $x(n)$ is approximated
by truncation in approximating the spectrum. Figure 4.1a and b shows the truncated
$x(n)$ and the samples of its DTFT spectrum, which is that of a lowpass filter. The
time-domain signal $x(n) = \frac{\sin(\frac{\pi n}{5})}{n\pi}$ is of infinite length and is truncated to 2001
samples. The peak value of the signal is 0.2 and the magnitude of the minimum
value of the truncated signal is 0.0003, while the minimum value approaches 0 as n
tends to infinity. Then, the frequency increment of the spectrum becomes $2\pi/2001$
and the frequency index $k = 200$ represents the frequency $\pi/5$ approximately. In
general, the signal is truncated so that the values of the discarded samples are
negligible. In this case, the signal is a sampled one. If the time-domain signal is a
continuous one, it has to be sampled with an appropriate sampling rate so that the
spectrum approximates the exact spectrum to a specified accuracy, both in terms of
the magnitudes of the spectral values and the density.

Gibbs Phenomenon
As always, at discontinuities, the Fourier reconstructed continuous waveform
converges to the average value at each discontinuity both in the frequency and
time domains. In addition, oscillations around the discontinuities, called Gibbs
phenomenon, also occur. No matter how many frequency components are used
for the reconstruction, the largest of the magnitude of the oscillations, as shown in
Fig. 4.1b, converges to the value of 1.0869 for a unit discontinuity with a relatively
small number of harmonics used. As the convergence is with respect to the least
squares error criterion, the area under the overshoots and undershoots tends to zero.
As the number of harmonics used in the reconstruction increases, the oscillations
are confined more closer to the discontinuity. There is no way a waveform with
discontinuity can be reconstructed as a linear combination of sinusoids that are
very smooth. The failure of the Fourier analysis with respect to providing the
uniform convergence at discontinuities and the requirement of an infinite number
of harmonics for a complete reconstruction of a waveform are only in theory. The
response of physical devices is never as sharp as the mathematical definition of a
discontinuity. Further, no physical device can generate a frequency component of
infinite order. Therefore, Fourier analysis is essential and widely used in practice

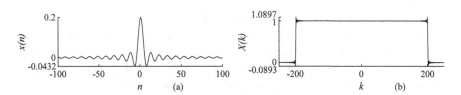

Fig. 4.1 Part of (**a**) the time-domain signal $x(n)$ truncated to $N = 2001$ samples and (**b**) the
samples of its DTFT spectrum, which is that of a lowpass filter

with disregard to these failures in theory. In applications, where the oscillations are objectionable, they can be removed, as shown in a later chapter, by modifying the Fourier spectrum. The modified spectrum is a representation of the signal by a trigonometric polynomial of a certain type, not a Fourier representation. For continuous waveforms in both the domains, the Fourier reconstruction provides uniform convergence. In conclusion, the finite numerical samples of a waveform, to be transformed by the DFT to the frequency domain, should be sufficiently close to the actual signal that is being approximated. It has to be ensured by trial and error that the record length and the number of samples are adequate to approximate the waveform with negligible truncation and aliasing errors. Further, the spectrum should be sufficiently denser so that all its essential features, such as a peak, are included. The DFT computation is always the same. The accuracy control lies in the data modeling. Just keep increasing the record length until there are no significant changes in two consecutive spectrums. If the record length is finite, then zero padding should be resorted to make the spectrum sufficiently denser.

When the time index is equal to 0 or the frequency variable $\omega = 0$ or $\omega = \pi$, the DTFT and the inverse DTFT become just summation or integral, making it easy to compute some specific values of the spectrum or time-domain signal. These values may be used to verify closed-form solutions. That is,

$$X(e^{j0}) = \sum_{n=-\infty}^{\infty} x(n), \qquad X(e^{j\pi}) = \sum_{n=-\infty}^{\infty} (-1)^n x(n),$$

$$x(0) = \frac{1}{2\pi} \int_{-\pi}^{\pi} X(e^{j\omega}) d\omega$$

For example, for the transform pair shown in Fig. 4.1,

$$x(0) = \frac{1}{2\pi} \int_{-\pi/5}^{\pi/5} 1 d\omega = 0.2 \quad \text{and} \quad X(e^{j0}) = \sum_{n=-\infty}^{\infty} \frac{\sin(\frac{\pi n}{5})}{n\pi} = 1$$

The summation can be verified numerically for large values of n. The Fourier series version of the Fourier analysis is the same as the DTFT with the roles of the time and frequency domains interchanged. In the Fourier series formulation, the periodic square wave spectrum of the DTFT becomes the periodic time-domain signal and the discrete samples in the time domain become the Fourier series.

DTFT Definitions with the Sampling Interval T_s

The DTFT and the inverse DTFT definitions were presented assuming that the sampling interval is 1 s. There are two ways to get the spectrum for other sampling intervals. One is that, we can rescale the frequency axis. For example, if the sampling interval $T_s \neq 1$, the frequency axis can be rescaled. If $T_s < 1$, we get a longer frequency range $\omega = 2\pi/T_s$. If $T_s > 1$, we get a shorter frequency range $\omega = 2\pi/T_s$. These are consistent with the sampling theorem, as the number of distinct frequency components is more with a shorter sampling interval. With $T_s = 1\,\text{s}$, $\omega = 2\pi$. Due to the redundancy of the spectrum by a factor of 2, the effective frequency range is from 0 to $\omega = \pi$ radians. For example, with $T_s = 2\,\text{ms}$, the frequency range is increased to 1000π rad/s. The effective frequency range is from 0 to $\omega = 500\pi$ rad.

Another way is to redefine the definitions by including the sampling interval as

$$X(e^{j\omega T_s}) = \sum_{n=-\infty}^{\infty} x(nT_s)e^{-jn\omega T_s} \tag{4.2}$$

$$x(nT_s) = \frac{1}{\omega_s} \int_{-\frac{\omega_s}{2}}^{\frac{\omega_s}{2}} X(e^{j\omega T_s})e^{jn\omega T_s}\, d\omega, \ \ n = 0, \pm 1, \pm 2, \ldots, \tag{4.3}$$

where $\omega_s = \frac{2\pi}{T_s}$. As the spectrum is conjugate-symmetric, specifying the spectrum over the half-period from $\omega = 0$ to $\omega = \pi/T_s$ is sufficient for real-valued signals.

Visualization of DFT and DTFT Spectra
In the case of DFT, the frequency components of a time-domain waveform are discrete with finite amplitudes. Therefore, we are able to easily visualize the reconstruction of a waveform. In the case of the DTFT, the spectrum is continuous, and the magnitude of each of the infinite frequency components is infinitesimal. The amplitudes of the components are given by

$$\frac{1}{2\pi}X(e^{j\omega})d\omega$$

As the amplitudes are immeasurably small, the spectral density $X(e^{j\omega})$ represents $x(n)$ in the frequency domain as a relative amplitude spectrum.

Convergence of the DTFT
The DTFT transform pair is defined as

$$X(e^{j\omega}) = \sum_{n=-\infty}^{\infty} x(n)e^{-j\omega n} \leftrightarrow x(n) = \frac{1}{2\pi}\int_{-\pi}^{\pi} X(e^{j\omega})e^{j\omega n}d\omega,$$

$$n = 0, \pm 1, \pm 2, \ldots$$

The Fourier reconstruction of a continuous waveform in time or frequency domains is with respect to the least squares error criterion. It includes all the energy signals with finite energy. However, a finite energy signal need not be absolutely summable. If the signal is continuous, the Fourier reconstructed signal absolutely converges to it. Since the spectrum is of finite duration, the power spectrum is integrable and the energy is finite. The energy of a signal can be expressed in both the time and frequency domains

$$E = \sum_{n=-\infty}^{\infty} |x(n)|^2 = \frac{1}{2\pi} \int_0^{2\pi} |X(e^{j\omega})|^2 d\omega$$

In the expression for the DTFT spectrum, the infinite summation converges in the least squares error sense, if $x(n)$ is square summable. That is,

$$\sum_{n=-\infty}^{\infty} |x(n)|^2 < \infty$$

The signals may be absolutely convergent or not. For signals that are not absolutely convergent, the reconstructed waveform has discontinuities. But, the waveform satisfies the least squares error criterion. Some signals are divergent, but square summable. For example,

$$\sum_{n=1}^{\infty} \frac{1}{n} \text{ is divergent} \quad \text{and} \quad \sum_{n=1}^{\infty} \frac{1}{n^2} \text{ is convergent}$$

Example 4.2 The nonzero samples of $x(n)$ are $\{x(-1) = 0.5, x(1) = -0.5\}$. Find the DTFT of $x(n)$.

Solution From the definition,

$$X(e^{j\omega}) = 0.5(e^{j\omega} - e^{-j\omega}) = j \sin(\omega)$$

The inverse DTFT is

$$x(n) = \frac{1}{2\pi} \int_{-\pi}^{\pi} (0.5e^{j\omega} - 0.5e^{-j\omega})e^{j\omega n} d\omega = \{0.5, -0.5\},$$

$$n = \{-1, 1\} \quad \text{and } 0 \quad \text{otherwise}$$

since

$$\int_{-\pi}^{\pi} e^{j\omega n} d\omega = \begin{cases} 1 \text{ for } n = 0 \\ 0 \text{ for } n = \pm 1, \pm 2, \ldots \end{cases}$$

One period of the magnitude spectrum $|X(e^{j\omega})|$ of $x(n)$ and the phase spectrum $\angle X(e^{j\omega})$ are shown in Fig. 4.2a and b, respectively. The samples of $x(n)$ are odd-symmetric. Therefore, its DTFT is imaginary and odd. As the basis function is the complex exponential, the DTFT spectrum is also complex. In the case the signal is real-valued, its DTFT spectrum will be conjugate-symmetric. That is, half the values are redundant, since two conjugate complex exponentials are required to reconstruct a real frequency component. For complex signals, there is no redundancy. While the signals in practical applications are mostly real-valued, the complex signals are also often used in applications such as Hilbert transform, single-sideband modulation in communication systems, and sampling of bandpass signals. Further, in order to reduce the redundancy of the spectrums of real-valued signals, two signals are combined in an appropriate way and used in fast numerical approximation of the Fourier spectra.

As the spectra of real-valued signals are also complex-valued, in general, two plots are required to represent a complex spectrum completely. Either frequency versus the magnitude of the DTFT, called the magnitude spectrum (Fig. 4.2a), and frequency versus phase, called the phase spectrum (Fig. 4.2b), or frequency versus the real part (Fig. 4.3a) and frequency versus the imaginary part (Fig. 4.3b) are adequate. Remember that a complex number can be represented either in Cartesian form or polar form. One period of the real part of the spectrum $X(e^{j\omega})$ of $x(n)$ and the imaginary part are shown in Fig. 4.3a and b, respectively.

∎

When a continuous signal is sampled, the range of the unique frequencies of its spectrum becomes limited, referred to as bandlimited. Unlike in the case of continuous signals with the highest frequency ∞, the highest distinct frequency is π for discrete signals. By making the sampling interval smaller, we can increase the frequency range to a desired value, but it will always be finite for discrete signals. As the frequency range is increased, the rate of change of the sample values increases. Therefore, the complex spectrum of discrete signals is always periodic, and it is periodic in ω with period 2π with a sampling interval of 1 s. Further, the spectrum of real-valued signals is conjugate-symmetric, as mentioned above. Therefore, the spectrum in the range $0 < \omega < \pi$ is adequate, and usually, that is the

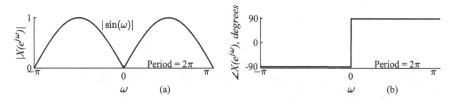

Fig. 4.2 (a) The magnitude spectrum $|X(e^{j\omega})|$ of $x(n)$ and (b) the phase spectrum

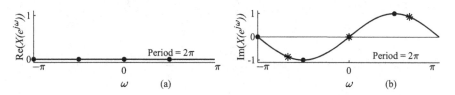

Fig. 4.3 (a) The real part of the spectrum $|X(e^{j\omega})|$ of $x(n)$ and (b) the imaginary part

way it is represented. In Figs. 4.2 and 4.3, we have shown the full spectrum just for completeness.

Relation Between the DFT and the DTFT

The four versions of Fourier versions are related. For example, DFT and DTFT are related. These relationships are important, since the DFT, which is to be used for the approximation of other versions, is discrete and finite in both the domains, while the other versions are either continuous or aperiodic at least in one or both of the domains. The DTFT $X(e^{j\omega})$ of $x(n)$, starting from $n = 0$, with all zeros outside of length N and the DFT $X(k)$ of the periodic extension of the N-point finite sequence are given as

$$X(e^{j\omega}) = \sum_{n=0}^{N-1} x(n)e^{-j\omega n}$$

and

$$X(k) = \sum_{n=0}^{N-1} x(n)e^{-jk\omega_0 n}, \quad \omega_0 = \frac{2\pi}{N}, \quad k = 0, 1, 2, \ldots, N-1$$

The periodic DTFT spectrum is continuous over a period of 2π rad. The spectrum is continuous, as the DTFT is defined for aperiodic sequences. The frequency increment tends to zero. The periodicity remains the same for both the DTFT and the DFT. Therefore, the DFT spectrum is the samples of the DTFT spectrum at equal intervals of $\omega_0 = \frac{2\pi}{N}$. That is,

$$X(k) = X(e^{j\omega})|_{\omega=k\omega_0} = X(e^{jk\omega_0})$$

Let the samples of a signal be

$$\{x(-1) = 0.5, x(0) = 0, x(1) = -0.5\}$$

and zero otherwise. The DTFT of $x(n)$ is $X(e^{j\omega}) = 0.5e^{j\omega} - 0.5e^{-j\omega} = j\sin(\omega)$. The set of samples of $X(e^{j\omega})$,

$$\{X(0) = 0, X(1) = j0.866, X(2) = -0.866\},$$

at

$$\omega = 0, \omega = \frac{2\pi}{3}, \omega = 2\frac{2\pi}{3}$$

is the DFT of $x(n)$. Remember that both have periodic spectra and the spectra can be defined over any interval of width one period. The DFT spectral values are shown by the symbol * in Fig. 4.3b.

Although all the defined values have been taken for computing the spectrum, we zero pad the signal to make the spectrum denser to ensure that all the essential features of the spectrum, such as a peak, are present. Let us append one zero to $x(n)$ to get

$$\{x(-2) = 0, x(-1) = -1, x(0) = 0, x(1) = 1\}$$

$X(e^{j\omega}) = 0.5e^{j\omega} - 0.5e^{-j\omega} = j\sin(\omega)$. The set of samples of $X(e^{j\omega})$,

$$\{X(0) = 0, X(1) = j1, X(2) = 0, X(3) = -j1\},$$

at

$$\omega = 0, \omega = \frac{2\pi}{4}, \omega = 2\frac{2\pi}{4}, \omega = 3\frac{2\pi}{4}$$

is the DFT of the zero-padded $x(n)$. The DFT spectral values are shown by dots in Fig. 4.3b. Now, the spectrum is denser and the peaks, missing with 3 samples, are shown in the spectrum. Therefore, proper sampling interval, adequate record length, and sufficient zero padding will ensure a good approximation of the spectrums of other versions of Fourier analysis using the DFT.

DTFT of the Unit-Impulse Signal
Example 4.3 Let $x(n) = \delta(n)$, the discrete unit impulse. Find its DTFT.

Solution Replacing $x(n)$ in the DTFT definition by $\delta(n)$, we get

$$X(e^{j\omega}) = \sum_{n=-\infty}^{\infty} \delta(n)e^{-j\omega n} = 1e^{-j\omega 0} = 1 \quad \text{and} \quad \delta(n) \leftrightarrow 1, \forall \omega$$

The impulse is an all-zero sequence except at $n = 0$. Therefore, for all values of ω, the periodic DTFT spectrum $X(e^{j\omega})$ is a constant with value 1. Figure 4.4a and b shows, respectively, part of the unit-impulse signal $\delta(n)$ and its constant DTFT spectrum over one period.

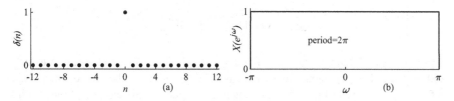

Fig. 4.4 (a) Part of the unit-impulse signal $\delta(n)$ and (b) its constant DTFT spectrum over one period

Consider the DTFT transform pair

$$\frac{\sin(\frac{\pi n}{5})}{n\pi}, \quad -\infty < n < \infty \leftrightarrow u\left(\omega + \frac{\pi}{5}\right) - u\left(\omega - \frac{\pi}{5}\right), \quad -\pi < \omega \leq \pi$$

The spectrum is composed of all complex exponentials in a limited portion of the period. If we include all complex exponentials over a period, we get the unit impulse $\delta(n)$ as its representation in the time domain. As the spectrum is even-symmetric,

$$x(n) = \frac{1}{\pi} \int_0^{\pi} \cos(\omega n)\, d\omega = \frac{\sin(\pi n)}{n\pi}, \quad -\infty < n < \infty$$

Replacing $\pi/5$ by π in the limit, we get the transform of the impulse as

$$\frac{\sin(\pi n)}{n\pi} = \delta(n), \quad -\infty < n < \infty \leftrightarrow 1, \quad -\pi < \omega \leq \pi$$

Therefore, the unit-impulse signal is composed of complex sinusoids of all the infinite frequencies from $\omega = -\pi$ to $\omega = \pi$ with equal strength.

■

DTFT of the DC Signal

Usually, any problem dealing with continuous impulse is approached through a limiting process. A continuous impulse is difficult to visualize, and it is usually considered as a finite width pulse with unit area of some functions as the width tends to zero. One of the functions that degenerates into the unit impulse is the unit-area rectangular pulse of width a and height $1/a$. As, $a \to 0$, the pulse degenerates into the unit impulse.

Some functions such as the DC and complex exponential, which are neither absolutely nor square summable, are often used in the analysis of signals and systems. The DTFT of these signals is obtained by applying a limiting process to appropriate summable signals that degenerate into these signals in the limit. Let us use this method to obtain the transform of the DC signal. Consider the DTFT transform pair

$$5\frac{\sin(\frac{\pi n}{5})}{n\pi}, \quad -\infty < n < \infty \leftrightarrow 5\left(u\left(\omega + \frac{\pi}{5}\right) - u\left(\omega - \frac{\pi}{5}\right)\right), \quad -\pi < \omega \leq \pi$$

shown in Fig. 4.5a and b. The sampling interval in (b) is $\pi/1000$ rad. Therefore, the 200 mark on the horizontal axis represents $0.2\pi = \pi/5$ rad. The area of the pulse is $5 \times 2\pi/5 = 2\pi$. In Fig. 4.5d, the height of the pulse is increased and the width reduced so that the area enclosed by the spectrum is still 2π. In Fig. 4.5c, the first zero crossings of the corresponding time-domain signal occurs farther than in (a). In Fig. 4.5f, the height of the pulse is further increased and the width reduced so that the area enclosed by the spectrum is still 2π. In Fig. 4.5e, the first zero crossings of

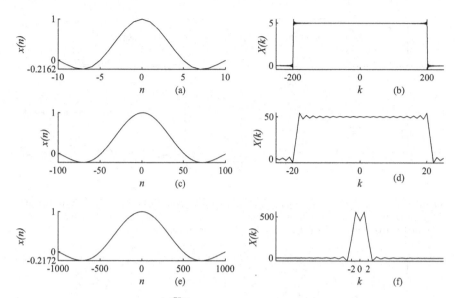

Fig. 4.5 (a) Part of the signal $5\frac{\sin(\frac{\pi n}{5})}{n\pi}$; (b) its DTFT spectrum over one period, which is a pulse of area 2π; (d) the width of the pulse in (b) is reduced and the height increased; (c) its inverse DTFT; (f) the width of the pulse in (d) is reduced and the height increased; (e) its inverse DTFT

the corresponding time-domain signal occurs farther than in (c). In the limit, when the first zero crossings occur at $n = \pm\infty$, the time-domain signal degenerates into a DC signal. Therefore, we get the transform pair for the DC signal as

$$1, \ -\infty < n < \infty \leftrightarrow 2\pi\delta(\omega), \ -\pi < \omega \leq \pi$$

From this result, we can also deduce the transform pair for the complex exponential, $e^{j\omega_0 n}$

$$e^{j\omega_0 n} - \infty < n < \infty \leftrightarrow 2\pi\delta(\omega - \omega_0), \ -\pi < \omega \leq \pi$$

The DC signal is $e^{j\omega n}$ with $\omega = 0$. For exponentials with nonzero frequency, the occurrence of their DTFT with a single impulse is delayed or advanced depending on their frequencies. One period of the complex exponential $e^{j0.1n}$ and its approximate triangular DTFT spectrum over one period are shown in Fig. 4.6a and b, respectively. The DTFT is approximated by taking 64 samples over one cycle. The frequency increment is $\pi/32$, and the height of the triangle is 63.9636. Multiplying these two values, the area of the impulse is approximately 2π at $\omega = \pi/32 \approx 0.1$ rad. This time a triangle approaches an impulse, as the width is decreased and keeping the area the same by increasing the number of time-domain samples. The transform pair for the complex exponential, $e^{j0.1n}$, is

$$e^{j0.1n} - \infty < n < \infty \leftrightarrow 2\pi\delta(\omega - 0.1), \ -\pi < \omega \leq \pi$$

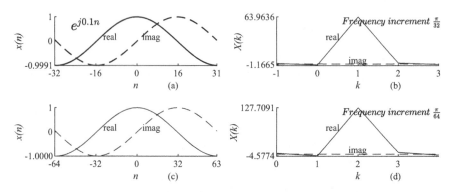

Fig. 4.6 (a) One period of the complex exponential $e^{j0.1n}$ with 64 samples; (b) its approximate triangular DTFT spectrum over one period; (c) one period of the complex exponential $e^{j0.1n}$ with 128 samples; (d) its approximate triangular DTFT spectrum over one period

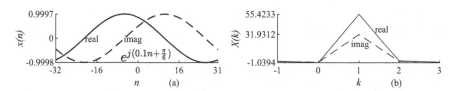

Fig. 4.7 (a) One period of the complex exponential $e^{j(0.1n+\frac{\pi}{6})}$; (b) its approximate triangular DTFT spectrum over one period

One period of the same complex exponential $e^{j0.1n}$ and its approximate triangular DTFT spectrum over one period are shown in Fig. 4.6c and d, respectively, with the samples doubled. The frequency increment is $\pi/64$, and the height of the triangle is 127.7. Multiplying these two values, the area of the impulse is again approximately 2π at $\omega = 0.1$ rad. In the limit, the transform value is exactly 2π.

Consider the transform pair for the complex exponential, $e^{j(0.1n+\frac{\pi}{6})}$.

$$e^{j(0.1n+\frac{\pi}{6})} \quad -\infty < n < \infty \leftrightarrow 2\pi(0.866 + j0.5)\delta(\omega - 0.1), \quad -\pi < \omega \le \pi$$

One period of the complex exponential $e^{j(0.1n+\frac{\pi}{6})}$ and its approximate triangular DTFT spectrum over one period are shown in Fig. 4.7a and b, respectively. Now, the coefficient of the complex exponential is complex, and its DTFT coefficient is complex with a complex scale factor. From the figure,

$$\frac{\pi}{32}(55.4233 + j31.9312) = 5.4412 + j3.1348 = 2\pi(0.8660 + j0.4989)$$

$$\approx 2\pi(0.866 + j0.5)$$

Note that, in numerical approximations, there will always be some error due to finite precision representation of the numbers. But, it can be reduced to a required accuracy with sufficient wordlength.

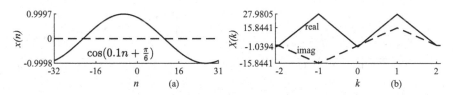

Fig. 4.8 (a) One period of the sinusoid $\cos(0.1n + \frac{\pi}{6})$; (b) its approximate triangular DTFT spectrum over one period with conjugate coefficients

As

$$\cos(\omega_0 n + \theta) = 0.5(e^{j\theta}e^{j\omega_0 n} + e^{-j\theta}e^{-j\omega_0 n})$$

$$\cos(\omega_0 n + \theta) \leftrightarrow e^{j\theta}\pi\delta(\omega - \omega_0) + e^{-j\theta}\pi\delta(\omega + \omega_0)$$

With $\theta = 0$ and $\theta = -\pi/2$, we get

$$\cos(\omega_0 n) \leftrightarrow \pi(\delta(\omega - \omega_0) + \delta(\omega + \omega_0))$$

$$\sin(\omega_0 n) \leftrightarrow j\pi(\delta(\omega + \omega_0) - \delta(\omega - \omega_0))$$

Consider the transform pair for the sinusoid, $\cos(0.1n + \frac{\pi}{6})$

$$\cos\left(0.1n + \frac{\pi}{6}\right) - \infty < n < \infty \leftrightarrow \pi(0.866 + j0.5)\delta(\omega - 0.1)$$

$$+ \pi(0.866 - j0.5)\delta(\omega + 0.1), \quad -\pi < \omega \leq \pi$$

The spectrum is complex conjugate. One period of the sinusoid $\cos(0.1n + \frac{\pi}{6})$ and its approximate triangular DTFT spectrum over one period are shown in Fig. 4.8a and b, respectively. Same coefficients as for the complex exponential, but with a conjugate pair and scaled by a factor of 2. Now, the coefficients of the complex exponentials composed of the sinusoid are complex, and its DTFT coefficients are complex with a complex scale factor. From these examples, we understand the fact that Fourier reconstruction can approximate any practical signal with an adequate accuracy with a finite number of samples in both the domains. Further, the availability of fast algorithms for the DFT makes the Fourier analysis dominant in the analysis of signals and systems.

Example 4.4 Find the DTFT of the signal $x(n) = a^n u(n)$, $|a| < 1$

$$X(e^{j\omega}) = \sum_{n=0}^{\infty} a^n e^{-j\omega n} = \sum_{n=0}^{\infty} (ae^{-j\omega})^n = \frac{1}{1 - ae^{-j\omega}}, \quad |a| < 1$$

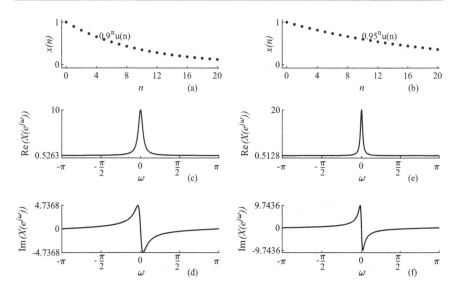

Fig. 4.9 (**a**) $x(n) = 0.9^n u(n)$; (**b**) $x(n) = 0.95^n u(n)$; (**c**) the real part of the DTFT spectrum of the signal in (**a**) and (**e**) its imaginary part; (**d**) the real part of the DTFT spectrum of the signal in (**b**) and (**f**) its imaginary part

The defining summation is a geometric progression with a common ratio of $(ae^{-j\omega})$, if $|(ae^{-j\omega})| < 1$. This implies $|a| < 1$, as $|(e^{-j\omega})| = 1$. The real exponentials with base 0.9 and 0.95 are shown, respectively, in Fig. 4.9a and b. The real and imaginary parts of the DTFT spectrum of the signal $0.9^n u(n)$, shown in Fig. 4.9a, are shown, respectively, in Fig. 4.9c and e. The real and imaginary parts of the DTFT spectrum of the signal $0.95^n u(n)$, shown in Fig. 4.9b, are shown, respectively, in Fig. 4.9d and f. The real parts of the spectrums are even-symmetric, and and the odd parts are odd-symmetric, as the spectrum of a real signal is conjugate-symmetric. As the base of the exponential approaches 1, the signal approximates the unit-step signal. Therefore, as we did earlier for the DC signal, we can derive the DTFT of the DC signal as the limiting case of the exponential with $a \to 1$.   ■

Example 4.5   Find the DTFT of the unit-step signal $x(n) = u(n)$.

Expressing the DTFT of the exponential signal in terms of its real and imaginary parts, as $a \to 1$, we get

$$X(e^{j\omega}) = \lim_{a \to 1} \frac{1}{1 - ae^{-j\omega}} = \lim_{a \to 1} \left(\frac{1 - a\cos(\omega)}{1 - 2a\cos(\omega) + a^2} - j\frac{a\sin(\omega)}{1 - 2a\cos(\omega) + a^2} \right)$$

With the imaginary part odd-symmetric and computing the inverse DTFT with $n = 0$, we get

$$x(0) = 1 = \frac{1}{2\pi} \int_{-\pi}^{\pi} \frac{1 - a\cos(\omega)}{1 - 2a\cos(\omega) + a^2} e^{j\omega 0} d\omega$$

$$= \frac{1}{2\pi} \int_{-\pi}^{\pi} \frac{1 - a\cos(\omega)}{1 - 2a\cos(\omega) + a^2} d\omega$$

Therefore, the area enclosed by the spectrum is 2π for any value of $a < 1$. In the limit, the area 2π is split by an impulse $\delta(\omega)$, at $\omega = 0$ with strength π and rest of the real part of the spectrum also contributing π. The strictly continuous component of the spectrum with $\omega \neq 0$ is

$$\frac{1}{1 - e^{-j\omega}}$$

Therefore, the transform pair for the unit-step signal becomes

$$u(n) \leftrightarrow \pi\delta(\omega) + \frac{1}{1 - e^{-j\omega}}$$

∎

Consider the transform pair

$$2u(n) - 1 \leftrightarrow 2\left(\pi\delta(\omega) + \frac{1}{1 - e^{-j\omega}}\right) - 2\pi\delta(\omega) = \frac{2}{1 - e^{-j\omega}}$$

$$2\{\ldots, 0, 0, 0, \breve{1}, 1, 1, 1, \ldots\} - \{\ldots, 1, 1, 1, \breve{1}, 1, 1, 1, \ldots\}$$

$$= \{\ldots, -1, -1, -1, \breve{1}, 1, 1, 1, \ldots\}$$

The sign or signum function $\text{sgn}(n)$ is defined as

$$\text{sgn}(n) = \begin{cases} 1 & \text{for } n \geq 0 \\ -1 & \text{for } n < 0 \end{cases}$$

There is an odd version of the sign or the signum function defined as

$$sgn(n) = 2u(n) - 1 - \delta(n) \leftrightarrow 2(\pi\delta(\omega) + \frac{1}{1 - e^{-j\omega}}) - 2\pi\delta(\omega) - 1 = \frac{2}{1 - e^{-j\omega}} - 1$$

$$2\{\ldots, 0, 0, 0, \breve{1}, 1, 1, 1, \ldots\} - \{\ldots, 1, 1, 1, \breve{1}, 1, 1, 1, \ldots\}$$

$$- \{\ldots, 0, 0, 0, \breve{1}, 0, 0, 0, \ldots\} = \{\ldots, -1, -1, -1, \breve{0}, 1, 1, 1, \ldots\}$$

4.1.1 The Relation Between the DTFT and DFT of a Discrete Periodic Signal

Let $x(n)$ be a periodic signal of period N with its DFT $X(k)$. The IDFT of $X(k)$ resulting in $x(n)$ is given by

$$x(n) = \frac{1}{N} \sum_{k=0}^{N-1} X(k) e^{jk\omega_0 n}, \quad \omega_0 = \frac{2\pi}{N}$$

The inverse DTFT of $X(e^{j\omega})$ resulting in $x(n)$ is given by

$$x(n) = \frac{1}{2\pi} \int_{-\pi}^{\pi} X(e^{j\omega}) e^{j\omega n} d\omega = \frac{1}{2\pi} \int_{-\pi}^{\pi} 2\pi \frac{X(k)}{N} \delta(\omega - k\omega_0) e^{j\omega n} d\omega,$$

$$n = 0, \pm 1, \pm 2, \ldots$$

Therefore, the DTFT of a periodic signal, with period 2π, is a set of impulses with strength $\frac{2\pi}{N} X(k)$ at $\omega = \frac{2\pi}{N} k$.

For example, the 4-point DFT of $2 \sin(\frac{2\pi}{4} n)$ is

$$\{X(0) = 0, X(1) = -j4, X(2) = 0, X(3) = j4\}$$

Its DTFT $X(e^{j\omega})$, over one period, is

$$\left\{ X(e^{j0}) = 0, \quad X(e^{j\frac{2\pi}{4}}) = -j2\pi\delta\left(\omega - \frac{2\pi}{4}\right), \right.$$

$$\left. X(e^{j2\frac{2\pi}{4}}) = 0, \quad X(e^{j3\frac{2\pi}{4}}) = j2\pi\delta\left(\omega - 3\frac{2\pi}{4}\right) \right\}$$

4.2 Properties of the Discrete-Time Fourier Transform

The effect of an operation or a characteristic of a signal in one domain can be found in the other domain through properties. For example, multiplication of the spectrums of two signals in the frequency domain corresponds to convolution of the two signals in the time domain. Further, the DTFT of signals can be easily obtained from the DTFT of simpler signals. For example, the DTFT of a sinusoid can be obtained as a linear combination of the DTFTs of a complex exponential. Of course, finding the inverse DTFT is also simplified.

4.2.1 Linearity

Let a signal be a linear combination of two signals. Then, the DTFT of the signal is the same linear combination of the DTFT of their components. That is,

$$x(n) \leftrightarrow X(e^{j\omega}), \quad y(n) \leftrightarrow Y(e^{j\omega}), \quad Cx(n) + Dy(n) \leftrightarrow CX(e^{j\omega}) + DY(e^{j\omega}),$$

where C and D are the arbitrary constants. The summation operation is linear and the DTFT definition is also a summation. Therefore, the DTFT operation is also linear. The inverse DTFT is also linear.

For example, let

$$X(e^{j\omega}) = \frac{e^{j2\omega}}{(e^{j\omega} - 0.5)(e^{j\omega} - 0.7)}$$

$$\frac{X(e^{j\omega})}{e^{j\omega}} = \frac{e^{j\omega}}{(e^{j\omega} - 0.5)(e^{j\omega} - 0.7)} = \frac{-2.5}{(e^{j\omega} - 0.5)} + \frac{3.5}{(e^{j\omega} - 0.7)}$$

$$X(e^{j\omega}) = \frac{-2.5e^{j\omega}}{(e^{j\omega} - 0.5)} + \frac{3.5e^{j\omega}}{(e^{j\omega} - 0.7)}$$

The inverse DTFT of the last equation yields

$$x(n) = (-2.5(0.5)^n + 3.5(0.7)^n)u(n)$$

4.2.2 Time Shifting

The shape of a signal is unaffected by a shift but is relocated. Consider delaying of a typical spectral component, $X(e^{j\omega_a})e^{j\omega_a n}$ by n_0 sample intervals. The shifted component is

$$X(e^{j\omega_a})e^{j\omega_a(n-n_0)} = e^{-j\omega_a n_0}X(e^{j\omega_a})e^{j\omega_a n}$$

As the term, $e^{-j\omega_a n_0}$, changes the phase only, the amplitude remains the same. Therefore, if

$$x(n) \leftrightarrow X(e^{j\omega})$$

then

$$x(n \pm n_0) \leftrightarrow e^{\pm j\omega n_0}X(e^{j\omega})$$

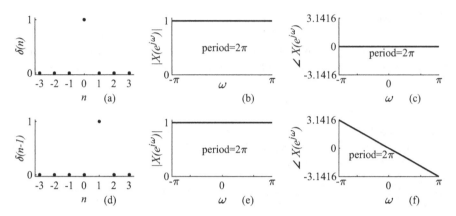

Fig. 4.10 (a) Unit impulse $\delta(n)$; (b) the magnitude of its DTFT over one period; (c) the phase of its DTFT over one period; (d) the delayed unit impulse $\delta(n-1)$; (e) the magnitude of its DTFT over one period; (f) the phase of its DTFT over one period

Time shifting of a signal results in adding linear phase to its spectrum. That is, delaying a signal by n_0 sample intervals results in the phase spectrum changed by $-n_0\omega$.

The unit impulse $\delta(n)$, the magnitude of its DTFT over one period, and the phase of its DTFT over one period are shown, respectively, in Fig. 4.10a–c. As $\delta(n)$ is an even-symmetric signal, its DTFT spectrum is real. Its reconstruction requires components with all the frequencies with equal strength from $\omega = -\pi$ and $\omega = \pi$.

The delayed unit impulse $\delta(n-1)$, the magnitude of its DTFT over one period, and the phase of its DTFT over one period are shown, respectively, in Fig. 4.10d–f. Therefore, the transform pair

$$\delta(n) \leftrightarrow 1, \ -\pi \leq \omega < \pi$$

becomes, due to this property, the transform pair

$$\delta(n-1) \leftrightarrow 1e^{-j\omega}, \ -\pi \leq \omega < \pi$$

While the periodic magnitude spectrum of the constituent frequency components remains the same, the phase spectrum undergoes, as the frequency increases, proportionally larger phase shifts.

4.2.3   Frequency Shifting

Multiplying a signal by a complex exponential $e^{\pm j\omega_0 n}$ shifts the spectrum of the signal by $\omega = \mp \omega_0$. The correlation of the modified signal occurs at a different frequency in the determination of its DTFT spectrum by the DTFT defining

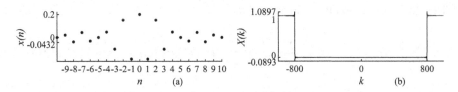

Fig. 4.11 (a) The impulse response of a highpass filter and (b) samples of its DTFT spectrum

equation. Therefore, if

$$x(n) \leftrightarrow X(e^{j\omega})$$

then

$$x(n)e^{\pm j\omega_0 n} \leftrightarrow X(e^{j(\omega \mp \omega_0)})$$

Figure 4.1 shows the frequency response of an ideal lowpass filter and its impulse response. The filter passes frequency components up to $\omega = \pi/5$ and rejects the rest. Therefore, it is a lowpass filter. The range of the frequencies from 0 to $\pi/5$ is called the passband, and the rest of the range is called the stopband. In a highpass filter, the passband and the stopband are the stopband and passband of the lowpass filter, respectively. The midpoints of the passbands of the lowpass and highpass filters are located at 0 and π, respectively. Therefore, by shifting the lowpass frequency response, shown in Fig. 4.1b, by π radians, we get the frequency response of an ideal highpass filter with cutoff frequency $\pi - \frac{\pi}{5} = \frac{4\pi}{5}$, as shown in Fig. 4.11b. The frequency response is shifted by π radians by multiplying that of the lowpass filter impulse response by $e^{j\pi n} = (-1)^n$. Therefore, impulse response of the highpass filter is $(-1)^n \frac{\sin(\frac{\pi n}{5})}{n\pi}$, shown in Fig. 4.11a.

4.2.4 Convolution in the Time Domain

The output of convolution of two complex exponentials with the same frequency is the product of their coefficients at the same frequency with a constant factor in Fourier analysis. If the frequencies are different, the output is zero. In Fourier analysis, an arbitrary signal is decomposed into its constituent complex exponentials and convolution is a linear operation. Therefore, convolution in the time domain becomes much simpler multiplication in the frequency domain. This result is a major reason for the dominance of the frequency-domain analysis of signals and systems, although most naturally occurring signals are with respect to time.

Let

$$x(n) \leftrightarrow X(e^{j\omega}) \quad \text{and} \quad h(n) \leftrightarrow H(e^{j\omega})$$

Then, the convolution theorem states that

$$\sum_{m=-\infty}^{\infty} x(m)h(n-m) = \frac{1}{2\pi} \int_{-\pi}^{\pi} X(e^{j\omega})H(e^{j\omega})e^{j\omega n} d\omega \leftrightarrow X(e^{j\omega})H(e^{j\omega})$$

That is, we find the individual DTFT of the two signals, multiply them, and find the inverse DTFT to find their convolution output.

Let us prove the convolution theorem formally. Taking the DTFT of the convolution equation in the time domain, we get

$$Y(e^{j\omega}) = \sum_{n=-\infty}^{\infty} \left(\sum_{m=-\infty}^{\infty} x(m)h(n-m) \right) e^{-j\omega n}$$

Substituting $k = n - m$, we get

$$Y(e^{j\omega}) = \sum_{k=-\infty}^{\infty} \sum_{m=-\infty}^{\infty} x(m)h(k)e^{-j\omega(k+m)}$$

$$= \sum_{m=-\infty}^{\infty} x(m) \left(\sum_{k=-\infty}^{\infty} h(k)e^{-j\omega k} \right) e^{-j\omega m} = X(e^{j\omega})H(e^{j\omega})$$

Example 4.6 Find the closed-form expression of the convolution of the sequences $x(n) = (0.14)^n u(n)$ and $h(n) = (0.05)^n u(n)$.

Solution Let us find the solution in the time domain

$$y(n) = \sum_{k=-\infty}^{\infty} x(k)h(n-k) = \sum_{k=0}^{n}(0.14)^k(0.05)^{n-k}, \ n \geq 0$$

$$= (0.05)^n \sum_{k=0}^{n} \left(\frac{0.14}{0.05} \right)^k = (0.05)^n \left(\frac{1 - \left(\frac{0.14}{0.05} \right)^{n+1}}{1 - \left(\frac{0.14}{0.05} \right)} \right)$$

$$= \left(\frac{10}{0.9} \right)((0.14)^{n+1} - (0.05)^{n+1})u(n)$$

The first four values of the convolution of $x(n)$ and $h(n)$ are

$$y(0) = 1 \times 1 = 1, \ y(1) = 1 \times 0.05 + 0.1400 \times 1 = 0.19,$$

$$y(2) = 0.0291, \ y(3) = 0.0042$$

Now, let us find the solution in the frequency domain. Consider the DTFT transform pair,

$$a^n u(n), \ |a| < 1 \leftrightarrow \frac{1}{1 - ae^{-j\omega}}$$

With $a = 0.14$ and $a = 0.05$, we get

$$(0.14)^n u(n) \leftrightarrow \frac{1}{1 - (0.14)e^{-j\omega}} \quad \text{and} \quad (0.05)^n u(n) \leftrightarrow \frac{1}{1 - (0.05)e^{-j\omega}}$$

Therefore, the DTFT of the transform of the convolution of the two signals is

$$(0.14)^n u(n) * (0.05)^n u(n) \leftrightarrow Y(e^{j\omega}) = \frac{1}{1 - (0.14)e^{-j\omega}} \frac{1}{1 - (0.05)e^{-j\omega}}$$

Expanding into partial fraction form, we get

$$Y(e^{j\omega}) = \frac{1.5556}{1 - (0.14)e^{-j\omega}} + \frac{-0.5556}{1 - (0.05)e^{-j\omega}}$$

The inverse DTFT yields the same convolution output

$$y(n) = (1.5556(0.14)^n - 0.5556(0.05)^n)u(n)$$

In practice, the DFT/IDFT, with fast algorithms, is used to implement the convolution operation. Let us compute the 4 samples at equal intervals of the continuous and periodic DTFT spectrum. With $T_s = 1$,

$$x(0) = 1, x(1) = 0.14, x(2) = 0.0196, x(3) = 0.0027$$

Although the signal is of infinite extent, the magnitude of the samples becomes negligibly small after $T = 3$ and the rest can be ignored. Its DFT is

$$X(0) = 1.1623, X(1) = 0.9804 - j0.1373,$$
$$X(2) = 0.8769, X(3) = 0.9804 + j0.1373$$

$$h(0) = 1, h(1) = 0.05, h(2) = 0.0025, h(3) = 0.0001$$

Its DFT is

$$H(0) = 1.0526, H(1) = 0.9975 - j0.0499,$$
$$H(2) = 0.9524, H(3) = 0.9975 + j0.0499$$

The pointwise product $X(k)H(k), k = 0, 1, 2, 3$ is

$$1.2235, 0.9711 - j0.1858, 0.8351, 0.9711 + j0.1858$$

These values are the 4 equally sampled values of the DTFT output spectrum. The IDFT of the product yields the first four time-domain convolution output samples with some error due to truncation.

$$\{1.0002, 0.1900, 0.0291, 0.0042\}$$

As the signals mostly occur in continuous form, it is often required to approximate the Fourier spectrum of signals adequately using their sampled version. Consider approximating the continuous signal $x(t) = (0.14)^t$. At any discontinuity of a signal, the sample value is the average of the values of the right and left limits. Therefore, the samples of the signal are, with $T_s = 1$,

$$x(0) = 0.5, x(1) = 0.14, x(2) = 0.0196, x(3) = 0.0027$$

Similarly,

$$h(0) = 0.5, h(1) = 0.05, h(2) = 0.0025, h(3) = 0.0001$$

The convolution output $y(n)$ is shown in Fig. 4.12b, and its DTFT spectrum is shown, using dots, in Fig. 4.12a. The continuous curve is the convolution of the continuous version of the signals. The first four values of the convolution of $x(n)$ and $h(n)$ are, with $T_s = 1$,

$$y(0) = 0.5 \times 0.5 = 0.25, y(1) = 0.5 \times 0.05 + 0.1400 \times 0.5 = 0.095,$$

$$y(2) = 0.0181, y(3) = 0.0028$$

The DTFT of the samples of the two signals are

$$(0.14)^n u(n) - 0.5\delta(n) \leftrightarrow \frac{1}{1 - ((0.14))e^{-j\omega}} - 0.5 \text{ and}$$

$$(0.05)^n u(n) - 0.5\delta(n) \leftrightarrow \frac{1}{1 - (0.05)e^{-j\omega}} - 0.5$$

Therefore, the DTFT of the transform of the convolution of the two signals is

$$(0.14)^n u(n) - 0.5\delta(n) * (0.05)^n u(n) - 0.5\delta(n) \leftrightarrow$$

$$Y(e^{j\omega}) = \left(\frac{1}{1 - (0.14)e^{-j\omega}} - 0.5 \right) \left(\frac{1}{1 - (0.05)e^{-j\omega}} - 0.5 \right)$$

The inverse DTFT yields the same convolution output.

The DFT of the first 4 samples of $x(n)$ is

$$X(0) = 0.6623, X(1) = 0.4808 - j0.1373,$$
$$X(2) = 0.3769, X(3) = 0.4808 + j0.1373$$

Similarly,

$$H(0) = 0.5526, H(1) = 0.4975 - j0.0499,$$
$$H(2) = 0.4524, H(3) = 0.4975 + j0.0499$$

The pointwise product $X(k)H(k), k = 0, 1, 2, 3$ is

$$0.3660, 0.2322 - j0.0922, 0.1705, 0.2322 + j0.0922$$

These values are the 4 equally sampled of the DTFT output spectrum. The IDFT of the product yields the first four time-domain samples we got earlier with a small error.

Usually, the sampling interval of $T_s = 1$ may be inadequate to represent adequately. Therefore, we are forced to take samples at shorter intervals. Let $T_s = 0.1$. Then, the samples are

$$((0.14)^{0.1})^n u(n) - 0.5\delta(n) \qquad ((0.05)^{0.1})^n u(n) - 0.5\delta(n)$$

and we get

$$((0.14)^{0.1})^n u(n) - 0.5\delta(n) \leftrightarrow \frac{1}{1 - ((0.14)^{0.1})e^{-j\omega}} - 0.5$$

and

$$((0.05)^{0.1})^n u(n) - 0.5\delta(n) \leftrightarrow \frac{1}{1 - ((0.05)^{0.1})e^{-j\omega}} - 0.5$$

Therefore, the DTFT of the transform of the convolution of the two signals is

$$((0.14)^{0.1})^n u(n) - 0.5\delta(n) * ((0.05)^{0.1})^n u(n) - 0.5\delta(n) \leftrightarrow$$

$$Y(e^{j\omega}) = \left(\frac{1}{1 - ((0.14)^{0.1})e^{-j\omega}} - 0.5 \right) \left(\frac{1}{1 - ((0.05)^{0.1})e^{-j\omega}} - 0.5 \right)$$

The convolution output $y(n)$ is shown in Fig. 4.12d, and its DTFT spectrum is shown, using dots, in Fig. 4.12c. The convolution output, with $T_s = 0.01$, $y(n)$ is shown in Fig. 4.12f, and its DTFT spectrum is shown, using dots, in Fig. 4.12e. In practice, neither the closed-form expression for the signal nor its transform is

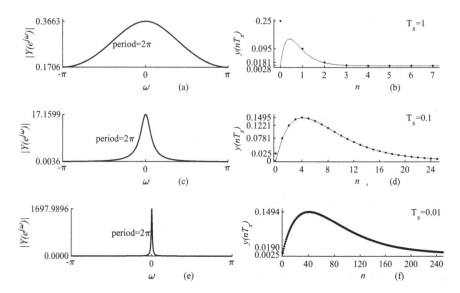

Fig. 4.12 (a) The magnitude of $Y(e^{j\omega}) = X(e^{j\omega})H(e^{j\omega})$ with $T_s = 1$; (b) the inverse DTFT of $Y(e^{j\omega})$, $y(nT_s)$, and the continuous curve shows the exact values of the convolution of the corresponding continuous signals; (c) the magnitude of $Y(e^{j\omega}) = X(e^{j\omega})H(e^{j\omega})$ with $T_s = 0.1$; (d) the inverse DTFT of $Y(e^{j\omega})$, $y(nT_s)$; (e) the magnitude of $Y(e^{j\omega}) = X(e^{j\omega})H(e^{j\omega})$ with $T_s = 0.01$; (f) the inverse DTFT of $Y(e^{j\omega})$, $y(nT_s)$

available. Therefore, we have used the DFT/IDFT with the trial and error method. Start with a signal length and number of samples. Keep increasing the length and reducing the sampling interval until two consecutive iterations yield almost the same output signal or the transform.

4.2.5 Convolution in the Frequency Domain

The convolution integral of the DTFT of two signals in the frequency domain corresponds to the multiplication of the inverse DTFT of the individual signals in the time domain with a scale factor. That is,

$$x(n)y(n) \leftrightarrow \sum_{n=-\infty}^{\infty} x(n)y(n)e^{-j\omega n} = \frac{1}{2\pi} \int_0^{2\pi} X(e^{ju})Y(e^{j(\omega-u)})du$$

Since the DTFT spectrum is periodic, this convolution is periodic.

A simple DFT/IDFT example will clarify this operation. Let

$$a = \{1, 2\}, \quad b = \{4, 3\}, \quad \text{and} \quad ab = \{4, 6\}$$

The respective DFTs are

$$A = \{3, -1\}, \quad B = \{7, 1\}, \quad \text{and} \quad AB = \{10, -2\}$$

The circular convolution of A and B is

$$A \circledast B = \{20, -4\}$$

which is a multiple of AB.

Let us prove this theorem. The problem is that if a signal $z(n)$ is expressed as a product of component signals as

$$z(n) = x(n)y(n),$$

then its DTFT is to be expressed in terms of the DTFT of $x(n)$ and $y(n)$. Consider the DTFT representations of $x(n)$ and $y(n)$

$$x(n) = \frac{1}{2\pi} \int_{-\pi}^{\pi} X(e^{ju})e^{jun} du \quad \text{and} \quad y(n) = \frac{1}{2\pi} \int_{-\pi}^{\pi} Y(e^{jv})e^{jvn} dv$$

Multiplying the two expressions, we get the DTFT representation of $z(n) = x(n)y(n)$ as

$$z(n) = x(n)y(n) = \frac{1}{2\pi} \int_{-\pi}^{\pi} \frac{1}{2\pi} \int_{-\pi}^{\pi} X(e^{ju})Y(e^{jv})e^{j(u+v)n} du\, dv$$

Letting $v = \omega - u$, we get $dv = d\omega$. Then,

$$z(n) = x(n)y(n) = \frac{1}{2\pi} \int_{-\pi}^{\pi} \left(\frac{1}{2\pi} \int_{-\pi}^{\pi} X(e^{ju})Y(e^{j(\omega-u)}) du \right) e^{j\omega n} d\omega$$

This is an inverse DTFT representation of $z(n) = x(n)y(n)$ with the DTFT

$$\left(\frac{1}{2\pi} \int_{-\pi}^{\pi} X(e^{ju})Y(e^{j(\omega-u)}) du \right)$$

That is,

$$z(n) = x(n)y(n) \leftrightarrow \sum_{n=-\infty}^{\infty} x(n)y(n)e^{-j\omega n} = \frac{1}{2\pi} \int_{-\pi}^{\pi} X(e^{ju})Y(e^{j(\omega-u)}) du$$

Consider finding the DTFT of the product of the signals

$$x(n) = (0.8)^n u(n) \quad \text{and} \quad h(n) = (0.9)^n u(n)$$

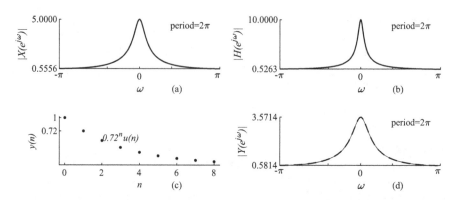

Fig. 4.13 (a) The magnitude of the DTFT of $x(n) = 0.8^n u(n)$; (b) the magnitude of the DTFT of $h(n) = 0.9^n u(n)$; (c) the exponential signal $y(n) = 0.9^n u(n) 0.8^n u(n) = 0.72^n u(n)$ and (d) its magnitude spectrum in continuous line. The same spectrum obtained by numerical convolution in dashed line

The individual DTFTs are

$$x(n) = (0.8)^n u(n)) \leftrightarrow X(e^{j\omega}) = \left(\frac{1}{1 - (0.8)e^{-j\omega}} \right) \text{ and}$$

$$h(n) = (0.9)^n u(n)) \leftrightarrow H(e^{j\omega}) = \left(\frac{1}{1 - (0.9)e^{-j\omega}} \right)$$

with period 2π, shown in Fig. 4.13a and b, respectively. The product and its DTFT are

$$y(n) = x(n)h(n) = (0.72)^n u(n)) \leftrightarrow Y(e^{j\omega}) = \left(\frac{1}{1 - (0.72)e^{-j\omega}} \right)$$

Signal $y(n)$ and its DTFT $Y(e^{j\omega})$ are shown in Fig. 4.13c and d (in continuous line), respectively.

While we can find the DTFT in this case by analytically, for an arbitrary signal, we have to use this property. The DTFT of $y(n) = x(n)h(n)$ is, from the theorem,

$$Y(e^{j\omega}) = \frac{1}{2\pi} \int_0^{2\pi} \frac{1}{1 - 0.9e^{-ju}} \frac{1}{1 - 0.8e^{-j(\omega-u)}} du$$

As it is difficult to find an expression for $Y(e^{j\omega})$, in practice, this convolution is computed numerically using the samples of $X(e^{j\omega})$ and $H(e^{j\omega})$. The numerically obtained DTFT of $x(n)y(n) = 0.72^n u(n)$ is also shown in Fig. 4.13d in dashed line. The results are the same, as it should be.

4.2.6 Time Reversal

Let $x(n) \leftrightarrow X(e^{j\omega})$. Then, $x(-n) \leftrightarrow X(e^{-j\omega})$. For example, with just one nonzero sample,

$$\{x(1) = 1\} \leftrightarrow e^{-j\omega} = \cos(\omega) - j\sin(\omega) \text{ and}$$

$$\{x(-1) = 1\} \leftrightarrow e^{j\omega} = \cos(\omega) + j\sin(\omega)$$

4.2.7 Conjugation

Let $x(n) \leftrightarrow X(e^{j\omega})$. Then,

$$x^*(-n) \leftrightarrow X^*(e^{j\omega}) \quad \text{and} \quad x^*(n) \leftrightarrow X^*(e^{-j\omega})$$

Conjugating both sides of DTFT definition, we get

$$X^*(e^{j\omega}) = \sum_{n=-\infty}^{\infty} x^*(n)e^{j\omega n} = \sum_{n=-\infty}^{\infty} x^*(-n)e^{-j\omega n}$$

Using the time reversal property, we get

$$X^*(e^{-j\omega}) = \sum_{n=-\infty}^{\infty} x^*(n)e^{-j\omega n}$$

For example,

$$\{x(1) = j\} \leftrightarrow je^{-j\omega} \quad \text{and} \quad \{x^*(1) = -j\} \leftrightarrow -je^{-j\omega}$$

4.2.8 Correlation

In summer season, the temperature increases and the sales of cold beverages also increase. Therefore, temperature and sales are positively correlated (mutually related) with a positive correlation coefficient. In apple season, supply of apples is more and the price comes down. Supply and price are negatively correlated with a negative correlation coefficient. The voltage across a linear resistor is positively correlated with the current through it. The input and output of an inverting amplifier are negatively correlated. The correlation coefficient of a random signal and another one is very small and close to zero. The output current in an electrical circuit is totally unrelated to another one in some other electrical circuit. The correlation coefficient between the two currents should be very small.

Correlation is a measure of the similarity relationship between two variables. Out of several applications of correlation, the most important and well-known example is Fourier analysis. Essentially, the correlation coefficient with zero lag is computed. As the frequency components are orthogonal, only the coefficient of the correlated frequency component is returned with the rest zero. In the frequency domain, the cross-spectrum is a measure of the spectral coincidence of the signals. In signal processing, signal detection and object identification are some of the applications of correlation.

The cross-correlation of two signals $x(n)$ and $h(n)$ is defined as

$$r_{xh}(m) = \sum_{n=-\infty}^{\infty} x(n)h^*(n-m), \quad m = -\infty \leq m \leq \infty \qquad (4.4)$$

An alternative cross-correlation definition is

$$r_{xh}(m) = \sum_{n=-\infty}^{\infty} x(n)h^*(n+m), \quad m = -\infty \leq m \leq \infty$$

where the symbol $*$ indicates complex conjugation. The outputs of the two definitions are the time-reversed version of the other, as shown in Fig. 4.14a and b. The correlation of $x(n)$ and $h(n)$ is a function of the delay time. That is the relative shift of one from the other. Our hungriness depends on the time elapsed after our last meal. Similarly, the independent variable in the correlation function is the time-lag or time-delay variable m that also has the dimension of time (delay time). The independent time variable n indicates the current time. Comparing the definitions, we find that the convolution operation without time reversal is the correlation operation. While the purpose of the two operations is different, for computational purposes, correlation is usually implemented with an algorithm for convolution operation.

The correlation of $h(n) = \{\check{2}, 4, 5, 7\}$ and $x(n) = \{\check{1}, 3, 4, 5\}$ is shown in Fig. 4.14a.

$$r_{xh}(m) = \{7, 26, 47, \check{69}, 47, 28, 10\}$$

corresponding to lags m

$$\{-3, -2, -1, 0, 1, 2, 3\}$$

Let us put $m = -3$ in Eq. (4.4). Then, we get

$$r_{xh}(-3) = \sum_{n=-\infty}^{\infty} x(n)h^*(n+3)$$

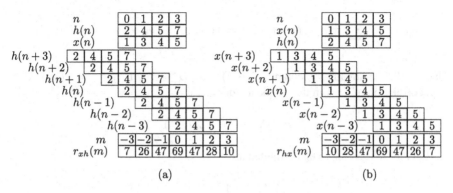

Fig. 4.14 (a) Linear correlation of $x(n)$ and $h(n)$ and (b) that of $h(n)$ and $x(n)$

That is, we are computing $r_{xh}(-3)$. Then, $h(n)$ is left shifted or advanced by 3 sample intervals, as shown in the figure. With different amount of shifts, we get all the outputs. The computation $r_{hx}(m)$ is shown in Fig. 4.14b. Obviously, $r_{xh}(m) = r_{hx}(-m)$. That is, $r_{xh}(m)$ is the time-reversed version of $r_{hx}(m)$. The correlation output we computed is called raw. Usually, this output is divided by a factor to get the normalized output, which is restricted to -1 to 1.

The correlation of signals $x(n)$ and $x(n)$, of a signal by itself, is called the autocorrelation. The autocorrelation of $x(n) = \{\overset{\vee}{1}, 3, 4, 5\}$ is

$$r_{xx}(m) = \{5, 19, 35, \overset{\vee}{51}, 35, 19, 5\}$$

corresponding to lags m

$$\{-3, -2, -1, 0, 1, 2, 3\}$$

The energy of the signal $x(n)$ is given by the autocorrelation with zero lag. That is,

$$r_{xx}(0) = \sum_{n=-\infty}^{\infty} x^2(n)$$

For real sequences, the autocorrelation output is even-symmetric.

The transform pair for the convolution of two signals $x(n)$ and $h(n)$ is given by

$$\sum_{n=-\infty}^{\infty} x(n)h(n-m) \leftrightarrow X(e^{j\omega})H(e^{j\omega})$$

For implementation purposes, correlation of two signals is usually implemented by using convolution of the two signals with one of them time-reversed. The transform pair for the correlation of signal $x(n)$ with $h(n)$ is given by

$$r_{xh}(m) = \sum_{n=-\infty}^{\infty} x(n)h^*(-(m-n)) = x(m) * h^*(-m) \leftrightarrow X(e^{j\omega})H^*(e^{j\omega})$$

due to the conjugation theorem $h^*(-m) \leftrightarrow H^*(e^{j\omega})$. If $x(n) = h(n)$, the autocorrelation of $x(n)$ in the frequency domain is

$$X(e^{j\omega})X^*(e^{j\omega}) = |X(e^{j\omega})|^2$$

Let the two signals to be correlated are

$$\{\check{2}, 1\} \quad \text{and} \quad \{\check{3}, -2\}$$

The corresponding DTFTs are

$$2 + e^{-j\omega} \quad \text{and} \quad 3 - 2e^{-j\omega}$$

By conjugating the DTFT of the second signal, we get $3 - 2e^{j\omega}$. The product of $2 + e^{-j\omega}$ and $3 - 2e^{j\omega}$ is $3e^{-j\omega} + 4 - 4e^{j\omega}$. By taking the inverse DTFT of the product, we get the correlation of $\{\check{2}, 1\}$ and $\{\check{3}, -2\}$ as $\{-4, \check{4}, 3\}$.

4.2.9 Symmetry

DTFT of Real-Valued Signals

A real frequency component is expressed in terms of two complex conjugate exponentials with the same magnitude and the same phase with opposite signs (Euler's formula) or, equivalently, the real part of its spectrum is even and the imaginary part is odd. Therefore, the DTFT of real signals is conjugate- or Hermitian symmetric. That is,

$$X^*(e^{-j\omega}) = X(e^{j\omega})$$

There are only N independent values, in a real-valued sequence of length N. Therefore, its DTFT, with $-\pi \leq \omega < \pi$, must be redundant by a factor 2. As the spectrum is conjugate-symmetric, the plotting of one-half of the spectrum, usually, in the range $0 \leq \omega < \pi$ is adequate.

For example, with just one nonzero sample,

$$\{x(1) = 1\} \leftrightarrow e^{-j\omega} = \cos(\omega) - j\sin(\omega)$$

The magnitude of the spectrum is 1 (even-symmetric) and the phase spectrum is odd, $-\omega$, $-\pi \leq \omega < \pi$ (odd-symmetric), since

$$\cos(\omega n) = \cos(-\omega n) \quad \text{and} \quad \sin(-\omega n) = -\sin(\omega n)$$

Let us reconstruct the signal. The inverse DTFT is

$$x(n) = \frac{1}{2\pi} \int_{-\pi}^{\pi} (e^{-j\omega})e^{j\omega n} d\omega = 1, \quad n = 1 \text{ and } 0 \text{ otherwise}$$

since

$$\int_{-\pi}^{\pi} e^{j\omega k} d\omega = \begin{cases} 1 \text{ for } k = 0 \\ 0 \text{ for } k = \pm 1, \pm 2, \ldots \end{cases}$$

DTFT of Real-Valued and Even-Symmetric Signals

If a signal is real and even, then its spectrum is also real and even, since only cosine components can be used to synthesize it.

Since $x(n)\cos(\omega n)$ is even and $x(n)\sin(\omega n)$ is odd,

$$X(e^{j\omega}) = x(0) + 2\sum_{n=1}^{\infty} x(n)\cos(\omega n) \quad \text{and} \quad x(n) = \frac{1}{\pi} \int_0^{\pi} X(e^{j\omega})\cos(\omega n)d\omega$$

The DTFT of cosine signal, which is even-symmetric, exhibits this symmetry.

DTFT of Real-Valued and Odd-Symmetric Signals

If a signal is real and odd, then its spectrum is imaginary and odd, since only sine components can be used to synthesize it. Since $x(n)\cos(\omega n)$ is odd and $x(n)\sin(\omega n)$ is even,

$$X(e^{j\omega}) = -j2\sum_{n=1}^{\infty} x(n)\sin(\omega n) \quad \text{and} \quad x(n) = \frac{j}{\pi} \int_0^{\pi} X(e^{j\omega})\sin(\omega n)d\omega$$

The DTFT of sine signal, which is odd-symmetric, exhibits this symmetry.

From the fact that the DTFT of a real and even signal is real and even and that of a real and odd is imaginary and odd, the real part of the DTFT, $\text{Re}(X(e^{j\omega}))$, of an arbitrary real signal $x(n)$ is the transform of its even component $x_e(n)$ and $j\,\text{Im}(X(e^{j\omega}))$ is that of its odd component $x_o(n)$, where $x(n) = x_e(n) + x_0(n)$.

4.2.10 Frequency Differentiation

By definition,

$$X(e^{j\omega}) = \sum_{n=-\infty}^{\infty} x(n)e^{-j\omega n}$$

If we differentiate this equation with respect to ω, the equation remains the same with the input and output variables differentiated. That is, $X(e^{j\omega})$ is replaced by $\frac{dX(e^{j\omega})}{d\omega}$ and $e^{-j\omega n}$ is replaced by $(-jn)e^{-j\omega n}$. That is,

$$(-jn)x(n) \leftrightarrow \frac{dX(e^{j\omega})}{d\omega} \quad \text{or} \quad (n)x(n) \leftrightarrow (j)\frac{dX(e^{j\omega})}{d\omega}$$

In general,

$$(-jn)^k x(n) \leftrightarrow \frac{d^k X(e^{j\omega})}{d\omega^k} \quad \text{or} \quad (n)^k x(n) \leftrightarrow (j)^k \frac{d^k X(e^{j\omega})}{d\omega^k}$$

This property is applicable only if the resulting signals still fulfill the conditions for DTFT representation.

For example,

$$\delta(n+3) \leftrightarrow e^{j3\omega}$$

and

$$-3\delta(n+3) \leftrightarrow (j)(j3)e^{j3\omega} = -3e^{j3\omega}$$

Note that $\delta(n+3)$ is nonzero only at $n = -3$.

Consider the transform pair,

$$x(n) = a^n u(n), \; |a| < 1 \leftrightarrow X(e^{j\omega}) = \frac{1}{1 - ae^{-j\omega}}$$

Then,

$$na^n u(n) \leftrightarrow j\frac{dX(e^{j\omega})}{d\omega} = j\frac{d\left(\frac{1}{1-ae^{-j\omega}}\right)}{d\omega} = \frac{ae^{-j\omega}}{(1 - ae^{-j\omega})^2}$$

From this transform pair, we can deduce

$$(n+1)a^n u(n) \leftrightarrow \frac{1}{(1 - ae^{-j\omega})^2}$$

4.2.11 Summation

Let the signal to be summed be

$$\{x(-1) = 3, x(0) = 5, x(1) = 2, x(2) = 4\} \text{ and } 0 \text{ otherwise}$$

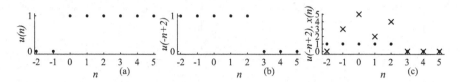

Fig. 4.15 (a) $u(n)$; (b) $u(-n + 2)$; (c) $u(-n + 2)$, $x(n)$

The summation from $n = -\infty$ to $n = 2$ is 14. The unit-step signal is shown in
Fig. 4.15a. Its time-reversed and delayed version $u(-n + 2)$ is shown in Fig. 4.15b.
The signals $x(n)$ and $u(-n + 2)$ are shown in Fig. 4.15c. Now, multiplying $x(n)$
by $u(-n + 2)$ yields the same samples of $x(n)$ up to $n = 2$. Therefore, the sum
remains the same. But, multiplying $x(n)$ by $u(-n + 2)$ and summing are the same
as convolution of $x(n)$ and $u(n)$ to find the convolution up to $n = 2$. Consequently,
in the frequency domain, the summation operation corresponds to $X(e^{j\omega})U(e^{j\omega})$.
The summation of a time-domain function, $x(n)$, can be expressed, in terms of its
DTFT $X(e^{j\omega})$, as

$$s(n) = \sum_{m=-\infty}^{n} x(m) \leftrightarrow S(e^{j\omega}) = \frac{X(e^{j\omega})}{(1 - e^{-j\omega})} + \pi X(e^{j0})\delta(\omega), \quad -\pi < \omega \leq \pi$$

If $X(e^{j0}) = 0$, the time-summation operation can be considered as the inverse of
the time-differencing operation. In addition to the strictly continuous component of
the spectrum, an impulsive component is required to represent the DC component
of $x(n)$. This property is applicable on the condition that the summation is bounded.
That is, its DTFT exists.
   With

$$X(e^{j\omega}) = 3e^{j\omega} + 5 + 2e^{-j\omega} + 4e^{-2j\omega},$$

the DTFT of the summation of $x(n)$ is

$$S(e^{j\omega}) = \frac{3e^{j\omega} + 5 + 2e^{-j\omega} + 4e^{-2j\omega}}{(1 - e^{-j\omega})} + 14\pi\delta(\omega), \quad -\pi < \omega \leq \pi$$

Using the transform pair

$$\begin{cases} 1 \text{ for } n \geq 0 \\ -1 \text{ for } n < 0 \end{cases} = u(n) - 0.5 \leftrightarrow= \frac{1}{1 - e^{-j\omega}},$$

we get the reconstructed $s(n)$ as shown in Fig. 4.16. The partial $s(n)$ corresponding
to the transforms

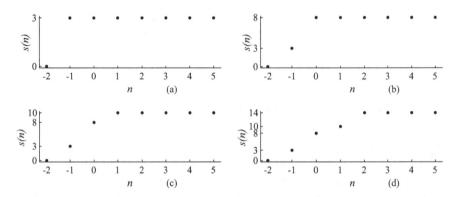

Fig. 4.16 (a)–(c) Partial $s(n)$; (d) $s(n)$

$$S(e^{j\omega}) = \frac{3e^{j\omega}}{(1 - e^{-j\omega})} + 3\pi\,\delta(\omega), \quad -\pi < \omega \leq \pi,$$

$$S(e^{j\omega}) = \frac{3e^{j\omega} + 5}{(1 - e^{-j\omega})} + 8\pi\,\delta(\omega), \quad -\pi < \omega \leq \pi,$$

$$S(e^{j\omega}) = \frac{3e^{j\omega} + 5 + 2e^{-j\omega}}{(1 - e^{-j\omega})} + 10\pi\,\delta(\omega), \quad -\pi < \omega \leq \pi$$

are shown, respectively, in Fig. 4.16a–c. The fully reconstructed $s(n)$ is shown in Fig. 4.16d.

We know that the summation of the delayed unit impulse is the delayed unit-step signal and

$$\delta(n - 1) \leftrightarrow e^{-j\omega}$$

Therefore, using this property, we get the DTFT of $u(n - 1)$, over one period, as

$$u(n - 1) = \sum_{m=-\infty}^{n} \delta(m) \leftrightarrow \frac{e^{-j\omega}}{(1 - e^{-j\omega})} + \pi\,\delta(\omega), \quad -\pi < \omega \leq \pi$$

For arbitrary signals, as usual, numerical approximation is required to implement the theorem.

4.2.12 Parseval's Theorem and the Energy Transfer Function

The energy of the signal $x(n)$ is defined as

$$E = r_{xx}(0) = \sum_{n=-\infty}^{\infty} |x(n)|^2,$$

which is the autocorrelation of $x(n)$ with lag 0. With $x(n) \leftrightarrow X(e^{j\omega})$, we get

$$E = \sum_{n=-\infty}^{\infty} |x(n)|^2 = \frac{1}{2\pi} \int_0^{2\pi} |X(e^{j\omega})|^2 d\omega$$

The energy of a signal can also be represented in the frequency-domain representation, as this representation is an equivalent representation. Since $\frac{1}{2\pi}|X(e^{j\omega})|^2 d\omega$ is the signal energy over the infinitesimal frequency band ω to $\omega + d\omega$, the quantity $|X(e^{j\omega})|^2$ is called the energy spectral density of the signal.

Consider the signal with only two nonzero values and its DTFT

$$\{x(-1) = -j, x(1) = j\} \leftrightarrow -j(e^{j\omega} - e^{-j\omega}) = 2\sin(\omega)$$

The energy of the signal is $1^2 + 1^2 = 2$ from its time-domain representation. The energy of the signal is

$$E = \frac{1}{2\pi} \int_0^{2\pi} |2\sin(\omega)|^2 d\omega = \frac{1}{\pi} \int_0^{2\pi} (1 - \cos(2\omega)) d\omega = 2$$

using its frequency-domain representation also.

The output of a LTI system is, in the frequency domain, the transfer function $H(e^{j\omega})$ times of that of the input. Therefore,

$$Y(e^{j\omega}) = H(e^{j\omega})X(e^{j\omega})$$

where $X(e^{j\omega})$, $Y(e^{j\omega})$, and $H(e^{j\omega})$ are the DTFT of the input $x(n)$, output $y(n)$, and impulse response of the system $h(n)$. Then, the output energy spectrum is given by

$$|Y(e^{j\omega})|^2 = Y(e^{j\omega})Y^*(e^{j\omega})$$
$$= H(e^{j\omega})X(e^{j\omega})H^*(e^{j\omega})X^*(e^{j\omega}) = |H(e^{j\omega})|^2|X(e^{j\omega})|^2$$

As it relates the input and output energy spectral densities of a system, the quantity $|H(e^{j\omega})|^2$ is called the energy transfer function of the system.

Let us do a simple problem using the DFT and IDFT to verify the theorem. The two sequences are specified by their nonzero samples as

$$\{x(0) = 3, x(1) = 2, x(2) = 1\} \quad \text{and} \quad \{h(0) = 1, h(1) = 3\}$$

and their convolution output is

$$\{y(0) = 3, y(1) = 11, y(2) = 7, y(3) = 3\}$$

The 4-point DFTs of the sequences are

$$X(k) = \{6, 2 - j2, 2, 2 + j2\}$$

$$H(k) = \{4, 1 - j3, -2, 1 + j3\}$$

$$Y(k) = X(k)H(k) = \{24, -4 - j8, -4, -4 + j8\}$$

Now,

$$|Y(k)|^2 = |H(k)|^2 |X(k)|^2 = \{576, 80, 16, 80\}$$

The energies of the signal $y(n)$ computed in the frequency and time domains are the same, 188.

4.3   System Response and the Transfer Function

In general, the main tasks in signal and system analysis and design are characterization and representation of signals and to find the relation between the input and output of systems. These tasks are extremely difficult to carry out with practical signals and systems, since signals have arbitrary amplitude in practice. Therefore, the basic procedure in the analysis of LTI systems is to use well-defined signals as intermediaries. In the time domain, arbitrary signals are represented in terms of scaled and shifted impulses. The impulse response represents a system in the time domain. The relationship between the input and output of systems is the convolution operation. In the frequency domain, signals are represented in terms of complex exponentials. The relationship between the input and output of systems is the ratio of the transforms of the output and input, called the transfer function. The transfer function represents a system in the frequency domain, which is the transform of the system impulse response. In both the domains, the input–output relationship is valid assuming that the system is initially relaxed. It is ironic that both the impulse and the complex exponential have no physical existence. However, both the signals are used as intermediaries in practical signal and system analysis.

4.3.1   The Transfer Function

With $x(n)$, $h(n)$, and $y(n)$, respectively, the system input, impulse response, and output in the time domain, and $X(e^{j\omega})$, $H(e^{j\omega})$, and $Y(e^{j\omega})$, their respective DTFT, the stable LTI system output is given, in the two domains, as

$$y(n) = \sum_{m=-\infty}^{\infty} x(m)h(n-m) \leftrightarrow Y(e^{j\omega}) = X(e^{j\omega})H(e^{j\omega})$$

Therefore, the transfer function, with respect to the DTFT, is given by

$$H(e^{j\omega}) = \frac{Y(e^{j\omega})}{X(e^{j\omega})},$$

provided $|X(e^{j\omega})| \neq 0$ for all frequencies and the system is initially relaxed. It is called the transfer function, since the input is transferred to output by multiplication with $H(e^{j\omega})$. The transfer function is also the frequency response, since it represents the filtering characteristic (response to real sinusoids or their equivalent complex exponentials) of the system.

Taking the DTFT of Equation (2.6), we get the DTFT version of the transfer function of a general Nth order causal LTI discrete system, with $M \leq N$ and the coefficient of $y(n)$ is a_0 instead of 1, is given by

$$H(e^{j\omega}) = \frac{(b_0 + b_1 e^{-j\omega} + \cdots + b_M e^{-j\omega M})}{(a_0 + a_1 e^{-j\omega} + \cdots + a_N e^{-j\omega N})} \tag{4.5}$$

4.3.2 System Analysis

Example 4.7 Find the output $y(n)$, using the DTFT, of the system characterized by the difference equation

$$y(n) = 2x(n) + 0.8y(n-1)$$

to the input $x(n) = 2\cos(\frac{2\pi}{8}n + \frac{\pi}{3})$.

Solution When a sinusoid or equivalent complex exponential is applied to a stable LTI system, the output signal is of the same form as the input with its amplitude and phase changed. These changes are the same for a sinusoid or equivalent complex exponential. Therefore, for easier analysis, the complex exponential $e^{j\omega n}$ is assumed to be the input. Taking the DTFT of the difference equation, we get

$$Y(e^{j\omega}) = 2X(e^{j\omega}) + 0.8e^{-j\omega}Y(e^{j\omega}) \text{ or } H(e^{j\omega}) = \frac{2}{1 - 0.8e^{-j\omega}} = \frac{2e^{j\omega}}{e^{j\omega} - 0.8}$$

Substituting $\omega = \frac{2\pi}{8}$, we get

$$H\left(e^{j\frac{2\pi}{8}}\right) = 2\frac{e^{j\frac{2\pi}{8}}}{e^{j\frac{2\pi}{8}} - 0.8} = 1.7078 - j2.2244 = 2.8043\angle(-0.9160)$$

The response of the system to the input $x(n) = 2\cos(\frac{2\pi}{8}n + \frac{\pi}{3})$ is $y(n) = (2)2.8043\cos(\frac{2\pi}{8}n + \frac{\pi}{3} - 0.9160)$.

The input is an everlasting signal. Therefore, the response of the system is the steady-state response. Since the input is applied at $n = -\infty$, the transient component of the output has died down for any finite n. The natural response of the system is of the form 0.8^n, which is a decaying exponential.

∎

Example 4.8 Find the impulse response $h(n)$, using the DTFT, of the system characterized by the difference equation

$$y(n) = 2x(n) + x(n-1) + x(n-2) + (2.1333)y(n-1)$$
$$- (1.5111)y(n-2) + (0.3556)y(n-3)$$

Verify the result by finding the first 4 output values by iterating the difference equation.

Solution

$$H(e^{j\omega}) = \frac{2 + e^{-j\omega} + e^{-j2\omega}}{(1 - 2.1333e^{-j\omega} + 1.5111e^{-j2\omega}) - 0.3556e^{-j3\omega})}$$

$$= \frac{2 + e^{-j\omega} + e^{-j2\omega}}{(1 - \frac{4}{5}e^{-j\omega})(1 - \frac{2}{3}(e^{-j\omega}))^2}$$

Expanding into partial fractions, we get

$$Y(e^{j\omega}) = \frac{173.25}{\left(1 - \frac{4}{5}e^{-j\omega}\right)} + \frac{-19.1667}{\left(1 - \frac{2}{3}e^{-j\omega}\right)^2} + \frac{-171.25}{\left(1 - \frac{2}{3}e^{-j\omega}\right)}$$

Taking the inverse DTFT, we get the impulse response

$$y(n) = (173.25)\left(\frac{4}{5}\right)^n + n(-19.1667)\left(\frac{2}{3}\right)^{(n-1)}$$

$$+ (-171.25)\left(\frac{2}{3}\right)^n, \quad n = 0, 1, \ldots$$

The first four values of the impulse response $h(n)$ are

$$h(0) = 2, \quad h(1) = 5.2666, \quad h(2) = 9.2133, \quad h(3) = 12.4077$$

∎

Example 4.9 Find the zero-state response, using the DTFT, of the system characterized by the difference equation

$$y(n) = x(n) + x(n-1) - 2x(n-2) + \frac{5}{4}y(n-1) - \frac{3}{8}y(n-2)$$

The input is $x(n) = (0.8)^n u(n)$. Verify the result by finding the first 4 output values by iterating the difference equation.

Solution

$$H(e^{j\omega}) = \frac{1 + e^{-j\omega} - 2e^{-j2\omega}}{(1 - \frac{5}{4}e^{-j\omega} + \frac{3}{8}e^{-j2\omega})} = \frac{1 + e^{-j\omega} - 2e^{-j2\omega}}{(1 - \frac{3}{4}e^{-j\omega})(1 - \frac{1}{2}e^{-j\omega})}$$

With $X(e^{j\omega}) = \frac{1}{(1 - 0.8e^{-j\omega})}$,

$$Y(e^{j\omega}) = H(e^{j\omega})X(e^{j\omega}) = \frac{1 + e^{-j\omega} - 2e^{-j2\omega}}{(1 - 0.8e^{-j\omega})(1 - \frac{3}{4}e^{-j\omega})(1 - \frac{1}{2}e^{-j\omega})}$$

The partial fraction expansion is

$$Y(e^{j\omega}) = -\frac{37.3333}{(1 - 0.8e^{-j\omega})} + \frac{55}{(1 - \frac{3}{4}e^{-j\omega})} - \frac{16.6667}{(1 - \frac{1}{2}e^{-j\omega})}$$

Taking the inverse DTFT, we get the zero-state response

$$y(n) = \left(-37.3333\,(0.8)^n + 55\left(\frac{3}{4}\right)^n - 16.6667\left(\frac{1}{2}\right)^n \right) u(n)$$

The first four values of the sequence $y(n)$ are

$$y(0) = 1, \quad y(1) = 3.05, \quad y(2) = 2.8775, \quad y(3) = 2.0051$$

∎

4.3.3 Digital Differentiator

Differentiation operation is fundamental to signal and system analysis. The frequency response of an ideal differentiator is linearly proportional to frequency. This operation can also be approximated numerically and implemented by digital systems adequately. As a discrete system, the digital differentiator is characterized by its discrete impulse response in the time domain and the transfer function in the frequency domain. We derive the impulse response of the digital differentiator

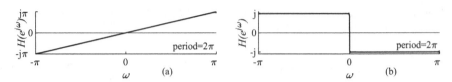

Fig. 4.17 (**a**) The frequency response of the ideal digital differentiator; (**b**) the frequency response of the ideal Hilbert transformer

from its frequency response. The input to the differentiator is the samples of the continuous signal $x(t)$, whose derivative is being approximated. The output is the samples of the derivative. That is,

$$y(t) = \frac{dx(t)}{dt}$$

in the continuous time domain. In the discrete-time domain,

$$y(n) = \frac{dx(t)}{dt}|_{t=n}$$

The sampling interval has to be fixed appropriately. As the sampling interval tends to 0, the approximation improves.

The periodic frequency response, shown in Fig. 4.17a over one period, of the ideal digital differentiator is defined as

$$H(e^{j\omega}) = j\omega, \qquad -\pi < \omega \le \pi$$

For example, the input and output of the differentiator are

$$\cos(\omega_0 n) \leftrightarrow \pi(\delta(\omega + \omega_0) + \delta(\omega - \omega_0))$$

$$j\omega_0\pi(-\delta(\omega + \omega_0) + \delta(\omega - \omega_0)) \leftrightarrow -\omega_0 \sin(\omega_0 n)$$

The impulse response of the ideal differentiator is obtained by finding the inverse DTFT of its frequency response

$$h(n) = \frac{1}{2\pi} \int_{-\pi}^{\pi} j\omega e^{j\omega n} d\omega = \frac{\cos(\pi n)}{n} = \begin{cases} \frac{(-1)^n}{n} & \text{for } n \ne 0 \\ 0 & \text{for } n = 0 \end{cases}, \qquad -\infty < n < \infty$$

As the frequency response of the differentiator is imaginary and odd-symmetric, the impulse response is real and odd-symmetric. An example of designing linear-phase FIR differentiating filter is presented later.

4.3.4   Hilbert Transform

An ideal transformer is an all-pass filter that imparts a $-90°$ phase shift on all the real frequency components of the input signal. For example, the Hilbert transform of $\cos(\omega n)$ is $\cos(\omega n - \frac{\pi}{2}) = \sin(\omega n)$. Obviously, the input and output are in the same domain, unlike other transforms with two domains. Let us form a complex signal with the given signal $\cos(\omega n)$ as the real part and its Hilbert transform as the imaginary part. We get

$$\cos(\omega n) + j\sin(\omega n) = e^{j\omega n}$$

the N-point DFT of which is nonzero, N, at ω only. In general, such signals with zero DFT coefficients in the negative part of the spectrum are called analytic signal. That type of signals is very useful in applications such as the sampling of bandpass signals and single-sideband amplitude modulation.

An arbitrary signal

$$x(n) = 0.5 + \cos\left(\frac{2\pi}{16}n\right) + \cos\left(3\frac{2\pi}{16}n + \frac{\pi}{6}\right) + \sin\left(5\frac{2\pi}{16}n\right) - 0.25\cos(\pi n)$$

and its DFT spectrum are shown, respectively, in Fig. 4.18a and b. The Hilbert transform of the signal in (a)

$$x_H(n) = \cos\left(\frac{2\pi}{16}n - \frac{\pi}{2}\right) + \cos\left(3\frac{2\pi}{16}n + \frac{\pi}{6} - \frac{\pi}{2}\right) + \sin\left(5\frac{2\pi}{16}n - \frac{\pi}{2}\right)$$

and its DFT spectrum are shown, respectively, in Fig. 4.18c and d. The DC component 0.5 and the component with frequency π, $-0.25\cos(\pi n)$ disappear, as all their samples become zero at these frequencies. The differences between the spectra in (b) and (d) are that the values at index $k = 0$ and at $k = \frac{N}{2} = 8$ are zero in (d), the values of the other positive frequency components in (b) are multiplied by $-j$, and those of the negative frequency components in (b) are multiplied by j. Therefore, the IDFT of the spectrum in (d) is the Hilbert transform of the signal.

Figure 4.18e and f shows, respectively, the signal $jx_H(n)$ and its DFT spectrum. Figure 4.18g and h shows, respectively, the complex signal $x(n) + jx_H(n)$ and its DFT spectrum. The spectral values in (h) with indices from 1 to 7 are twice of those in the first half of (b). Values with indices 0 and 8 are the same, and the rest of the values are zero, which is a one-sided spectrum.

The periodic frequency response, shown in Fig. 4.17b over one period, of the ideal Hilbert transformer is defined as

$$H(e^{j\omega}) = \begin{cases} -j & \text{for } 0 < \omega \leq \pi \\ j & \text{for } -\pi < \omega < 0 \end{cases}$$

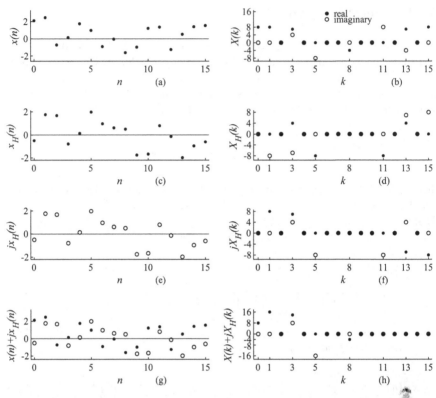

Fig. 4.18 (**a**) An arbitrary signal and (**b**) its DFT; (**c**) the Hilbert transform of signal in (**a**) and (**d**) its DFT; (**e**) the signal in (**c**) multiplied by j and (**f**) its DFT; (**g**) the sum of signals in (**a**) and (**e**), and (**h**) its one-sided DFT spectrum

The impulse response of the ideal Hilbert transformer is obtained by finding the inverse DTFT of its frequency response.

$$h(n) = \frac{1}{2\pi} \int_0^\pi -je^{j\omega n} d\omega + \frac{1}{2\pi} \int_{-\pi}^0 je^{j\omega n} d\omega$$

$$= \begin{cases} \frac{2\sin^2(\frac{\pi n}{2})}{\pi n} & \text{for } n \neq 0 \\ 0 & \text{for } n = 0 \end{cases}, \quad -\infty < n < \infty$$

4.4   Approximation of the DTFT

In engineering design and analysis of signals and systems, usually, there will be some tolerances given in the specifications of systems. First, we start studying ideal systems. They set a bound in the performance. Unfortunately, ideal systems cannot be realized in practice due to the limitations in the response of physical components of a system. Fortunately, it is adequate to approximate the ideal systems with some tolerances. Therefore, we start studying the ideal case and then approximate it to the required accuracy.

The signals and the system responses usually have arbitrary amplitude profiles and aperiodic. The DFT with fast algorithms for its computation is the workhorse in design and analysis. The DFT is finite and discrete in both the domains. In practice, signals are not so. First, we study the transforms with well-specified signals, such as impulse, step, and sinusoids, to understand the theory. Then, we study the approximation of signals using the DFT. Therefore, the approximation of the analysis is very important in practice.

The spectra of both the DFT and DTFT are periodic with the same limits. The DFT spectrum is the uniform samples of the DTFT spectrum. The corresponding time-domain signals are periodic and aperiodic. Due to this difference, the time-domain signal has to be truncated and assumed to be periodic. The result is that the DTFT spectrum is not exact. The solution is to truncate the signal with sufficient window length so that the distortion of the DTFT spectrum is negligible. Fortunately, practical signals have samples with significant magnitude within a finite range.

4.4.1 Approximating the DTFT Spectrum by the DFT

As the input signal to the DTFT is aperiodic, while that of the DFT is periodic, inevitably, truncation is required. If no truncation is necessary, we approximate the samples of the continuous DTFT spectrum by its exact samples using the DFT. We can increase the number of samples by zero padding the input data. Let

$$x(n) = \{x(-2) = 3, x(-1) = -3, x(0) = 2, x(1) = 1\}$$

and zero otherwise. The DTFT of $x(n)$ is

$$X(e^{j\omega}) = 3e^{j2\omega} - 3e^{j\omega} + 2 + e^{-j\omega}$$

The input samples $x(n)$ can be specified anywhere in the infinite range. The DFT algorithms usually assume that the N-point data are in the range from 0 to $N - 1$. Using the assumed periodicity of the DFT, the data can be specified in that range by periodic extension, and we get

$$xp(n) = \{x(0) = 2, x(1) = 1, x(2) = 3, x(3) = -3\}$$

The fast and practically used DFT algorithms are of length that is an integer power of 2. This requirement can be fulfilled by sufficient zero padding of the data and, at the same time, to satisfy the required number of samples of the spectrum. For example, with data length 3 and spectral samples 5, we have to zero pad the data to make the length 8. The DFT of $xp(n)$, $XP(k)$, is

$$\{XP(0) = 3, XP(1) = -1 - j4, XP(2) = 7, XP(3) = -1 + j4\}$$

The set of samples of the DTFT $X(e^{j\omega})$ at

$$\omega = 0, \omega = \frac{2\pi}{4}, \omega = \frac{2\pi}{4}, \omega = 3\frac{2\pi}{4}$$

are the same as the DFT $XP(k)$. Let us zero pad the signal $xp(n)$ to make its length 8. Then,

$$xp(n) = \{x(0) = 2, x(1) = 1, x(2) = 3, x(3) = -3, x(4) = 0, x(5) = 0,$$
$$x(6) = 0, x(7) = 0\}$$

The DFT is

$$\{3, 4.8284 - j1.5858, -1 - j4, -0.8284 + j4.4142,$$
$$7, -0.8284 - j4.4142, -1 + j4, 4.8284 + j1.5858\}$$

The even-indexed samples are the same as those obtained for the $xp(n)$ with 4 samples.

Let us consider the effect of data truncation. The criterion for data truncation is that most of the energy of the signal is retained. Let

$$x(n) = \{x(-3) = 3, x(-2) = -3, x(-1) = 1, x(0) = 2, x(1) = 1\}$$

and zero otherwise. Then,

$$xp(n) = \{x(0) = 2, x(1) = 1, x(2) = 3, x(3) = -3, x(4) = 1\}$$

The DFT of $xp(n)$ is

$$XP(k) = \{4, 2.6180 - j3.5267, 0.3820 + j5.7063, 0.3820 - j5.7063,$$
$$2.6180 + j3.5267\}$$

The truncation operation can be considered as multiplying the signal with a window, which reduces the effective length of the data. The DFT of window

$$w(n) = \{1, 1, 1, 1, 0\}$$

is

$$W(k) = \{4, -0.3090 - j0.9511, 0.8090 - j0.5878, 0.8090 + j0.5878i$$

$$-0.3090 + j0.9511\}$$

If we multiply $xp(n)$ point by point by the window $w(n)$, we get the truncated signal $xp\_t(n)$. The DFT of the truncated signal $xp\_t(n)$

$$xp\_t(n) = \{2, 1, 3, -3, 0\}$$

is

$$XP\_t(k) = \{3, 2.3090 - j4.4778, 1.1910 + j5.1186, 1.1910 - j5.1186,$$

$$2.3090 + j4.4778\}$$

This is the DFT of the circular convolution of a rectangular window $W(k)$ and that of $xp(n)$, $XP(k)$, divided by 5, the convolution theorem in the frequency domain. Let us also compute the spectrum of the truncated signal using the truncation model. The circular convolution of the two spectra can be obtained using the DFT and IDFT. The DFT of $W(k)$ is

$$\{5, 0, 5, 5, 5\}$$

The DFT of $XP(k)$ is

$$\{10, 5, -15, 15, 5\}$$

The point-by-point multiplication of the two spectra divided by 5 is

$$\{10, 0, -15, 15, 5\}$$

The IDFT of this spectrum is the spectrum of $xp\_t(n)$, $XP\_t(k)$, which is the same as that we found already. In summary, to approximate the DTFT of an arbitrary signal, start with some data length and find its DFT. Repeat this operation by doubling the data length until the difference between the energy of the signal of two consecutive iterations is negligible.

4.4.2 Approximating the Inverse DTFT by the IDFT

Let

$$x(n) = \{x(0) = 3, x(1) = 1, x(2) = 2, x(3) = 4\}$$

The DTFT of $x(n)$ is

$$X(e^{j\omega}) = 3 + e^{-j\omega} + 2e^{-j2\omega} + 4e^{-j3\omega}$$

The samples of the spectrum at $\omega = 0, \pi/2, \pi, 3\pi/2$ are

$$\{10, 1 + j3, 0, 1 - j3\}$$

The 4-point DFT of $x(n)$ is also the same. The IDFT of these samples yields $x(n)$ exactly, since the number of samples of the spectrum that represents the data is sufficient. However, if we take just two samples of the spectrum $\{10, 0\}$, and taking the IDFT, we get $\{5, 5\}$. Time-domain aliasing has occurred. That is, the time samples are

$$\{3 + 2 = 5, 1 + 4 = 5\}$$

due to insufficient number of bins to hold the exact data. Aliasing is unavoidable in practice due to the finite and infinite natures, respectively, of the DFT and the DTFT in the time domain. If we take more number of samples of the DTFT spectrum, we get the exact data with some zeros appended. With 6 spectral samples

$$\{10, -1.5000 - j2.5981, 5.5000 + j0.8660, 0, 5.5000 - 0.8660, -1.5000 + j2.5981\},$$

the IDFT yields

$$\{3, 1, 2, 4, 0, 0\}$$

With time-limited data, the exact time-domain samples can be obtained from adequate samples of the DTFT spectrum. Otherwise, aliasing is unavoidable. Then, it has to be ensured that time-domain aliasing is negligible by taking sufficient number of spectral samples. In summary, to approximate the inverse DTFT of an arbitrary signal, start with some number of spectral samples and find its IDFT. Repeat this operation by doubling the data length until the difference between the energy of the signal of two consecutive iterations is negligible.

4.5   Summary

- The DTFT finds the frequency content of aperiodic discrete signals. As the period of the time-domain signal is infinite, the frequency increment of the spectrum tends to zero and the spectrum is continuous. A sampled signal can only have a periodic spectrum, as the effective frequency range is finite. Therefore, the DTFT spectrum is continuous and periodic.
- As always, at discontinuities, the Fourier reconstructed continuous waveform converges to the average value at each discontinuity both in the frequency and time domains. In addition, oscillations around the discontinuities, called Gibbs

phenomenon, also occur. No matter how many frequency components are used for the reconstruction, the largest of the magnitude of the oscillations converges to the value of 1.0869 for a unit discontinuity with a relatively small number of harmonics used.

- In the case of the DTFT, the spectrum is continuous and the magnitude of each of the infinite frequency components is infinitesimal. The amplitudes of the components are given by

$$\frac{1}{2\pi} X(e^{j\omega}) d\omega$$

As the amplitudes are immeasurably small, the spectral density $X(e^{j\omega})$ represents $x(n)$ in the frequency domain as a relative amplitude spectrum.

- The DFT spectrum is the samples of the DTFT spectrum at equal intervals of $\frac{2\pi}{N}$. That is,

$$X(k) = X(e^{j\omega})|_{\omega=k\omega_0} = X(e^{jk\omega_0})$$

- The DTFT of a periodic signal, with period 2π, is a set of impulses with strength $\frac{2\pi}{N} X(k)$ at $\omega = \frac{2\pi}{N} k$.
- Convolution in the time domain becomes much simpler multiplication in the frequency domain.
- The energy of a signal can be represented either in the frequency domain or time domain, as the two representations are equivalent.
- The DTFT and the inverse DTFT can be adequately approximated by the DFT and the IDFT in practice.

Exercises

4.1
   (a) Find the DTFT of

$$x(n) = \begin{cases} 1 \text{ for } -N \leq n \leq N \\ 0 \text{ otherwise} \end{cases}$$

With $N = 4$, deduce the samples of $X(e^{j\omega})$ in polar form at $\omega = 0$ and $\omega = \pi$. From the DTFT, deduce the samples of $x(0)$.
   (b) Using the time-domain shift theorem, deduce the DTFT of

$$x(n) = \begin{cases} 1 \text{ for } 0 \leq n \leq N \\ 0 \text{ otherwise} \end{cases}$$

With $N = 4$, deduce the samples of $X(e^{j\omega})$ in polar form at $\omega = 0$ and $\omega = \pi$. From the DTFT, deduce the value of the sample $x(0)$.

4.2 The samples over a period of a periodic sequence $x(n)$ are given. Find its DTFT using the DFT.

4.2.1 $\{x(0) = 1, x(1) = 1, x(2) = 2, x(3) = -3\}$.

* 4.2.2. $\{x(0) = 2, x(1) = -1, x(2) = 4, x(3) = 3\}$.

4.2.3. $\{x(0) = -2, x(1) = 3, x(2) = -2, x(3) = 4\}$.

4.3 Find the DFT, $X(k)$, of $x(n)$ with its only 4 nonzero values given. Find the samples of the DTFT, $X(e^{j\omega})$, of $x(n)$ at $\omega = \{0, \frac{\pi}{2}, \pi, \frac{3\pi}{2}\}$. Verify that they are the same as those of $X(k)$.

* 4.3.1. $\{x(n), n = 0, 1, 2, 3\} = \{4, 3, -1, 2\}$ and $x(n) = 0$ otherwise.

4.3.2. $\{x(n), n = 0, 1, 2, 3\} = \{1, 2, 3, 4\}$ and $x(n) = 0$ otherwise.

4.3.3 $\{x(n), n = 0, 1, 2, 3\} = \{0, 0, 0, 1\}$ and $x(n) = 0$ otherwise.

4.4 Find the convolution $y(n)$ of $x(n)$ and $h(n)$, using the time-domain convolution property of the DTFT. Find the first 4 values of $y(n)$ by time-domain convolution and verify that they are the same as those obtained using the DTFT.

4.4.1. $x(n) = (0.8)^n u(n)$ and $h(n) = x(n)$.

4.4.2. $x(n) = (0.9)^n u(n)$ and $h(n) = u(n)$.

* 4.4.3. $x(n) = (0.8)^n u(n)$ and $h(n) = (0.6)^n u(n)$.

4.5 Using the frequency-domain convolution property, find the DTFT of the product of $x(n)$ and $h(n)$ and then find the inverse of the DTFT to verify that it is the same as the product of $x(n)$ and $h(n)$. Use the DFT to approximate the DTFT.

4.5.1. The nonzero samples of $x(n)$ are $\{x(n), n = 0, 1, 2, 3\} = \{1, 2, 3, 4\}$. The nonzero samples of $h(n)$ are $\{x(n), n = 0, 1, 2, 3\} = \{3, -1, 2, 4\}$.

* 4.5.2. The nonzero samples of $x(n)$ are $\{x(n), n = 0, 1, 2, 3, 4\} = \{3, 2, 1, 4, -2]\}$. The nonzero samples of $h(n)$ are $\{x(n), n = 0, 1, 2, 3, 4\} = \{1 - 1, 3, 4, -1\}$.

4.5.3. $x(n) = 2\sin(n)$ and $h(n) = \sin(n)$.

4.6 A sequence $x(n)$ is given. Find the summation

$$s(n) = \sum_{n=-\infty}^{n} x(n)$$

using the time-summation property of the DTFT.

4.6.1 $x(n) = n^2 u(n)$.

4.6.2 $x(n) = \cos(\frac{2\pi}{4}n)$, $n = 0, 1, 2, 3, 4$.

* 4.6.3 $x(n) = (0.8)^n u(n)$.

4.6.4 $x(n) = (1.4286)^n u(-n)$.

4.7 The difference equation governing a system is

$$y(n) = x(n) - x(n-1) + x(n-2) + \frac{7}{6}y(n-1) - \frac{1}{3}y(n-2)$$

with input $x(n)$ and output $y(n)$. Use the DTFT to find the impulse response $h(n)$ of the system. List the first four values of $h(n)$.

4.8 The difference equation governing a system is

$$y(n) = x(n) + 2x(n-1) - x(n-2) + \frac{17}{12}y(n-1) - \frac{1}{2}y(n-2)$$

with input $x(n)$ and output $y(n)$. Use the DTFT to find the zero-state response $y(n)$ of the system with the input $x(n) = u(n)$, the unit-step function. List the first four values of $y(n)$.

* **4.9** The difference equation governing a system is

$$y(n) = x(n) - 2x(n-1) - x(n-2) + \frac{4}{3}y(n-1) - \frac{4}{9}y(n-2)$$

with input $x(n)$ and output $y(n)$. Use the DTFT to find the zero-state response $y(n)$ of the system with the input $x(n) = (\frac{4}{5})^n u(n)$. List the first four values of $y(n)$.

4.10 Use the DFT and IDFT to construct an analytic signal with one-sided spectrum.

4.10.1 $\{x(n), \ n = 0, 1, 2, 3\} = \{2, 1, 3, 4\}$

* **4.10.2** $\{x(n), \ n = 0, 1, 2, 3\} = \{3, -1, 2, 2\}$

The z-Transform

5

In such applications as stability analysis of systems, convolution output may be unbounded. However, convolution is defined only for signals with bounded output. But the unbounded signals can be converted to bounded ones by multiplying with an appropriate exponential. For example, let the signals to be convolved be

$$x(n) = 0.8^n u(n) \quad \text{and} \quad h(n) = 0.6^n u(n)$$

The convolution output is

$$y(n) = x(n) * h(n) = \left(\frac{0.8^{n+1} - 0.6^{n+1}}{0.8 - 0.6} \right) u(n)$$

which is bounded. Let the signals to be convolved be

$$x(n) = 1.4^n u(n) \quad \text{and} \quad h(n) = 0.8^n u(n)$$

The convolution output is

$$y(n) = x(n) * h(n) = \left(\frac{1.4^{n+1} - 0.8^{n+1}}{1.4 - 0.8} \right) u(n)$$

which is unbounded. Let both the signals to be convolved be multiplied by the exponential $0.4^n u(n)$. Then, we get

$$x(n) = 0.56^n u(n) \quad \text{and} \quad h(n) = 0.32^n u(n)$$

Now, the convolution output is

$$y(n) = x(n) * h(n) = \left(\frac{0.56^{n+1} - 0.32^{n+1}}{0.56 - 0.32} \right) u(n)$$

which is bounded. This convolution operation can be implemented faster in the frequency domain, as shown for other transforms. By multiplying $y(n)$ with the exponential $0.4^{-n}u(n)$, we get the output as

$$y(n) = x(n) * h(n) = 0.4^{-n}u(n) \left(\frac{0.56^{n+1} - 0.32^{n+1}}{0.56 - 0.32} \right) u(n)$$

which is unbounded. Therefore, by multiplying signals with appropriate exponential as a preprocessing, we are able to extend the range of signals to be convolved. This is possible due to commutability of convolution and multiplication by an exponential. This procedure is basically an extension of the Fourier analysis. The complex basis function z can have magnitude other than 1 also, in contrast to Fourier analysis. This extension is called the z-transform, which is extensively used in system analysis. While the DTFT changes a sequence of numbers into a function of the purely imaginary complex variable $e^{j\omega}$, the z-transform changes a sequence of numbers into a function of the complex variable z with an expanded set of basis functions. That is, the DTFT of a signal is its z-transform with $z = e^{j\omega}$, if the DTFT exists. The variable z is restricted to the unit-circle in the z-plane.

As pointed out earlier, by this generalization of the Fourier analysis, a wider class of signals is included in the frequency-domain analysis. This transform, as in the case of other transforms, is a transform of a signal into another form so that analysis of signals and systems is simplified. In addition, it provides insight into system behavior such as filtering. Further, for real-valued signals, the z-transform becomes a real-rational function of the complex variable z. The properties of these functions are well-established and can be used in system analysis. By designing the z-transform only for causal signals, the systems with initial conditions can be easily analyzed. The zero-input response is important in applications, such as control systems. A short list of z-transform pairs is adequate for most of the analysis. In addition, the use of variable z instead of $e^{j\omega}$ makes it simpler to manipulate the z-transform expressions. The analysis of discrete systems using z-transform is similar to that of other transforms. Application of this transform reduces a difference equation characterizing a system into an algebraic equation, thereby simplifying system analysis.

5.1   The z-Transform

In this chapter, we introduce the commonly used one-sided version of the z-transform, called the unilateral z- transform, that is suitable for analyzing causal systems with causal signals. Therefore, in this chapter, signals are assumed to be causal, $x(n) = 0$, $n < 0$, unless otherwise stated. If a causal signal $x(n)$ is unbounded, for example, $(2)^n u(n)$ with first few values

$$\{1, 2, 4, 8, 16, 32, \ldots, \}$$

it can be made bounded by multiplying it by another causal exponential $(r)^{-n} u(n)$ and $|r| > 2$, say 2.1, with initial values

$$\{1, 0.4762, 0.2268, 0.1080, 0.0514, 0.0245, \ldots\}$$

The product is the sequence $(0.9524)^n u(n)$

$$\{1, 0.9524, 0.9070, 0.8638, 0.8227, \ldots\}$$

and it has a DTFT $1/(1 - 0.9524e^{-j\omega})$. The sequence $(0.9524)^n u(n)$ multiplied by the exponential $(r)^n u(n)$ gets back the original sequence $(2)^n u(n)$. In essence, the given sequence is appropriately modified to make it convergent and a Fourier forward or inverse representation is found. If a signal $x(n)u(n)$ is not bounded and, hence, not Fourier transformable, then its exponentially weighted version, $(x(n)r^{-n})$, may be Fourier transformable for certain range of r. Therefore, the DTFT gets modified as

$$\sum_{n=0}^{\infty} (x(n)r^{-n}) e^{-j\omega n}$$

By combining the exponential factors, we get

$$X\left(re^{j\omega}\right) = \sum_{n=0}^{\infty} x(n)\left(re^{j\omega}\right)^{-n}$$

This definition is a generalized version of Fourier analysis, with the basis functions becoming complex exponentials with varying amplitudes. Denoting $z = re^{j\omega}$, the defining equation of the one-sided or unilateral z-transform of $x(n)$ becomes

$$X(z) = \sum_{n=0}^{\infty} x(n)z^{-n} \tag{5.1}$$

Explicitly writing the first few terms of the infinite summation, we get

$$X(z) = x(0) + x(1)z^{-1} + x(2)z^{-2} + x(3)z^{-3} + \cdots$$

where z is a complex variable.

Writing the basis functions of the z-transform in various forms of the complex variable, we get

$$z^n = e^{(\sigma+j\omega)n} = r^n e^{j\omega n} = (a + jb)^n = r^n(\cos(\omega n) + j\sin(\omega n))$$

$X(z) = X(re^{j\omega})$ is the DTFT of $x(n)r^{-n}$ for all values of r for which $\sum_{n=0}^{\infty} |x(n)r^{-n}| < \infty$. Of course, when $r = 1$ is one of such values, then the z-transform reduces to the DTFT. That is, replacing z by $e^{j\omega}$ in the z-transform of a signal yields its DTFT in such cases. We suitably damp an unbounded signal by multiplying it by an exponential to get its z-transform. Even then, if the signal does not get bounded, there is no z-transform representation for the signal as it grows faster than an exponential. For example, $x(n) = b^{n^2}$ does not have a z-transform. All practical signals satisfy the convergence condition and are, therefore, have z-transform representation. The absolute value of a complex number is a constant defined as $|a + jb| = \sqrt{a^2 + b^2}$. The area in the z-plane, where the z-transform of a function is defined, is the area outside of a circle with this absolute value as the radius. This area is called the region of convergence (ROC) of the z-transform. A circle in the z-plane with center at the origin and radius c is the border between the ROC and region of divergence. The condition such as $|z| > r$ for ROC specifies the region outside the circle with center at the origin in the z-plane with radius r.

Example 5.1 Determine the z-transform of the unit-impulse signal, $\delta(n)$.

Solution Since the only nonzero sample of the signal occurs at $n = 0$, from the definition, we get

$$X(z) = 1, \text{ for all } z \qquad \text{and} \qquad \delta(n) \leftrightarrow 1, \text{ for all } z$$

Since the transform is independent of z, it is a valid representation of the impulse all over the z-plane.

The transform pair for a delayed impulse by k samples, $\delta(n - k)$, is

$$\delta(n - k) \leftrightarrow z^{-k}, \ |z| > 0,$$

since the terms of inverse power of z appear in the definition of the z-transform, the ROC is the entire z-plane except the origin.   ∎

Example 5.2 Determine the z-transform of the finite sequence with its only nonzero samples specified as

(a)

$$\{x(0) = 1, x(2) = -3, x(5) = 4\}$$

(b)

$$\{x(2) = 1, x(3) = -3, x(5) = 4\}$$

(c)

$$\{x(-1) = 1, x(0) = -3, x(5) = 4\}$$

(d)

$$\{x(-5) = 2, x(-3) = -3, x(-1) = 4\}$$

(e)

$$a^n, \ 0 \le n \le N \text{ and } 0 \text{ otherwise}$$

with $|a| < \infty$, $z \ne 0$, and N finite.

Solution Since the sequences are finite, all are convergent except at the origin. From the definition, we get

(a)

$$X(z) = 1 - 3z^{-2} + 4z^{-5} = \frac{1z^5 - 3z^3 + 4}{z^5}$$

(b)

$$X(z) = z^{-2} - 3z^{-3} + 4z^{-5} = \frac{1z^3 - 3z^2 + 4}{z^5}$$

(c)

$$X(z) = -3 + 4z^{-5} = \frac{-3z^5 + 4}{z^5}$$

(d)

$$X(z) = 0$$

(e)

$$X(z) = \sum_{n=0}^{N} a^n z^{-n} = \sum_{n=0}^{N} (az^{-1})^n = \left(\frac{z^{N+1} - a^{N+1}}{z^N(z - a)} \right)$$

■

The geometric sequence, $a^n u(n)$, is of fundamental importance in the study of linear systems, as the natural response of systems is in that form, and also as the basis functions of transforms.

Example 5.3 Determine the z-transform of the geometric sequence, $a^n u(n)$.

Solution Substituting $x(n) = a^n$ in the defining equation of the z-transform and writing $a^n z^{-n}$ as $(az^{-1})^n$ yields

$$X(z) = \sum_{n=0}^{\infty} (az^{-1})^n = 1 + (az^{-1})^1 + (az^{-1})^2 + (az^{-1})^3 + \cdots$$

$$= \frac{1}{1 - (az^{-1})} = \frac{z}{z-a}, \quad |z| > |a|$$

It is known that the geometric series $1 + r + r^2 + \cdots$ converges to $\frac{1}{1-r}$, if $|r| < 1$. If $|z| > |a|$, the common ratio of the series $r = \frac{a}{z}$ has magnitude that is less than one. That is, the magnitude of the complex number $(a^{-1}z)^{-n}$ approaches zero as n tends to ∞. Therefore, the ROC of the z-transform is given as $|z| > |a|$ and we get the transform pair

$$a^n u(n) \leftrightarrow \frac{z}{z-a}, \quad |z| > |a|$$

∎

The form of the transform $z/(z - a)$ is called a rational function. A rational function is a function of a variable z, if it can be expressed as a fraction of numerator and denominator polynomials in z. The z-transform of most of the signals of interest can be expressed as a rational function in the variable z. One of the effective ways of characterization of rational functions is the z-plane representation in terms of its of frequency factors or poles and zeros (pole-zero plot). At roots of the denominator polynomial of a rational function, its value tends to ∞. These roots are called *poles* of the rational function. Similarly, its *zeros* are the roots of its numerator polynomial, where its value tends to zero. The pole-zero plot and the magnitude in dB of $H(z) = \frac{z}{z-0.7}$ of the signal $0.7^n u(n)$ are shown in Fig. 5.1a and b, respectively. The pole at $z = 0.7$ of $X(z)$ is shown by a cross (the peak in in the magnitude plot Fig. 5.1b). It is called a pole, since it looks like a tent supported by a pole of infinite height at $z = 0.7$. The tent is fastened down at $z = 0$. The pole-zero plot depicts all the essential characteristics of $X(z)$. The zero at $z = 0$ of $X(z)$ is shown by a small circle (the valley in Fig. 5.1b). The transform $X(z)$ is specified in the pole-zero plot except for a constant factor. In the region outside the circle with radius 0.7, $X(z)$ exists and it is a valid frequency domain representation of $x(n)$. With several poles, the ROC is the region in the z-plane that is exterior to the smallest circle, centered at the origin, enclosing all its poles. The z-transform of the unit-step signal can be obtained by replacing a by 1 in the transform of $a^n u(n)$.

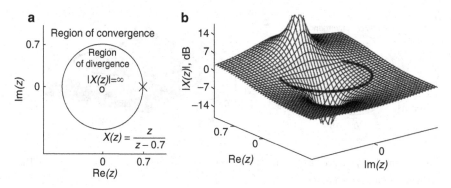

Fig. 5.1 (a) The pole-zero plot of the z-transform $\frac{z}{z-0.7}$ of $(0.7)^n u(n)$; (b) the magnitude in dB of the z-transform

$$u(n) \leftrightarrow \frac{z}{z-1}, \quad |z| > |1|$$

Example 5.4 Determine the z-transform of the signal $e^{j\omega_0 n} u(n)$. From the z-transform obtained, deduce the z-transform of the causal sinusoidal sequence $\sin(\omega_0 n) u(n)$.

Solution Replacing a by $e^{j\omega_0}$ in the transform of $a^n u(n)$, we get

$$e^{j\omega_0 n} u(n) \leftrightarrow \frac{z}{z - e^{j\omega_0}}, \quad |z| > 1$$

Expressing the sinusoid in terms of complex exponentials using Euler identity, we get

$$j2 \sin(\omega_0 n) = \left(e^{j\omega_0 n} - e^{-j\omega_0 n} \right)$$

Taking the transform, we get

$$j2X(z) = \frac{z}{z - e^{j\omega_0}} - \frac{z}{z - e^{-j\omega_0}}, \quad |z| = |e^{j\omega_0}| > 1$$

$$\sin(\omega_0 n) u(n) \leftrightarrow \frac{z \sin(\omega_0)}{(z - e^{j\omega_0})(z - e^{-j\omega_0})} = \frac{z \sin(\omega_0)}{z^2 - 2z \cos(\omega_0) + 1}, \quad |z| > 1$$

■

Figure 5.2a shows the pole-zero plot and Fig. 5.2b shows the magnitude of the z-transform $\frac{z \sin(\frac{\pi}{3})}{z^2 - 2z \cos(\frac{\pi}{3}) + 1}$ of the signal $\sin(\frac{\pi}{3} n) u(n)$. There is a zero at $z = 0$ and

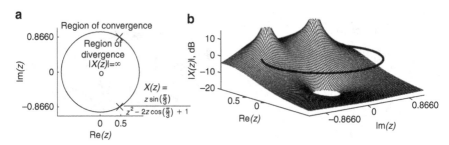

Fig. 5.2 (a) The pole-zero plot of the z-transform $\dfrac{z\sin(\frac{\pi}{3})}{(z-e^{j\frac{\pi}{3}})(z-e^{-j\frac{\pi}{3}})}$ of $\sin(\frac{\pi}{3}n)u(n)$; (b) The magnitude of the z-transform

poles at $z = e^{j\frac{\pi}{3}}$ and $z = e^{-j\frac{\pi}{3}}$, a pair of complex conjugate poles at $z = e^{j\frac{\pi}{3}}$ and $z = e^{-j\frac{\pi}{3}}$. As expected, the poles pass through the circular boundary with center at origin separating the regions of convergence (exterior of the circle) and divergence (interior of the circle). Following the same procedure, we get the z-transform $\dfrac{z(z-\cos(\frac{\pi}{3}))}{z^2-2z\cos(\frac{\pi}{3})+1}$ of the signal $\cos(\frac{\pi}{3}n)u(n)$.

Example 5.5 Determine the z-transform of the signal $x(n)$ composed of pseudo-geometric sequences

$$x(n) = \begin{cases} (0.3)^n \text{ for } 0 \le n \le 4 \\ (0.6)^n \text{ for } 5 \le n < \infty \end{cases}$$

Solution From the definition of the z-transform, we get

$$X(z) = \sum_{n=0}^{4}(0.3)^n z^{-n} + \sum_{n=0}^{\infty}(0.6)^n z^{-n} - \sum_{n=0}^{4}(0.6)^n z^{-n}$$

$$= \frac{z^5 - (0.3)^5}{z^4(z-0.3)} + \frac{z}{z-0.6} - \frac{z^5 - (0.6)^5}{z^4(z-0.6)}, \quad |z| > 0.6$$

∎

5.2   The Inverse z-Transform

In using the transforms to process signals, we find the forward transform, do the required processing in the frequency domain and find the inverse transform to get its time-domain version. Therefore, in z-transform applications also, invariably, finding the inverse z-transform is an important step. As the z-transform is a generalization of the DTFT, we can start from the DTFT definitions and derive its inverse definition. The DTFT of damped $x(n)u(n)$, $x(n)r^{-n}$, is given by

$$X\left(re^{j\omega}\right) = \sum_{n=0}^{\infty} x(n)\left(re^{j\omega}\right)^{-n}$$

Taking the inverse DTFT of $X(re^{j\omega})$, we get

$$x(n)r^{-n} = \frac{1}{2\pi} \int_{-\pi}^{\pi} X\left(re^{j\omega}\right)e^{j\omega n}d\omega$$

As we premultiplied $x(n)$ by r^{-n}, we have to multiply both sides by r^n to get

$$x(n) = \frac{1}{2\pi} \int_{-\pi}^{\pi} X\left(re^{j\omega}\right)\left(re^{j\omega}\right)^n d\omega$$

To get the inverse transform in terms of the variable z, let us replace $re^{j\omega}$ by z. Then, we get $dz = jre^{j\omega}d\omega = jzd\omega$. Substituting for $d\omega$ yields the inverse z-transform of $X(z)$ as

$$x(n) = \frac{1}{2\pi j} \oint_C X(z)z^{n-1}dz \tag{5.2}$$

with the integral evaluated, in the counterclockwise direction, along any simply connected closed contour C, encircling the origin, that lies in the ROC of $X(z)$. As ω varies from $-\pi$ to π in the DTFT, the variable z traverses the circle of radius r in the counterclockwise direction once. The contour of integration in evaluating the inverse z-transform could lie anywhere in the z-plane as long it is completely in the ROC satisfying the requirements. There are infinite choices for the contour of integration.

5.2.1 Inverse z-Transform by Partial-Fraction Expansion

As finding the inverse z-transform by evaluating the contour integral Eq. (5.2) is relatively difficult, for most practical purposes, two relatively simpler methods, the partial-fraction method and the long-division method, are commonly used.

Most of the z-transforms of practical interest are rational functions (a ratio of two polynomials in z). The denominator polynomial can be factored into a product of first- or second-order terms. This type of z-transforms can be expressed as the sum of partial fractions with each denominator forming a factor. The inverse z-transforms of the individual fractions can be easily found from a short table of z-transforms, such as those of $\delta(n)$, $a^n u(n)$, and $na^n u(n)$, shown in the Appendix. The sum of the individual inverses is the inverse of the given z-transform.

Two rational functions are added by converting them to a common denominator, add and then simplify. For example, the sum of the two rational functions is

$$\frac{2}{(z-1)} + \frac{3}{(z-2)} = \frac{2(z-2)+3(z-1)}{(z-1)(z-2)} = \frac{(5z-7)}{(z^2-3z+2)}$$

Usually, we are given the $H(z)$ to be inverted in the form on the right. The task is to find an expression like that on the left. The numerator polynomial of the rightmost expression is of order 1, whereas that of the denominator is of order 2. Partial fraction expansion of a rational function expresses it as a sum of appropriate fractions with the coefficient of each fraction to be found. For the time being, we assume that the denominator is of higher degree than the numerator.

Example 5.6 The difference equation of a causal discrete system is

$$y(n) = 5x(n) - 7x(n-1) + 3y(n-1) - 2y(n-2), \quad n = 0, 1, \ldots$$

Using the z-transform, find a closed-form expression for the time-domain response $y(n)$. Assume that the system is initially relaxed and the input is the unit-impulse, $x(n) = \delta(n)$. Verify that the first 4 values of the response of the system obtained using the expression for $y(n)$ and by iterating the difference equation are the same.

Solution Taking the z-transform of both sides, we get, with zero initial conditions,

$$Y(z) = X(z)\left(5 - 7z^{-1}\right) + Y(z)\left(3z^{-1} - 2z^{-2}\right) \text{ or } Y(z) = \frac{z(5z-7)}{(z^2-3z+2)}$$

with $X(z) = 1$. Decomposing $Y(z)$ into partial fractions, there are two poles contributing to two terms and we get

$$\frac{Y(z)}{z} = \frac{(5z-7)}{(z^2-3z+2)} = \frac{A}{(z-1)} + \frac{B}{(z-2)}$$

Since the z-transform of functions, such as $a^n u(n)$, have a factor z in the numerator, it is convenient to expand $Y(z)/z$ into partial fractions and then multiply both sides by z, after decomposing. To find A, multiply the rightmost terms by $(z-1)$ to isolate A. Then, we get

$$\frac{(5z-7)}{(z-2)} = A + \frac{B(z-1)}{(z-2)}$$

Replacing z by 1, we get $A = 2$. That is,

$$A = \frac{(5z-7)}{(z-2)}\Big|_{z=1} = 2$$

To find B, multiply the rightmost terms by $(z-2)$ to isolate B. Then, we get

$$\frac{(5z-7)}{(z-1)} = \frac{A(z-2)}{(z-1)} + B$$

Replacing z by 2, we get $B = 3$. That is,

$$B = \left.\frac{(5z-7)}{(z-1)}\right|_{z=2} = 3$$

The required partial fraction is

$$Y(z) = \frac{z(5z-7)}{(z^2-3z+2)} = \frac{2z}{(z-1)} + \frac{3z}{(z-2)}$$

By simplifying the right side, we get the left side proving that the expansion is correct. Finding the inverse z-transform of each term, we get the inverse of $Y(z)$, that is, the time-domain sequence $y(n)$ corresponding to $Y(z)$, as

$$y(n) = \left(2\,(1)^n + 3\,(2)^n\right) u(n)$$

The first four values of the sequence $y(n)$ are

$$\{y(0) = 5, \quad y(1) = 8, \quad y(2) = 14, \quad y(3) = 26\}$$

Note that the roots may be real and/or complex. ∎

Example 5.7 Find the z-transform $Y(z)$ of the difference equation of a causal discrete system

$$y(n) = 5x(n-1) - 7x(n-2) + 3y(n-1) - 2y(n-2),$$
$$n = 0, 1, \ldots$$

Find the inverse z-transform of $Y(z)$ getting a closed-form expression for the time-domain response $y(n)$. Assume that the system is initially relaxed and the input is the unit-impulse, $x(n) = \delta(n)$. Verify that the first 4 values of the response of the system obtained using the expression for $y(n)$ and iterating the difference equation are the same.

Solution Taking the z-transform of both sides, we get with zero initial conditions

$$Y(z) = X(z)\left(5z^{-1} - 7z^{-2}\right) + Y(z)\left(3z^{-1} - 2z^{-2}\right) \text{ or } Y(z) = \frac{(5z-7)}{(z^2-3z+2)}$$

with $X(z) = 1$. Dividing both sides by z, we get the modified problem as

$$\frac{Y(z)}{z} = \frac{(5z - 7)}{z(z^2 - 3z + 2)} = \frac{A}{z} + \frac{B}{(z - 1)} + \frac{C}{(z - 2)}$$

To find A, multiply the rightmost terms by z to isolate A. Then, we get

$$\frac{(5z - 7)}{(z - 1)(z - 2)} = A + \frac{B(z)}{(z - 1)} + \frac{C(z)}{(z - 2)}$$

Replacing z by 0, we get $A = -3.5$. To find B, multiply the rightmost terms by $(z - 1)$ to isolate B. Then, we get

$$\frac{(5z - 7)}{(z - 2)z} = \frac{A(z - 1)}{(z)} + B + \frac{C(z - 1)}{(z - 2)}$$

Replacing z by 1, we get $B = 2$. To find C, multiply the rightmost terms by $(z - 2)$ to isolate C. Then, we get

$$\frac{(5z - 7)}{z(z - 1)} = \frac{A(z - 2)}{(z)} + \frac{B(z - 2)}{(z - 1)} + C$$

Replacing z by 2, we get $C = 1.5$. The required partial fraction is

$$Y(z) = \frac{(5z - 7)}{(z^2 - 3z + 2)} = -3.5 + \frac{2z}{(z - 1)} + \frac{1.5z}{(z - 2)}$$

By simplifying the right side, we get the left side proving that the expansion is correct. Finding the inverse z-transform of each term, we get the inverse of $Y(z)$, that is, the time-domain sequence $y(n)$ corresponding to $Y(z)$, as

$$y(n) = -3.5\delta(n) + \left(2\,(1)^n + 1.5\,(2)^n\right) u(n)$$

The first four values of the sequence $y(n)$ are

$$\{y(0) = 0, \quad y(1) = 5, \quad y(2) = 8, \quad y(3) = 14\}$$

∎

Multiple-Order Poles
Each repeated linear factor $(z - a)^m$ contributes a sum of the form

$$\frac{A_m}{(z - a)^m} + \frac{A_{m-1}}{(z - a)^{m-1}} + \cdots + \frac{A_1}{(z - a)}$$

For a second-order pole, the transform pair is

$$na^n u(n) \leftrightarrow \frac{az}{(z-a)^2} \quad \text{and} \quad na^{n-1}u(n) \leftrightarrow \frac{z}{(z-a)^2}, \quad |z| > |a|$$

Example 5.8 Find the z-transform $Y(z)$ of the difference equation of a causal discrete system

$$y(n) = 3x(n) + 3y(n-2) - 2y(n-3), \quad n = 0, 1, \ldots$$

Find the inverse z-transform of $Y(z)$ getting a closed-form expression for the time-domain response $y(n)$. Assume that the system is initially relaxed and the input is the unit-impulse, $x(n) = \delta(n)$. Verify that the first 4 values of the impulse, $x(n) = \delta(n)$, response of the system obtained using the expression for $y(n)$ and iterating the difference equation are the same.

Solution Taking the z-transform of both sides, we get with zero initial conditions

$$Y(z) = 3X(z) + Y(z)(3z^{-2} - 2z^{-3}) \text{ or } Y(z) = \frac{3z^3}{(z^3 - 3z + 2)}$$

with $X(z) = 1$.

$$\frac{Y(z)}{z} = \left(\frac{3z^2}{(z-1)^2(z+2)} \right) = \left(\frac{A}{(z-1)^2} + \frac{B}{(z-1)} + \frac{C}{(z+2)} \right)$$

To find A, multiply the rightmost terms by $(z-1)^2$ to isolate A. Then, we get

$$\frac{(3z^2)}{(z+2)} = A + \frac{B(z-1)^2}{(z-1)} + \frac{C(z-1)^2}{(z+2)}$$

Replacing z by 1, we get $A = 1$. To find C, multiply the rightmost terms by $(z+2)$ to isolate C. Then, we get

$$\frac{(3z^2)}{(z-1)^2} = \frac{A(z+2)}{(z-1)^2} + \frac{B(z+2)}{(z-1)} + C$$

Replacing z by -2 we get $C = 4/3$. Now, we have

$$\left(\frac{3z^2}{(z-1)^2(z+2)} \right) = \left(\frac{1}{(z-1)^2} + \frac{B}{(z-1)} + \frac{\frac{4}{3}}{(z+2)} \right)$$

To find B, one method is to replace z by any value other then the roots. For example, by replacing z by 0, we get $B = 5/3$.

Another method is to subtract the term $\frac{1}{(z-1)^2}$ from the expression $\left(\frac{3z^2}{(z-1)^2(z+2)}\right)$ to get $\frac{3z+2}{(z-1)(z+2)}$. Substituting $z = 1$ in the expression $\frac{3z+2}{(z+2)}$, we get $B = 5/3$. Therefore, the partial-fraction expansion is

$$Y(z) = \left(\frac{z}{(z-1)^2} + \frac{\frac{5}{3}z}{(z-1)} + \frac{\frac{4}{3}z}{(z+2)}\right)$$

$$y(n) = \left(n(1)^{n-1} + \frac{5}{3}(1)^n + \frac{4}{3}(-2)^n\right)u(n)$$

The first four values of the sequence $y(n)$ are

$$\{y(0) = 3, \quad y(1) = 0, \quad y(2) = 9, \quad y(3) = -6\}$$

∎

For a third-order pole, the transform pair is

$$0.5n(n-1)a^{n-2}u(n) \leftrightarrow \frac{z}{(z-a)^3}, \quad |z| > |a|$$

In general,

$$\frac{n(n-1)(n-2)\cdots(n-k+1)}{k!}a^{n-k}u(n) \leftrightarrow \frac{z}{(z-a)^{k+1}}, \quad |z| > |a|$$

Example 5.9 Find the z-transform $Y(z)$ of the difference equation of a causal discrete system

$$y(n) = x(n) + 3.4y(n-1) - 4.32y(n-2) + 2.432y(n-3) - 0.512y(n-4),$$

$$n = 0, 1, \ldots$$

Find the inverse z-transform of $Y(z)$ getting a closed-form expression for the time-domain response $y(n)$. Assume that the system is initially relaxed and the input is the unit-impulse, $x(n) = \delta(n)$. Verify that the first 4 values of the impulse, $x(n) = \delta(n)$, response of the system obtained using the expression for $y(n)$ and iterating the difference equation are the same.

Solution Taking the z-transform of both sides, we get with zero initial conditions

$$Y(z) = X(z) + Y(z)\left(-3.4z^{-1} + 4.32z^{-2} - 2.432z^{-3} + 0.512z^{-4}\right)$$

That is,

$$Y(z) = \frac{z^4}{(z^4 - 3.4z^3 + 4.32z^2 - 2.432z + 0.512)}$$

with $X(z) = 1$.

$$\frac{Y(z)}{z} = \left(\frac{z^3}{(z - 0.8)^3 (z - 1)} \right)$$

$$= \left(\frac{A}{(z - 0.8)^3} + \frac{B}{(z - 0.8)^2} + \frac{C}{(z - 0.8)} + \frac{D}{(z - 1)} \right)$$

Finding the constants, we get

$$Y(z) = \left(\frac{-2.56z}{(z - 0.8)^3} - \frac{22.4z}{(z - 0.8)^2} - \frac{124z}{(z - 0.8)} + \frac{125z}{(z - 1)} \right)$$

$$y(n) = \left(-2.56(0.5)n(n - 1)(0.8)^{n-2} - 22.4(n)(0.8)^{n-1} - 124(0.8)^n + 125 \right) u(n)$$

The first four values of the sequence $y(n)$ are

$$\{ y(0) = 1, \quad y(1) = 3.4, \quad y(2) = 7.24, \quad y(3) = 12.36 \}$$

■

Complex Poles

For complex poles also, the same procedure is applicable. With the coefficients of the rational function real-valued, pairs of complex conjugate poles will always have complex conjugate partial-fraction coefficients. The terms combine to produce an oscillating response. Therefore, only one of the complex coefficients needs to be computed for each pair.

Example 5.10 Find the expression for the closed-form output of the system characterized by the difference equation

$$y(n) = x(n) - 2y(n - 1) - 2y(n - 2)$$

using the z-transform. Find also the first 4 values of the output using iteration. Assume that the system is initially relaxed and the input is the unit-impulse, $x(n) = \delta(n)$. Verify that the outputs are the same.

Solution Taking the z-transform, we get

$$Y(z) = \frac{z^2}{(z^2 + 2z + 2)}$$

Factorizing the denominator of $Y(z)$ and finding the partial-fraction expansion, we get

$$\frac{Y(z)}{z} = \left(\frac{z}{(z - (-1 + j1))(z - (-1 - j1))} \right)$$

$$= \left(\frac{0.5 + j0.5}{(z - (-1 + j1))} + \frac{0.5 - j0.5}{(z - (-1 - j1))} \right)$$

$$y(n) = (0.5 + j0.5)(-1 + j1)^n + (0.5 - j0.5)(-1 - j1)^n, \quad n = 0, 1, \ldots$$

As expected, $y(n)$ is composed of a complex conjugate pair. The sum of a complex number and its conjugate is twice the real part. Let one of the complex conjugate poles be $(a + jb)$ and the corresponding partial-fraction coefficient is $(c + jd)$. Then, the magnitude and phase of the poles determine the damping factor r and the frequency ω. The magnitude and phase of the coefficient determine the amplitude A and the phase θ. The time-domain response is

$$y(n) = 2A(r)^n \cos(\omega n + \theta)u(n), \text{ where } r = \sqrt{a^2 + b^2}, \ \omega = \tan^{-1}\left(\frac{b}{a}\right),$$

$$A = \sqrt{c^2 + d^2}, \ \theta = \tan^{-1}\left(\frac{d}{c}\right)$$

For the example, twice the real part of either of the terms forming the output is

$$y(n) = (2)\left(\frac{1}{\sqrt{2}}\right)(\sqrt{2})^n \cos\left(\frac{3\pi}{4}n + \frac{\pi}{4}\right)u(n)$$

The first four values of the sequence $y(n)$ are

$$\{y(0) = 1, \quad y(1) = -2, \quad y(2) = 2, \quad y(3) = 0\}$$

■

Rational Function with the Same Order of Numerator and Denominator
A rational function is called a proper function, if the degree of the numerator polynomial is less than that of the denominator. Only proper functions can be expanded into partial-fraction expansion. Other functions, called improper functions, can be expressed as a sum of a polynomial and a proper function by division. Then, the proper function part can be expanded into partial fractions. In practical system analysis, we are interested in rational functions with the degree of the numerator polynomial less than or equal to that of the denominator. In the only case with both the degrees equal, we divide both sides by z and make it proper.

Example 5.11 Find the z-transform $Y(z)$ of the difference equation of a causal discrete system

$$y(n) = 3x(n) - 5x(n-1) - 3x(n-2) + 5y(n-1) - 6y(n-2), \quad n = 0, 1, \dots$$

Find the inverse z-transform of $Y(z)$ getting a closed-form expression for the time-domain response $y(n)$. Assume that the system is initially relaxed and the input is the unit-impulse, $x(n) = \delta(n)$. Verify that the first 4 values of the impulse, $x(n) = \delta(n)$, response of the system obtained using the expression for $y(n)$ and by iterating the given difference equation are the same.

Solution Taking the z-transform of both sides, we get with zero initial conditions

$$Y(z) = X(z)\left(3 - 5z^{-1} - 3z^{-2}\right) + Y(z)\left(-5z^{-1} + 6z^{-2}\right) \text{ or } Y(z) = \frac{(3z^2 - 5z - 3)}{(z^2 - 5z + 6)}$$

with $X(z) = 1$. Dividing both sides by z, we get the modified problem as

$$\frac{Y(z)}{z} = \frac{(3z^2 - 5z - 3)}{z(z^2 - 5z + 6)} = \frac{A}{z} + \frac{B}{(z-3)} + \frac{C}{(z-2)}$$

To find A, multiply the rightmost terms by z to isolate A. Then, we get

$$\frac{(3z^2 - 5z - 3)}{(z^2 - 5z + 6)} = A + \frac{B(z)}{(z-3)} + \frac{C(z)}{(z-2)}$$

Replacing z by 0, we get $A = -0.5$. To find B, multiply the rightmost terms by $(z-3)$ to isolate B. Then, we get

$$\frac{(3z^2 - 5z - 3)}{(z(z-2))} = \frac{A(z-3)}{(z)} + B + \frac{C(z-3)}{(z-2)}$$

Replacing z by 3, we get $B = 3$. To find C, multiply the rightmost terms by $(z-2)$ to isolate C. Then, we get

$$\frac{(3z^2 - 5z - 3)}{z(z-3)} = \frac{A(z-2)}{(z)} + \frac{B(z-2)}{(z-3)} + C$$

Replacing z by 2, we get $C = 0.5$. The required partial-fraction expansion is

$$Y(z) = \frac{(3z^2 - 5z - 3)}{z(z^2 - 5z + 6)} = (-0.5)z + \frac{3z}{(z-3)} + \frac{0.5z}{(z-2)}$$

By simplifying the right side, we get the left side proving that the expansion is correct. Finding the inverse of each term, we get the inverse z-transform of $Y(z)$, that is, the time-domain sequence $y(n)$ corresponding to $Y(z)$, as

$$y(n) = -0.5\delta(n) + \left(3\,(3)^n + 0.5\,(2)^n\right)u(n)$$

The first four values of the sequence $x(n)$ are

$$\{y(0) = 3, \quad y(1) = 10, \quad y(2) = 29, \quad y(3) = 85\}$$

■

The Long-Division Method

The z-transform of causal sequences is a power series in the variable z^{-1}. For a causal and rational $X(z)$, we simply divide the numerator polynomial by the denominator polynomial. We get a series that is the defining series of the z-transform. Then, the sequence $x(n)$ corresponding to the $X(z)$ are the coefficients of the power series. The inverse z-transform of $X(z) = \frac{z}{z-0.5}$, for example, is obtained dividing z by $z - 0.5$. The quotient is

$$X(z) = 1 + 0.5z^{-1} + 0.25z^{-2} + 0.125z^{-3} + 0.0625z^{-4}$$
$$+ 0.0313z^{-5} + 0.0156z^{-6} + 0.0078z^{-7} + \cdots$$

Inverse z-transform by long division is shown in Fig. 5.3. Comparing with the definition of the z-transform, the time-domain values are $x(0) = 1$, $x(1) = 0.5$, $x(2) = 0.25$, $x(3) = 0.125$, and so on. Using the closed-form solution of the inverse z-transform $x(n) = (0.5)^n u(n)$, these values can be verified. This method of finding the first few values of the time-domain sequence is particularly useful to verify the closed-loop solution and when the first few values are adequate.

$$
\begin{array}{r}
1 + 0.5z^{-1} + 0.25z^{-2} + 0.125z^{-3} + 0.0625z^{-4} + \cdots \\
z - 0.5 \overline{)\,z} \\
\underline{z - 0.5} \\
0.5 \\
\underline{0.5 - 0.25z^{-1}} \\
0.25z^{-1} \\
\underline{0.25z^{-1} - 0.125z^{-2}} \\
0.125z^{-2} \\
\underline{0.125z^{-2} - 0.0625z^{-3}} \\
0.0625z^{-3}
\end{array}
$$

Fig. 5.3 Inverse z-transform by long division

5.3   Properties of the z-Transform

Properties greatly simplify signal and system analysis. For example, based on the symmetry of a signal, its processing can be made more efficient in the time domain or frequency domain. Implementing the convolution operation is, in general, faster in the frequency domain. The visualization of the effect of operations may be easier in one of the domains. Further, the forward and inverse transforms can be derived easily using properties.

5.3.1   Linearity

The linearity property is extensively used in the analysis of LTI systems. Transform analysis is based on linearity. Linearity is very useful in finding the forward and inverse transforms of signals. The difference equation characterizing a system is linear, if the input $x(n)$, the output $y(n)$, and their differences are not raised to any power other than unity or no terms involving the product of input and output terms or a constant term. One implication of the property is that if $y(n)$ is the response to input $x(n)$, then $a\, y(n)$ is the response to input $a\, x(n)$, where a is an arbitrary constant. This type of response is called the scaling property. Another property is called additive or superposable property. If y_1 is the output of a system to the input x_1 and y_2 is the output of the system to the input x_2, then the system is linear if $y_1 + y_2$ is the output of the system to the input $x_1 + x_2$.

Combining the two properties, we get the linearity property with respect to the z-transform. If $x(n) \leftrightarrow X(z)$ and $y(n) \leftrightarrow Y(z)$, then

$$cx(n) + dy(n) \leftrightarrow cX(z) + dY(z),$$

where c and d are arbitrary constants. In finding the inverse transform in the last section, we used the linearity property several times. That is, the sequence corresponding to complicated z-transforms is found by first expanding z-transforms into a linear combination of simpler z-transforms and summing the corresponding individual sequences in the same linear combination.

5.3.2   Right Shift of a Sequence

The causal version of signal $x(n)$ is $x(n)u(n)$. Let $x(n)u(n) \leftrightarrow X(z)$ and m is a positive integer. Let the signal $x(n)u(n)$ is passed through a set of m unit-delay units to get $x(n - m)u(n)$. Due to the right shifting, some samples are gained and they have to be taken care of in expressing its transform in terms of $X(z)$. That is,

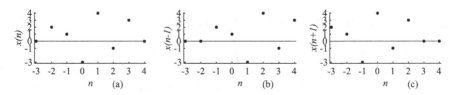

Fig. 5.4 (a) Sequence $x(n)$; (b) delayed sequence $x_d(n) = x(n-1)$; (c) advanced sequence $x_a(n) = x(n+1)$

$$x(n-m)u(n) \leftrightarrow z^{-m}X(z) + z^{-m}\sum_{n=1}^{m}x(-n)z^n$$

For example, with $m = 1$ and $m = 2$, we get

$$x(n-1)u(n) \leftrightarrow z^{-1}X(z) + x(-1)$$
$$x(n-2)u(n) \leftrightarrow z^{-2}X(z) + z^{-1}x(-1) + x(-2)$$

Consider the sequence $x(n)$ with its only nonzero values

$$\{x(-2) = 2, x(-1) = 1, x(0) = -3, x(1) = 4, x(2) = -1, x(3) = 3, \}$$

shown in Fig. 5.4a. The z-transform of $x(n)u(n)$ is

$$X(z) = -3 + 4z^{-1} - z^{-2} + 3z^{-3}$$

The delayed sequence, shown in Fig. 5.4b, is the right shifted sequence $x(n-1)$.

$$x_d(n) = \{x(-1) = 2, x(0) = 1, x(1) = -3, x(2) = 4, x(3) = -1, x(4) = 3, \}$$

The z-transform of $x(n-1)u(n)$ is

$$X_d(z) = 1 - 3z^{-1} + 4z^{-2} - z^{-3} + 3z^{-4}$$
$$= 1 + z^{-1}\left(-3 + 4z^{-1} - z^{-2} + 3z^{-3}\right) = x(-1) + z^{-1}X(z)$$

The input–output relationship of a discrete system can be characterized by a difference equation. While DTFT also reduces a difference equation into an algebraic equation making the solution easier, however, it is the z-transform that is more convenient for system analysis with nonzero initial conditions. The output is usually required for $n \geq 0$. In determining the output, in addition to the component due to input, the component due to the initial conditions, such as $y(-1)$ and $y(-2)$, must also be taken into account. The total output can be obtained using the z-

transform. When the difference equation is in delay-operator form, which is more suitable for causal systems, the right shift property is used.

Consider solving the difference equation

$$y(n) = 0.8y(n-1)$$

with the initial condition $y(-1) = 2$. By iteration, the first few terms of the output are

$$\{y(0) = (0.8)(y(-1)) = (0.8)(2) = 1.6, \ y(1) = (0.8)(1.6) = 1.28,$$

$$y(2) = (0.8)(1.28) = 1.024\}$$

That is,

$$y(n) = 1.6(0.8)^n u(n)$$

using the time-domain method.

Taking the z-transform of the difference equation, we get

$$Y(z) = 0.8\left(z^{-1}Y(z) + 2\right) = 0.8\left(z^{-1}Y(z)\right) + 1.6 = Y(z)\left(1 - 0.8\left(z^{-1}\right)\right) = 1.6$$

Rearranging, we get

$$Y(z) = \frac{1.6z}{z - 0.8}$$

Taking the inverse transform, we get

$$y(n) = 1.6\,(0.8)^n\,u(n)$$

as obtained by the time-domain method earlier.

System Response

Example 5.12 A discrete system is governed by the difference equation

$$y(n) = x(n) + 2x(n-1) - x(n-2) + \frac{5}{4}y(n-1) - \frac{3}{8}y(n-2)$$

Using the z-transform, find the zero-input, zero-state, transient, steady-state, and complete responses of the system for the input $x(n) = u(n)$, the unit-step function. The initial conditions are $y(-1) = -2$ and $y(-2) = 3$. List the first four values of the complete response of the system. Verify your answer by finding the first few values of the response by iterating the difference equation. Verify that the zero-input response satisfies the initial conditions. Verify that the total response in the z-transform domain satisfies the initial and final value theorems.

Solution Taking the z-transforms of the terms of the difference equation, we get

$$x(n) \leftrightarrow \frac{z}{z-1}, \qquad x(n-1) \leftrightarrow \frac{1}{z-1}, \qquad x(n-2) \leftrightarrow \frac{1}{z(z-1)}$$

$$y(n) \leftrightarrow Y(z), \qquad y(n-1) \leftrightarrow y(-1) + z^{-1}Y(z) = z^{-1}Y(z) - 2$$

$$y(n-2) \leftrightarrow y(-2) + z^{-1}y(-1) + z^{-2}Y(z) = z^{-2}Y(z) - 2z^{-1} + 3$$

Substituting the corresponding transform for each term in the difference equation, factoring and simplifying, we get

$$\frac{Y(z)}{z} = \frac{z^2 + 2z - 1}{(z-1)\left(z - \frac{3}{4}\right)\left(z - \frac{1}{2}\right)} + \frac{\left(-\frac{29}{8}z + \frac{3}{4}\right)}{\left(z - \frac{3}{4}\right)\left(z - \frac{1}{2}\right)}$$

The first and second terms on the right-hand side correspond, respectively, to the zero-state response and zero-input response. The numerator coefficients of the zero-input terms are obtained as

$$\frac{5}{4}y(n-1) - \frac{3}{8}y(n-2) = -\frac{29}{8} \quad \text{and} \quad -\frac{3}{8}y(-1) = \frac{3}{4}$$

The corresponding partial-fraction expansion is

$$\frac{Y(z)}{z} = \frac{16}{(z-1)} - \frac{17}{\left(z - \frac{3}{4}\right)} + \frac{2}{\left(z - \frac{1}{2}\right)} - \frac{7.8750}{\left(z - \frac{3}{4}\right)} + \frac{4.25}{\left(z - \frac{1}{2}\right)}$$

The inverse z-transform yields the complete response.

$$y(n) = \overbrace{16 - 17\left(\frac{3}{4}\right)^n + 2\left(\frac{1}{2}\right)^n}^{\text{zero-state}} \overbrace{-7.8750\left(\frac{3}{4}\right)^n + 4.25\left(\frac{1}{2}\right)^n}^{\text{zero-input}}, \; n = 0, 1, \ldots$$

$$y(n) = \underbrace{16}_{\text{steady-state}} \underbrace{-24.8750\left(\frac{3}{4}\right)^n + 6.25\left(\frac{1}{2}\right)^n}_{\text{transient}}, \; n = 0, 1, \ldots$$

The first four values of $y(n)$ are

$$\{y(0) = -2.6250, \quad y(1) = 0.4688, \quad y(2) = 3.5703, \quad y(3) = 6.2871\}$$

The responses are shown in Fig. 5.5. The initial and final values of the total response $y(n)$ are -2.6250 and 16, respectively, which can be verified by applying the initial

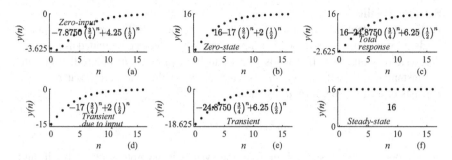

Fig. 5.5 Various components of the response of the system in Example 5.12

and final value properties to $Y(z)$. In addition, the initial conditions at $n = -1$ and at $n = -2$ are satisfied by the zero-input component of the response. ∎

The Transfer Function

The transfer function is defined, as in the case of DTFT, as the ratio of the transforms of the output and input, assuming that the system is initially relaxed. That is,

$$H(z) = \frac{Y(z)}{X(z)}$$

Replacing $e^{j\omega}$ by z in Equation (4.5), we get the z-transform version of the transfer function of a general Nth order causal LTI discrete system, with $M \le N$, is given by

$$H(z) = \frac{(b_0 + b_1 z^{-1} + \cdots + b_M z^{-M})}{(a_0 + a_1 z^{-1} + \cdots + a_N z^{-N})} \tag{5.3}$$

In terms of positive powers of z, we get

$$H(z) = \frac{z^{N-M}(b_0 z^M + b_1 z^{(M-1)} + \cdots + b_M)}{(a_0 z^N + a_1 z^{(N-1)} + \cdots + a_N)} \tag{5.4}$$

By multiplying $X(z)$ with $H(z)$, we get the z-transform of the zero-state response. Taking the inverse z-transform, we get the zero-state response in the time domain. For the above example,

$$H(z) = \frac{z^2 + 2z - 1}{\left(z - \frac{3}{4}\right)\left(z - \frac{1}{2}\right)}$$

System Stability

A discrete system is asymptotically stable if, and only if, its zero-input response tends to zero as $n \to \infty$. If the zero-input response becomes unbounded as $n \to \infty$, the system is unstable. If it tends to a finite limit (a constant or oscillating with a constant amplitude), the system is marginally stable. The zero-input response consists of terms of the form a_k^n, where

$$\{a_k\}, \quad k = 1, 2, \ldots, N$$

are the poles of the system. Therefore, the system is asymptotically stable if, and only if,

$$|a_k| < 1, \quad k = 1, 2, \ldots, N$$

In summary,

1. A LTI discrete system is asymptotically stable if, and only if, all the roots (simple or repeated) of the denominator polynomial of the transfer function are located inside the unit-circle.
2. A LTI discrete system is unstable if, and only if, one or more of the roots of the denominator polynomial of the transfer function are located outside the unit-circle or repeated roots are located on the unit-circle.
3. A LTI discrete system is marginally stable if, and only if, no roots of the denominator polynomial of the transfer function are located outside the unit-circle and some unrepeated roots are located on the unit-circle.

5.3.3   Convolution

Let $x(n)u(n) \leftrightarrow X(z)$ and $h(n)u(n) \leftrightarrow H(z)$. Then, the convolution operation in the time domain becomes the much simpler multiplication of the transforms of the two signals to be convolved in the z-domain. That is,

$$y(n) = \sum_{m=0}^{\infty} h(m)x(n - m) = h(n) * x(n) \leftrightarrow Y(z) = H(z)X(z)$$

The z-transform $Y(z)$ of $y(n)$, by definition, is

$$Y(z) = \sum_{n=0}^{\infty} y(n)z^{-n}$$

In terms of convolution, we get

$$Y(z) = \sum_{n=0}^{\infty} (h(0)x(n) + h(1)x(n-1) + h(2)x(n-2) + \cdots)z^{-n}$$

$$= \sum_{n=0}^{\infty} h(0)x(n)z^{-n} + \sum_{n=0}^{\infty} h(1)x(n-1)z^{-n} + \sum_{n=0}^{\infty} h(2)x(n-2)z^{-n} + \cdots$$

$$= h(0) \sum_{n=0}^{\infty} x(n)z^{-n} + h(1) \sum_{n=0}^{\infty} x(n-1)z^{-n} + h(2) \sum_{n=0}^{\infty} x(n-2)z^{-n} + \cdots$$

$$= h(0)X(z) + h(1)X(z)z^{-1} + h(2)X(z)z^{-2} + \cdots$$

$$= (h(0) + h(1)z^{-1} + h(2)z^{-2} + \cdots)X(z) = H(z)X(z)$$

Let the two transform pairs of the two sequences $x(n)$ and $h(n)$ be

$$x(n) = \left(\frac{1}{4}\right)^n u(n) \leftrightarrow X(z) = \frac{z}{z - \frac{1}{4}} \quad \text{and} \quad h(n) = \left(\frac{1}{5}\right)^n u(n) \leftrightarrow H(z) = \frac{z}{z - \frac{1}{5}}$$

The convolution of the sequences in the transform domain is the product of their z-transforms $Y(z) = X(z)H(z)$. Expanding $Y(z)$ into partial-fraction expansion, we get

$$Y(z) = X(z)H(z) = \left(\frac{z}{z - \frac{1}{4}}\right)\left(\frac{z}{z - \frac{1}{5}}\right) = \frac{5z}{z - \frac{1}{4}} - \frac{4z}{z - \frac{1}{5}}$$

Taking the inverse transform of $Y(z)$, we get the convolution of the sequences in the time domain as

$$y(n) = \left(5\left(\frac{1}{4}\right)^n - 4\left(\frac{1}{5}\right)^n\right) u(n)$$

The first 4 values of $\{y(n)$ are $1, 0.45, 0.1525, 0.0461\}$, which can be verified using time-domain convolution.

5.3.4 Left Shift of a Sequence

Let $x(n)u(n) \leftrightarrow X(z)$ and m is a positive integer. That is, the signal $x(n)$ is passed through a set of m unit-advance units to get $x(n+m)u(n)$. Due to the left shifting, some samples are lost and they have to be taken care of in expressing its transform in terms of $X(z)$. That is,

$$x(n+m)u(n) \leftrightarrow z^m X(z) - z^m \sum_{n=0}^{m-1} x(n)z^{-n}$$

For example, with $m = 1$ and $m = 2$, we get

$$x(n+1)u(n) \leftrightarrow z^1 X(z) - zx(0)$$
$$x(n+2)u(n) \leftrightarrow z^2 X(z) - z^2 x(0) - zx(1)$$

Let the z-transform of the sequence $x(n+m)u(n)$ be $Y(z)$. Then, after left shifting, the z-transform of the remaining samples is

$$Y(z) = x(m) + x(m+1)z^{-1} + x(m+2)z^{-2} + \cdots$$

Multiplying both sides by z^{-m}, we get

$$z^{-m} Y(z) = x(m)z^{-m} + x(m+1)z^{-m-1} + \cdots$$

By adding m terms, $\sum_{n=0}^{m-1} x(n)z^{-n}$, to both sides of the equation, we get

$$z^{-m} Y(z) + x(m-1)z^{-m+1} + x(m-2)z^{-m+2} + \cdots + x(0) = X(z)$$

Multiplying both sides by z^m and rearranging, we get

$$Y(z) = z^m X(z) - z^m \sum_{n=0}^{m-1} x(n)z^{-n}$$

The z-transform of $x(n)u(n)$ is

$$X(z) = -3 + 4z^{-1} - z^{-2} + 3z^{-3}$$

The advanced sequence, shown in Fig. 5.4c, is the left shifted sequence $x(n+1)$.

$$x_a(n) = \{x(-3) = 2, x(-2) = 1, x(-1) = -3, x(0) = 4, x(1) = -1, x(2) = 3, \}$$

The z-transform of $x(n+1)u(n)$ is

$$X_a(z) = 4 - z^{-1} + 3z^{-2}$$
$$= z^1(-3 + 4z^{-1} - z^{-2} + 3z^{-3}) - zx(0) = z^1 X(z) - zx(0)$$

The left shift property is more convenient for solving difference equations in advance operator form and in state-space analysis of systems.

5.3.5 Multiplication by n

Let $x(n)u(n) \leftrightarrow X(z)$. Then,

$$nx(n)u(n) \leftrightarrow -z\frac{d(X(z))}{dz}$$

Differentiating the defining expression for $X(z)$ with respect to z and multiplying it by $-z$, we get

$$-z\frac{d(X(z))}{dz} = -z\frac{d}{dz}\sum_{n=0}^{\infty} x(n)z^{-n} = \sum_{n=0}^{\infty} nx(n)z^{-n} = \sum_{n=0}^{\infty}(nx(n))z^{-n}$$

For example,

$$a^n u(n) \leftrightarrow \frac{z}{z-a}, \ z > |a| \quad \text{and} \quad na^n u(n) \leftrightarrow \frac{az}{(z-a)^2}, \ z > |a|$$

5.3.6 Multiplication by a^n

This property allows us to find the transforms of signals $a^n x(n)$ those are the product of $x(n)$ and a^n in terms of the transform $X(z)$ of $x(n)$, generating a large number of z-transforms. Let $x(n)u(n) \leftrightarrow X(z)$. Then,

$$a^n x(n)u(n) \leftrightarrow X\left(\frac{z}{a}\right)$$

for any constant a, real or complex. From the defining equation of the z-transform, we get the transform of $a^n x(n)u(n)$ as

$$\sum_{n=0}^{\infty} a^n x(n)z^{-n} = \sum_{n=0}^{\infty} x(n)\left(\frac{z}{a}\right)^{-n} = X\left(\frac{z}{a}\right)$$

The variable z is replaced by z/a, which amounts to scaling the frequency variable z. For example,

$$(0.5)^n u(n) \leftrightarrow \frac{z}{z-0.5}, \ z > 0.5 \quad \text{and}$$

$$(2)^n (0.5)^n u(n) \leftrightarrow \frac{\frac{z}{2}}{(\frac{z}{2}-0.5)} = \frac{z}{z-1} \ z > 1$$

The pole at $z = 0.5$ in the transform of $(0.5)^n u(n)$ is shifted to the point $z = 1$ in the transform of $(2)^n (0.5)^n u(n)$.

5.3.7 Summation

Let $x(n)u(n) \leftrightarrow X(z)$. Then, by summation property,

$$s(n) = \sum_{m=0}^{n} x(m) \leftrightarrow S(z) = \frac{z}{z-1} X(z)$$

The convolution of $x(n)$ with the unit-step function $u(n)$ is the cumulative sum $s(n)$ of $x(n)$ at n. In the frequency domain, it corresponds to the product of $X(z)$ and $U(z) = \frac{z}{z-1}$.

For example, $x(n) = \delta(n-p) \leftrightarrow z^{-p}$. Then,

$$S(z) = \frac{z}{z-1} z^{-p}$$

Taking the inverse z-transform, we get $s(n) = u(n-p)$, the delayed unit-step function. If $p = 0$, then $s(n) = u(n)$, as it is known that the sum of unit-impulse is the unit-step.

Similarly, the cumulative sum of delayed unit-step signal $u(n-1)$ is the unit-ramp signal $r(n)$. Using the theorem, we get

$$S(z) = \frac{z}{z-1} \frac{z(z^{-1})}{z-1} = \frac{z}{(z-1)^2}$$

which is the z-transform of $r(n) = nu(n)$.

5.3.8 Initial Value

Without finding the inverse z-transform, we can determine the initial term $x(0)$ of a sequence $x(n)$ from its z-transform $X(z)$ directly. Let $x(n)u(n) \leftrightarrow X(z)$. Then, the identity

$$x(0) = \lim_{|z| \to \infty} X(z)$$

is called initial value property. Extending this procedure, we get

$$x(1) = \lim_{|z| \to \infty} (z(X(z) - x(0)))$$

The definition of the z-transform is a power series. Therefore, we get

$$\lim_{|z| \to \infty} X(z) = \lim_{|z| \to \infty} \left(x(0) + x(1)z^{-1} + x(2)z^{-2} + x(3)z^{-3} + \cdots \right) = x(0)$$

Letting the variable z^{-1} approach zero, all the terms, except $x(0)$, tend to zero. This property is useful to verify the closed-form expression for the time-domain sequence obtained from its transform.

For example, let $X(z) = \frac{(3z^2-5z+2)}{(z^2+z-3)}$. Then,

$$x(0) = \lim_{|z| \to \infty} \frac{(3z^2 - 5z + 2)}{(z^2 + z - 3)} = 3$$

Only the terms of the highest power are significant with $z \to \infty$.

5.3.9 Final Value

Without finding the inverse z-transform, we can determine the limit of a sequence $x(n)$, as n tends to ∞ from its z-transform $X(z)$ directly. Let $x(n)u(n) \leftrightarrow X(z)$. Then, the identity

$$\lim_{n \to \infty} x(n) = \lim_{z \to 1} (z - 1)X(z)$$

is called the final value property, if the ROC of $(z - 1)X(z)$ includes the unit-circle. This condition excludes all the cases with no limit as $n \to \infty$. As the contribution of all other poles inside the unit-circle tends to zero as $\lim_{n\to\infty} x(n)$, the final value is due the pole at $z = 1$. That is, this value is the coefficient of the partial-fraction expansion of $X(z)$. That coefficient is obtained by multiplying $X(z)$ by $(z - 1)$ and then setting $z = 1$.

For example, let

$$X(z) = \frac{(z^2 - 2z + 5)}{(z^2 - 5z + 6)}$$

The property does not apply as the poles are located at $z = 2$ and $z = 3$, and, hence, the ROC of $(z - 1)X(z)$ does not include the unit-circle. Let

$$X(z) = \frac{(z^2 - 3z + 1)}{(z^2 - 1.8z + 0.8)}$$

Then,

$$\lim_{n \to \infty} x(n) = \lim_{z \to 1} (z - 1)\frac{(z^2 - 3z + 1)}{(z^2 - 1.8z + 0.8)} = \lim_{z \to 1} \frac{(z^2 - 3z + 1)}{(z - 0.8)} = -5$$

5.3.10 Transform of Semiperiodic Functions

As the input signal is restricted to be causal, the z-transform cannot handle periodic signals. However, it can handle semiperiodic signals with period N and 0 for $n < 0$ defined as $x(n + N) = x(n)$, $n \geq 0$. The z-transform of the first period of the semiperiodic function is

$$x_0(n) = x(n)u(n) - x(n - N)u(n - N) \leftrightarrow X_0(z), \quad z \neq 0$$

That is, $x_0(n)$ is the same as $x(n)u(n)$ over its first period with zero elsewhere. Then, due to periodicity,

$$x(n)u(n) = x_0(n) + x_0(n - N) + x_0(n - 2N) + \cdots$$

From the right shift property, the transform $X(z)$ of $x(n)u(n)$ is

$$X(z) = X_0(z)\left(1 + z^{-N} + z^{-2N} + \cdots\right) = \frac{X_0(z)}{1 - z^{-N}} = \left(\frac{z^N}{z^N - 1}\right) X_0(z), \quad |z| > 1$$

For example, let $x_0(n) = nu(n)$, $n = 0, 1, 2$ with period $N = 3$. The transform of the first period samples is

$$X_0(z) = z^{-1} + 2z^{-2} = \frac{z + 2}{z^2}$$

From the property,

$$X(z) = \frac{z^3}{(z^3 - 1)} \frac{(z + 2)}{z^2} = \frac{z(z + 2)}{(z^3 - 1)}$$

Let us find the inverse z-transform of $X(z)$.

$$X(z)/z = \frac{(z + 2)}{(z^3 - 1)} = \frac{A}{z - 1} + \frac{B}{z + (0.5 - j0.866)} + \frac{C}{z + (0.5 + j0.866)}$$

$$X(z) = \frac{z}{z - 1} + \frac{z(-0.5000 + j0.2887)}{z + (0.5 - j0.866)} + \frac{z(-0.5000 - j0.2887)}{z + (0.5 + j0.866)}$$

Taking the inverse, we get

$$x(n)u(n) = 1 + 2(0.5774) \cos\left(\frac{2\pi}{3}n + (2.6180)\right) = \{\hat{0}, 1, 2, 0, 1, 2, \ldots\}$$

5.4   Summary

- While the DTFT changes a sequence of numbers into a function of the purely imaginary complex variable $e^{j\omega}$, the z-transform changes a sequence of numbers into a function of the complex variable z with an expanded set of basis functions. That is, the DTFT of a signal is its z-transform with $z = e^{j\omega}$, if the DTFT exists.
- For real-valued signals, the z-transform becomes a real-rational function of the complex variable z. The properties of these functions are well-established and can be used in system analysis. By designing the z-transform only for causal signals, the systems with initial conditions can be easily analyzed.
- A short list of z-transform pairs is adequate for most of the analysis. In addition, the use of variable z instead of $e^{j\omega}$ makes it simpler to manipulate the z-transform expressions.
- There is no z-transform representation for a signal, which grows faster than an exponential.
- All practical signals satisfy the convergence condition and are, therefore, have z-transform representation.
- The ROC of a z-transform is the region in the z-plane that is exterior to the smallest circle, centered at the origin, enclosing all its poles.
- At roots of the denominator polynomial of a rational function, its value tends to ∞. These roots are called *poles* of the rational function. Similarly, its *zeros* are the roots of its numerator polynomial, where its value tends to zero.
- As finding the inverse z-transform by evaluating a contour integral is relatively difficult, for most practical purposes, two relatively simpler methods, the partial-fraction method and the long-division method are commonly used.
- Partial-fraction expansion of a rational function expresses it as a sum of appropriate fractions with the coefficient of each fraction to be found.
- The convolution operation in the time domain becomes the much simpler multiplication of the z-transforms of the two signals to be convolved in the frequency domain.

Exercises

5.1 The nonzero values of the sequence $x(n)$ are given as

$$\{x(n) = x(-3) = 3, x(0) = -1, x(2) = -2\}$$

Find the unilateral z-transform of $x(n)$.
5.1.1. $x(n)$.
* 5.1.2. $2x(n-1)$.
5.1.3. $x(-n)$.
5.1.4. $3x(n+1)$.

5.1.5. $-2x(n+3)$.

5.2 Find the nonzero values of the inverse of the given unilateral z-transform.
5.2.1. $X(z) = 1 + 2z^{-1} + 4z^{-2}$.
5.2.2. $X(z) = 2 + z^{-3} - z^{-5}$.
* **5.2.3.** $X(z) = z^{-7} + 2z^{-9}$.
5.2.4. $X(z) = 2(1 + z^{-3} + 2z^{-5})$.
5.2.5. $X(z) = -z^{-1} - z^{-4}$.

5.3 The nonzero values of two sequences $x(n)$ and $h(n)$ are given. Using the z-transform, find the convolution of the sequences $y(n) = x(n) * h(n)$.
* **5.3.1.** $\{x(0) = 1, x(1) = 2, x(2) = -4\}$ and $\{h(0) = 3, h(1) = 2\}$.
5.3.2. $\{x(1) = 1, x(3) = 3\}$ and $\{h(2) = -3, h(4) = 2\}$.
5.3.3. $\{x(2) = 1, x(3) = -3\}$ and $\{h(1) = 3, h(2) = 2\}$.
5.3.4. $\{x(0) = 2, x(4) = -1\}$ and $\{h(1) = -2, h(3) = 2\}$.
5.3.5. $\{x(1) = -2, x(3) = -1\}$ and $\{h(2) = 3, h(3) = 1\}$.

5.4
5.4.1. Use the multiplication by n property to find the z-transform of $x(n)$. Verify that the inverse of the resulting transform gets back the given signal.
5.4.1. $x(n) = n\delta(n-4)u(n)$.
5.4.2. $x(n) = n(-0.8)^n u(n)$.
* **5.4.3.** $x(n) = nu(n-1)$.

5.5 Given the samples $x_0(n)$ of the first period of the semiperiodic signal $x(n)$, find its z-transform $X(z)$.
5.5.1.

$$x_0(n) = \{0.5\hat{0}00, -0.8660, -0.5000, 0.8660\}.$$

5.5.2.

$$x_0(n) = \{\hat{1}, 2, 1\}.$$

* **5.5.3.**

$$x_0(n) = \{\hat{3}, 2\}.$$

5.6 The difference equation of a causal discrete system is given. Using the z-transform, find a closed-form expression for the time-domain response $y(n)$. Assume that the system is initially relaxed and the input is the unit-impulse, $x(n) = \delta(n)$. Verify that the first 4 values of the response of the system obtained using the expression for $y(n)$ and by iterating the difference equation are the same.

* 5.6.1

$$y(n) = x(n) - 2y(n-1) - y(n-2) + 2y(n-3).$$

5.6.2

$$y(n) = x(n) + 2y(n-1) + y(n-2) - 2y(n-3).$$

5.6.3

$$y(n) = x(n) + x(n-1) - x(n-2) + 2y(n-1) + y(n-2) - 2y(n-3).$$

5.6.4

$$y(n) = x(n) + 5y(n-1) - 8y(n-2) + 4y(n-3).$$

5.7 A discrete system is governed by the difference equation

$$y(n) = 2x(n) - x(n-1) + x(n-2) + \frac{3}{2}y(n-1) - \frac{9}{16}y(n-2)$$

Using the z-transform, find the complete response of the system for the input $x(n) = u(n)$, the unit-step function. The initial conditions are $y(-1) = 1$ and $y(-2) = 2$. List the first four values of the complete response of the system. Verify your answer by finding the first few values of the response by iterating the difference equation. Verify that the zero-input response satisfies the initial conditions. Verify that the total response in the z-transform domain satisfies the initial and final value theorems

5.8 A discrete system is governed by the difference equation

$$y(n) = x(n) - 2x(n-1) + x(n-2) + \frac{5}{6}y(n-1) - \frac{1}{6}y(n-2)$$

Using the z-transform, find the complete response of the system for the input $x(n) = 0.8^n u(n)$. The initial conditions are $y(-1) = 0$ and $y(-2) = 1$. List the first four values of the complete response of the system. Verify your answer by finding the first few values of the response by iterating the difference equation. Verify that the zero-input response satisfies the initial conditions. Verify that the total response in the z-transform domain satisfies the initial and final value theorems

* 5.9 A discrete system is governed by the difference equation

$$y(n) = x(n) - 2x(n-1) + \frac{5}{6}y(n-1)$$

Using the z-transform, find the complete response of the system for the input $x(n) = \cos(\frac{2\pi}{8}n + \frac{\pi}{3})u(n)$. The initial condition is $y(-1) = 3$. List the first four values of the complete response of the system. Verify your answer by finding the first few values of the response by iterating the difference equation. Verify that the zero-input response satisfies the initial conditions. Verify that the total response in the z-transform domain satisfies the initial and final value theorems.

5.10 A discrete system is governed by the difference equation

$$y(n) = 2x(n) - 2x(n-1) + x(n-2) + 1.6y(n-1) - 0.64y(n-2)$$

Using the z-transform, find the complete response of the system for the input $x(n) = 0.8^n u(n)$. The initial conditions are $y(-1) = 2$ and $y(-2) = 1$. List the first four values of the complete response of the system. Verify your answer by finding the first few values of the response by iterating the difference equation. Verify that the zero-input response satisfies the initial conditions. Verify that the total response in the z-transform domain satisfies the initial and final value theorems.

Finite Impulse Response Filters

6

Filter means that removes something that passes through it. A coffee filter removes coffee grounds from the coffee extract, allowing the decoction. Water filters are used to remove salts and destroy bacteria. An electrical filter modifies the spectrum of signals passing through it in a desired way. Filters are commonly used for several purposes. One common example of the usage of electrical filters is to reduce the noise mixed up with the signals. Another common example is to separate the desired signal from the intermediate signal frequency in receiving systems.

Natural electrical filters made of components, such as resistors, capacitors, and inductors, are analog devices. Using such filters or simulating them will result in a filter with infinite impulse response. This type of filters is called recursive or IIR (infinite impulse response) filter. The major disadvantages of this type are that the components are analog and when implemented results in inferior characteristics and the stability problem. While the digital version of IIR filters have wide applications, the other type with finite impulse response is also often used. This type of filters is called nonrecursive or FIR (finite impulse response) filters, although it can also be implemented recursively. This type has the potential disadvantage of requiring more computation with more number of coefficients. However, fast DFT algorithms, implemented in hardware and/or software, make it up. Further, linear-phase filters, which are frequently required in applications, can be designed easily. In this chapter, we study the FIR filter. One of the important differences with digital filter compared with the analog filter is that the frequency response is circular or periodic. While the highest frequency can be anything finite, it can be never be infinity as in the case of analog filter due to the distinguishability of only a finite number of discrete sinusoids due to sampling.

The impulse response of FIR filters is of finite length, and, therefore, implemented with finite number of coefficients. The output of FIR filters is an exclusive function of past and present input samples only, without feedback. That means no inherent stability problem. The reactance of the inductor increases with increasing

D. Sundararajan, *Digital Signal Processing*,
https://doi.org/10.1007/978-3-030-62368-5_6

frequency and that of a capacitor decreases. These components, along with resistors, enable us to realize a filter and visualize its action. The frequency response of inductors and capacitors are approximately simulated by the coefficients in the case of digital filters. The coefficients of the filters weigh the input samples appropriately to provide the filter action. For example, if the impulse response of the filter is slowly varying, then the filter will match with low frequency signals better and pass them more readily compared with high frequency signals. Similarly, a filter with a fast varying impulse response will more readily pass high frequency signals. While this is the essential point, several methods of filter design are available giving choices of simplicity and optimality. These methods provide appropriate filter coefficients, so that the filter's response satisfies the given specifications.

Overview of FIR Filter Design Using Windows
The infinite extent impulse response of the ideal filter that meets the specification is first found by taking the inverse DTFT of the frequency response. The impulse response is made realizable by shifting and truncation of the ideal impulse response. The truncation is carried out by multiplying the impulse response by a window of suitable length and characteristics. Several windows are available and the one suitable has to be selected. The filter order is approximately found for each window. Filters designed using each window have different transition bandwidths and sidelobe characteristics. Trial and error is required to find the best possible filter coefficients.

6.1   Characterization and Realization

In the time domain, the causal FIR filter is characterized by a difference equation that is a linear combination of past and present input samples $x(n)$ multiplied by the coefficients $h(n)$. For example, the difference equation of a second-order filter is

$$y(n) = h(0)x(n) + h(1)x(n-1) + h(2)x(n-2)$$

Samples $\{x(n), x(n-1), x(n-2)\}$ are the present and the past two input samples and $\{h(0), h(1), h(2)\}$ are the respective filter coefficients. The output is $y(n)$. The filter length M is 3, which is the number of terms of the impulse response. The order N of a FIR filter is the filter length minus one ($N = M - 1$) or the longest delay of the input terms in number of samples. In general, the difference equation is of the form

$$y(n) = \sum_{k=0}^{N} h(k)x(n-k) = \sum_{k=0}^{N} x(k)h(n-k)$$

where $y(n)$, $h(n)$, and $x(n)$ are, respectively, the output samples, coefficients or the values of the impulse response, and the input samples to the filter. The difference equation is the convolution of the impulse response of the filter and the input samples. The filter realization requires multipliers, adders, and unit delays. Subtraction operation is carried out by the adder by complementing the subtrahend and then adding with the minuend. The delay units create the necessary delay of the input samples. Then, the operation is sum of products of the input samples with their corresponding coefficients. The coefficients of the filter are the terms of its impulse response. The desired modification of the spectrum of the input signal (filtering) is carried out by using appropriate coefficients. Figure 6.1 shows the direct-form realization of a second-order FIR filter. There are two delay units in cascade with the input applied producing the necessary delayed samples of the input. The input and delayed samples are multiplied by their corresponding coefficients and all the three products are summed to produce the present output sample. This is a straightforward realization of the equation defining the filter. As most FIR filters are designed with linear phase, their coefficients are either symmetrical or asymmetrical, the number of multipliers can be reduced to about one-half.

With zero-initial conditions and the unit-impulse input, the response of the filter is its impulse response.

$$y(n) = \sum_{k=0}^{N} h(k)\delta(n-k) = h(n)$$

In the case of FIR filters, the numerator coefficients of the transfer function are the same as the impulse response. With

$$b(n) = \begin{cases} h(n) \text{ for } 0 \le n \le N \\ 0 \quad \text{otherwise} \end{cases}$$

the difference equation is usually written as

$$y(n) = \sum_{k=0}^{N} b(k)x(n-k) = \sum_{k=0}^{N} x(k)b(n-k)$$

in digital filter descriptions.

As impulse is composed of all frequency components, its transform is both the transfer function and the frequency response with respect to the transform used. In general, the transfer function is more useful to design the filter and visualize its operation. Let the z-transforms of the input $x(n)$, output $y(n)$, and the impulse response $h(n)$ be, respectively, $X(z)$, $Y(z)$, and $H(z)$. Then, the output of the filter, as convolution in the time domain becomes multiplication in the transform domain, is given by

Fig. 6.1 Direct-form
realization of a second-order
FIR filter

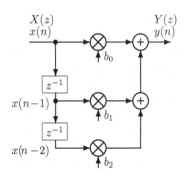

$$Y(z) = X(z)H(z)$$

in the z-domain. Replacing z by $e^{j\omega}$, we get in the DTFT-domain

$$Y\left(e^{j\omega}\right) = X\left(e^{j\omega}\right)H\left(e^{j\omega}\right)$$

As the transform of the unit-impulse $X(z) = 1$, the z-transform transfer function of a second-order FIR filter is given as

$$H(z) = \frac{Y(z)}{X(z)} = h(0) + h(1)z^{-1} + h(2)z^{-2}$$

As the transfer function is a numerator polynomial only in z, the transfer function has no poles except at $z = 0$. Hence, FIR filters are inherently stable. The transfer function with respect to the DTFT or the frequency response of a second-order FIR filter is obtained replacing z by $e^{j\omega}$ in $H(z)$ as

$$H\left(e^{j\omega}\right) = \frac{Y\left(e^{j\omega}\right)}{X\left(e^{j\omega}\right)} = h(0) + h(1)e^{-j\omega} + h(2)e^{-j2\omega}$$

6.1.1 Ideal Lowpass Filters

While FIR filters must necessarily have finite impulse response, we usually start with the study of ideal filters with infinite impulse response. Ideal filters set a maximum bound for performance. Further, practical filters are often designed approximating the impulse response of ideal filters to satisfy the given specifications. Some of the basic types of filters are lowpass, highpass, bandpass, and bandreject. Lowpass filters are often used as a starting point in the design of other types of filters also, as they can be expressed as a linear combination of lowpass filters or by some transformation. Therefore, a thorough study of lowpass filter design is required.

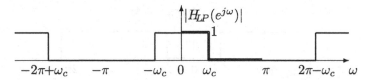

Fig. 6.2 Magnitude of the periodic frequency response of an ideal digital lowpass filter

The normalized range of the periodic frequency response of a digital filter is 0 to 2π radians. The reason that is periodic is due to sampled representation of the signals. While the upper limit can be varied with different sampling intervals to any desired finite limit, it can never be ∞. Since the coefficients are real-valued, the frequency response is conjugate symmetric. Therefore, due to the redundancy, the frequency response is usually specified in the range $\omega = 0$ to π. In the positive normalized range $\omega = 0$ to π, the passband is near the frequency $\omega = 0$ where the frequency response has a value of 1 and the stopband is away from $\omega = 0$ where the frequency response has a value of 0. That is, the frequency components of the signal with frequencies in the passband are passed through the filter and the rest rejected. The magnitude of the frequency response of an ideal digital lowpass filter is shown in Fig. 6.2. It is even-symmetric and periodic with period 2π. The response is specified as

$$|H_{LP}(e^{j\omega})| = \begin{cases} 1 \text{ for } 0 \leq \omega \leq \omega_c \\ 0 \text{ for } \omega_c < \omega \leq \pi \end{cases}$$

The response is even-symmetric, if the impulse response is real-valued. The frequency response is the transform of the impulse response. That is, if the coefficients of the filter are real-valued, the magnitude of the frequency response is even-symmetric. The lowpass filter has its passband centered on low frequencies. The range of frequencies from 0 to ω_c is called the passband and that from ω_c to π is called the stopband. The ideal filter model is physically unrealizable, since it requires a noncausal system. It should be ensured that the input range of frequencies is within the range $\omega = 0$ to π only. Otherwise, the signal will be distorted due to aliasing effect.

6.1.2 Ideal Highpass Filters

The magnitude of the frequency response of an ideal digital highpass filter is shown in Fig. 6.3. It is even-symmetric and periodic with period 2π. The response is specified as

$$|H_{HP}(e^{j\omega})| = \begin{cases} 0 \text{ for } 0 \leq \omega < \omega_c \\ 1 \text{ for } \omega_c \leq \omega \leq \pi \end{cases}$$

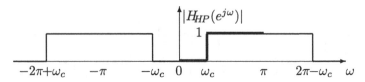

Fig. 6.3 Magnitude of the periodic frequency response of an ideal digital highpass filter

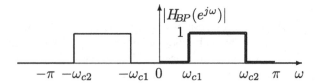

Fig. 6.4 Magnitude of the periodic frequency response of an ideal digital bandpass filter

For the same cutoff frequency ω_c, the difference between the frequency response of an allpass filter and that of the lowpass filter is the frequency response of the highpass filter. That is,

$$H_{HP}\left(e^{j\omega}\right) = 1 - H_{LP}\left(e^{j\omega}\right)$$

and, in the time domain,

$$h_{HP}(n) = \delta(n) - h_{LP}(n)$$

In the normalized range $\omega = 0$ to π, the stopband is near the frequency $\omega = 0$ where the frequency response has a value of 0 and the passband is away from $\omega = 0$ where the frequency response has a value of 1. That is, the frequency components of the signal with frequencies in the passband are passed through the filter and the rest rejected. The highpass filter has its stopband centered on low frequencies.

6.1.3   Ideal Bandpass Filters

The magnitude of the periodic frequency response of an ideal digital bandpass filter is shown in Fig. 6.4. The response is specified as

$$|H_{BP}\left(e^{j\omega}\right)| = \begin{cases} 1 \text{ for } \omega_{c1} \leq \omega \leq \omega_{c2} \\ 0 \text{ for } 0 \leq \omega < \omega_{c1} \text{ and } \omega_{c2} < \omega \leq \pi \end{cases}$$

The passband is centered in the middle of the spectrum, unlike at the origin or end of the spectrum as in the case of lowpass and highpass filters. Consequently, there are two stopbands. The filter passes frequency components with unity gain in the passband from ω_{c1} to ω_{c2} and rejects everything else in the two stopbands.

Fig. 6.5 Magnitude of the
periodic frequency response
of an ideal digital bandstop
filter

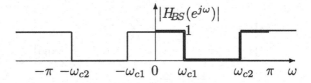

6.1.4 Ideal Bandstop Filters

The magnitude of the periodic frequency response of an ideal digital bandstop filter
is shown in Fig. 6.5. The response is specified as

$$|H_{BS}(e^{j\omega})| = \begin{cases} 0 \text{ for } \omega_{c1} \leq \omega \leq \omega_{c2} \\ 1 \text{ for } 0 \leq \omega < \omega_{c1} \text{ and } \omega_{c2} < \omega \leq \pi \end{cases}$$

The stopband is centered in the middle of the spectrum, unlike at the end or horizon
of the spectrum as in the case of lowpass and highpass filters. Consequently, there
are two passbands. The filter passes frequency components with unity gain in the
two passbands from

$$\text{for } 0 \leq \omega < \omega_{c1} \text{ and } \omega_{c2} < \omega \leq \pi$$

and rejects everything else in the stopband.
   Bandpass filters can be realized using two lowpass filters as

$$h_{bp}(k) = lh_c(k) - ll_c(k)$$

with filter $lh_c(k, l)$ having a higher cutoff frequency. Bandstop filters can be realized
using a bandpass filter as

$$h_{bs}(k) = \delta(k) - h_{bp}(k)$$

Frequency Response

Type I Filter
Figure 6.6a shows a typical 4-th order FIR filter with even-symmetric impulse
response and odd number of coefficients $NC = N + 1 = 5$, where N is the order
of the filter. This type of filters is called Type I filters. Depending on the symmetry
(odd or even) and number of coefficients (odd or even), the linear-phase FIR filters
are classified into four types. Each type is suitable for the design of some filters. The
impulse response values are

$$\{h(-2) = 0.1592, h(-1) = 0.2251, h(0) = 0.25, h(1) = 0.2251, h(2) = 0.1592\}$$

The frequency response, which is the transform of its impulse response, is

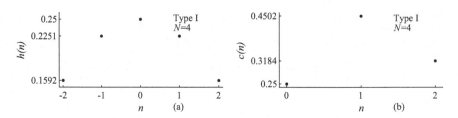

Fig. 6.6 (a) Typical FIR filter with even-symmetric impulse response and odd number of coefficients (Type I); (b) filter coefficients using the even symmetry

$$H\left(e^{j\omega}\right) = h(-2)e^{j2\omega} + h(-1)e^{j\omega} + h(0) + h(1)e^{-j\omega} + h(2)e^{-j2\omega}$$

$$= h(0) + 2(h(1)e^{-j\omega} + h(-1)e^{j\omega}) + 2(h(2)e^{-j2\omega} + h(2)e^{j2\omega})$$

using the even symmetry of the impulse response. Since the impulse response has to be a causal sequence for practical filters, we have to delay the response by $N/2$ samples. Taking this into account and further simplifying, we get the frequency response of the filter as

$$H\left(e^{j\omega}\right) = e^{-j\frac{N}{2}\omega}(c(0) + c(1)\cos(\omega) + c(2)\cos(2\omega))$$

where

$$c(0) = h(0) = 0.25, c(1) = h(1) + h(-1) = 0.4502 \text{ and}$$

$$c(2) = h(2) + h(-2) = 0.3184$$

The filter coefficients using the even symmetry are shown in Fig. 6.6b. For example filter,

$$H\left(e^{j\omega}\right) = e^{-j2\omega}(0.25 + 0.4502\cos(\omega) + 0.3184\cos(2\omega))$$

In general, the frequency response of Type I filter is given by

$$H\left(e^{j\omega}\right) = e^{-j\omega\left(\frac{N}{2}\right)} \sum_{n=0}^{\frac{N}{2}} c(n)\cos(\omega n),$$

$$c(n) = \begin{cases} hs(\frac{N}{2}) & \text{for } n = 0 \\ 2hs\left(\frac{N}{2} - n\right) & \text{for } n = 1, 2, \dots, \frac{N}{2} \end{cases} \qquad (6.1)$$

assuming the impulse response $h(n)$ is shifted, starting at index 0, $hs(n)$. While the spectrum is continuous, in practice, frequency response is usually approximated by sufficient number of spectral samples using the DFT. In this example, there are

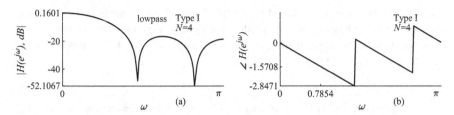

Fig. 6.7 (a) The magnitude of the frequency response of the filter; (b) the phase response

5 impulse response values. By taking the 5-point DFT, we get 5 samples of the continuous DTFT spectrum. The density of the spectral samples is not adequate. Therefore, it is a necessity to zero pad the impulse response values and then compute the DFT. If we zero pad the impulse response of the example filter by 3 zeros, we get

$$\{0.1592, 0.2251, 0.2500, 0.2251, 0.1592, 0, 0, 0\}$$

Computing the DFT, we get 8 spectral samples. Typically, the impulse response values are padded up by zeros to make the length of the sequence to 512. The length of the zero-padded sequence has to be decided in each case.

The magnitude of the frequency response of the example filter is

$$|H(e^{j\omega})| = (0.25 + 0.4502\cos(\omega) + 0.3184\cos(2\omega))$$

The phase response is

$$\angle H(e^{j\omega}) = -2\omega \text{ radians}$$

The magnitude of the frequency response is shown in Fig. 6.7a. The linear-phase response of the filter is shown in Fig. 6.7b. At $\omega = \pi/4$, the phase is $-2\pi/4 = -\pi/2$, as shown in Fig. 6.7b. The phase of the frequency response of the filter, proportional to an integer, is a linear function of ω as the summation value is real (with a phase of 0 or π radians).

Phase and Group Delay

The frequency components of the input signal in the passband are delayed when it passes through practical filters. If the phase response of the filter in the passband is linear (of the form $-n_0\omega$), then the frequency components of the output signal are a delayed version of the input frequency components in the passband. Such a phase delay results in no phase distortion. Differentiating the phase delay with respective to frequency yields the group delay given by

$$-\frac{d(n_0\omega)}{d\omega} = -n_0$$

The group delay gives the delay of the input frequency components in the passband are subjected to in passing through a filter. If the phase response of the filter is linear, then the shape of the signal formed by the frequency components remains the same, subject to an amplitude scaling.

Type II Filter

Figure 6.8a shows a typical 5-th order FIR filter with even-symmetric impulse response and even number of coefficients $NC = N + 1 = 6$, where N is the order of the filter. Figure 6.8b shows the filter coefficients $NC = N + 1 = 6$ using the even symmetry. The shifted impulse response values are

$$\{h(0) = 0.1176, h(1) = 0.1961, h(2) = 0.2436, h(3) = 0.2436,$$
$$h(4) = 0.1961, h(5) = 0.1176\}$$

The frequency response, which is the transform of its impulse response, is

$$H\left(e^{j\omega}\right) = h(0) + h(1)e^{-j\omega} + h(2)e^{-j2\omega} + h(2)e^{-j3\omega} + h(1)e^{-j4\omega} + h(0)e^{-j5\omega}$$

$$= e^{-j\omega\frac{N}{2}} \sum_{n=0}^{\frac{N-1}{2}} c(n)\cos(\omega(n+0.5))$$

where

$$c(n) = 2h\left(\frac{N-1}{2} - n\right) \text{ for } n = 0, 1, \ldots, \frac{N-1}{2}$$

and $h(n)$ are the shifted values starting with index 0. For example filter, the magnitude of the frequency response is

$$|H\left(e^{j\omega}\right)| = 2(0.2436\cos(0.5\omega) + 0.1961\cos(1.5\omega) + 0.1176\cos(2.5\omega))$$

The phase response is

$$\angle H\left(e^{j\omega}\right) = -2.5\omega \text{ radians}$$

The magnitude of the frequency response is shown in Fig. 6.9a. The linear-phase response of the filter is shown in Fig. 6.9b. The phase of the frequency response of the filter, proportional to a rational factor, is a linear function of ω as the summation value is real (with a phase of 0 or π radians).

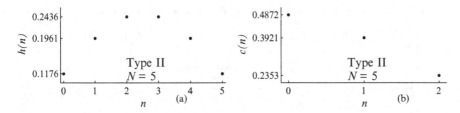

Fig. 6.8 (a) Typical 5-th order FIR filter with even-symmetric impulse response (Type II); (b) filter coefficients using the even symmetry

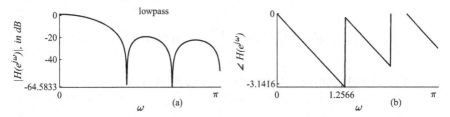

Fig. 6.9 (a) The magnitude of the frequency response of the filter; (b) the phase response

6.1.5 FIR Filters

The inverse DTFT of a periodic rectangular spectrum, centered at origin, yields the infinite-length real and even-symmetric coefficients of a lowpass filter. The coefficients are truncated to the required order of the filter. Then, the coefficients are shifted to make the filter causal, which adds linear phase to the frequency components. Therefore, the filtered output is delayed but not phase distorted.

Now, the truncated response can be of four types. One type, designated Type I, is even-symmetric with the number of coefficients $N + 1$ odd, where the order of the filter N is even. That is,

$$h(n) = \left\{ h\left(-\frac{N}{2}\right), \ldots, h(-1), h(0), h(1), \ldots, h\left(\frac{N}{2}\right) \right\}$$

Using Euler's identity and the even symmetry, the DTFT of $h(n)$, which is the frequency response is

$$H\left(e^{j\omega}\right) = h(0) + \sum_{n=1}^{\frac{N}{2}} 2h(n) \cos(\omega n)$$

For example, with the order $N = 6$, we get

$$h(n) = \{h(-3), h(-2), h(-1), h(0), h(1), h(2), h(3)\}$$

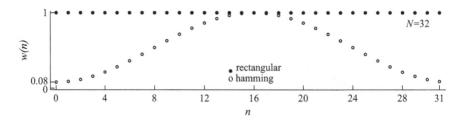

Fig. 6.10 Rectangular and Hamming window functions in the time domain

$$H\left(e^{j\omega}\right) = h(0) + \sum_{n=1}^{3} 2h(n)\cos(\omega n)$$

Then, with the right shift of $N/2$ samples to make it causal, we get

$$H\left(e^{j\omega}\right) = e^{-j\omega\left(\frac{N}{2}\right)} \sum_{n=0}^{\frac{N}{2}} c(n)\cos(\omega n), \quad c(n) = \begin{cases} h(0) & \text{for } n = 0 \\ 2h(n) & \text{for } n = 1, 2, \ldots, \frac{N}{2} \end{cases}$$

$$(6.2)$$

The DTFT spectrum of the impulse response of an ideal lowpass filter with a cutoff frequency ω_{ci} is a rectangular waveform, the inverse DTFT of which is shown to be

$$h_i(n) = \frac{\sin(\omega_{ci}n)}{\pi n}, \quad -\infty < n < \infty$$

in Chap. 4. As it is of infinite length and noncausal, it is practically unrealizable. Therefore, it has to be appropriately truncated and shifted. Truncation by rectangular window is very abrupt and results in big sidelobes in the spectrum. Therefore, other windows are used to reduce the sidelobes to the required level. Rectangular and Hamming window functions in the time domain are shown in Fig. 6.10. While all the samples of the rectangular window are 1s, the samples of the other windows are carefully designed to reduce the sidelobes to various levels. The price that is paid is the creation of the transition band at the cost of attenuation in the stopband.

6.1.6   Rectangular Window

The first window, called the rectangular window, is defined as

$$rect_w(n) = \begin{cases} 1 & \text{for } n = 0, 1, \ldots, N \\ 0 & \text{otherwise} \end{cases}$$

As all the samples of this window are 1s, the signal multiplied by it is the same as the truncated version of itself. The DTFT is the same as the Fourier series with role of the time and frequency variables interchanged. Therefore, the inability to converge pointwise in the neighborhood of discontinuities presents in the DTFT spectrum, called the Gibbs phenomenon. The use of this window provides the shortest transition band with smallest attenuation in the stopband.

The magnitude of the largest ripple in the stopband is 0.0895 and it determines the maximum stopband attenuation possible using this window to be $-20 \log_{10} 0.0895 = 20.96\,\text{dB}$. As this attenuation is fixed irrespective of the window length, only the width of the transition band gets shorter and the ripples get shorter in duration as the window length increases. The approximate average slope of the transition band this window provides is given as $0.9 \frac{2\pi}{N}$. Some trial and error may be required to make the frequency response suit the requirements. Despite the method of design, the frequency response of the designed filter must be verified that it meets the given specifications. Given the passband edge frequency ω_c and stopband edge frequency ω_s of a lowpass filter, the order of the filter is found using the formula

$$N \geq 0.9 \frac{2\pi}{\omega_s - \omega_c}$$

Usually, the resulting N is real-valued. We take the nearest higher integer as the filter order. If N is an even integer, we can design the appropriate type of filter and vice versa. If N is not suitable for a certain type, then increase the order by 1 and take $N + 1$ as the order.

Then, the shifted impulse response is specified, for both odd and even N, as

$$h_s(n) = \frac{\sin\left(\omega_{ci}\left(\frac{N}{2} - n\right)\right)}{\pi\left(\frac{N}{2} - n\right)}, \quad n = 0, 1, \ldots, N, \quad \omega_{ci} = \frac{\omega_s + \omega_c}{2}$$

Example 6.1 (Type I Lowpass) The passband and stopband edge frequencies of a lowpass filter are specified, respectively, as $\omega_c = 0.3\pi$ radians and $\omega_s = 0.45\pi$ radians, respectively. The minimum attenuation required in the stopband is 18 dB. Design the lowpass filter using the rectangular window. The sampling frequency is $f_s = 1024\,\text{Hz}$.

Solution Frequency 0.3π corresponds to $\frac{0.3\pi}{2\pi} 1024 = 153.6\,\text{Hz}$. Frequency 0.45π corresponds to $\frac{0.45\pi}{2\pi} 1024 = 230.4\,\text{Hz}$. The maximum attenuation provided by filters using the rectangular window is about 21 dB. Therefore, the specification of 18 dB is realizable. Now, we find the order of the filter as

$$N \geq 0.9 \frac{2\pi}{0.45\pi - 0.3\pi} = 12$$

The cutoff frequency of the corresponding ideal filter is computed as

$$\omega_{ci} = \frac{\omega_s + \omega_c}{2} = \frac{0.3\pi + 0.45\pi}{2} = 0.375\pi$$

The shifted impulse response of the filter is given by

$$h(n) = \frac{\sin\left(\omega_{ci}\left(\frac{N}{2} - n\right)\right)}{\pi\left(\frac{N}{2} - n\right)} = \frac{\sin(0.375\pi(6 - n))}{\pi(6 - n)}, \; n = 0, 1, \ldots, 12$$

The shifted impulse response values, with a precision of four digits after the decimal point, are

$$\{h_s(n), n = 0, 1, \ldots, 12\} = \{0.0375, -0.0244, -0.0796, -0.0406,$$

$$0.1125, 0.2941, 0.3750, 0.2941,$$

$$0.1125, -0.0406, -0.0796, -0.0244, 0.0375\}$$

The impulse response, and the magnitude and phase of the frequency response of the filter are shown in Fig. 6.11a, b, and d, respectively. The passband in (b) is shown in expanded linear scale in Fig. 6.11c. ∎

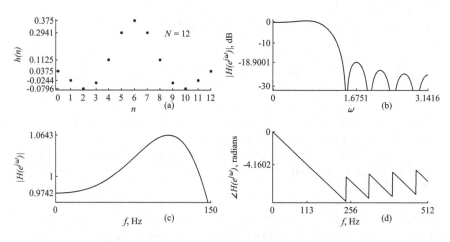

Fig. 6.11 (a) Causal filter impulse response; (b) the magnitude of the frequency response; (c) the passband in (b) shown in expanded linear scale; (d) the phase response

The frequency response is

$$H\left(e^{j\omega}\right)=e^{-j6\omega}\sum_{n=0}^{6}c(n)\cos(\omega n),\quad c(n)=\begin{cases}h\left(\frac{N}{2}\right)&\text{for }n=0\\2h\left(\frac{N}{2}-n\right)&\text{for }n=1,2,\ldots,6\end{cases}$$

The phase response in the passband is linear. For example, at the frequency $f = 113\,\text{Hz}$, the phase is

$$(113)(-6)\frac{\pi}{512}=-4.1602\text{ radians}$$

as shown in Fig. 6.11d. The specifications have been met. ■

Example 6.2 (Type I Highpass) The passband and stopband edge frequencies of a highpass filter are specified, respectively, as $\omega_c = 0.67\pi$ radians and $\omega_s = 0.3\pi$ radians, respectively. The minimum attenuation required in the stopband is 40 dB. Design the highpass filter using the Hamming window. The sampling frequency is $f_s = 1024\,\text{Hz}$.

Solution Frequency 0.3π corresponds to $\frac{0.3\pi}{2\pi}1024 = 153.6\,\text{Hz}$. Frequency 0.67π corresponds to $\frac{0.67\pi}{2\pi}1024 = 343.04\,\text{Hz}$. The maximum attenuation provided by filters using the Hamming window is about 41 dB. Therefore, the specification of 40 dB is realizable. The approximate average slope of the transition band of the Hamming window provides is given as $3.3\frac{2\pi}{N}$.

Now, we find the order of the filter as

$$N \geq 3.3\frac{2\pi}{0.67\pi - 0.3\pi} = 17.8378 \rightarrow 18$$

The cutoff frequency of the corresponding ideal filter is computed as

$$\omega_{ci} = \frac{\omega_s + \omega_c}{2} = \frac{0.3\pi + 0.67\pi}{2} = 0.485\pi$$

The shifted impulse response of the lowpass filter is given by

$$h'(n) = \frac{\sin\left(\omega_{ci}\left(\frac{N}{2}-n\right)\right)}{\pi\left(\frac{N}{2}-n\right)} = \frac{\sin(0.485\pi(9-n))}{\pi(9-n)},\quad n = 0,1,\ldots,18$$

The values, for this example, are

$$\{h'_s(n), n = 0, 1, \ldots, 18\} = \{0.0322, -0.0146, -0.0430, 0.0148, 0.0619,$$

$$-0.0149, -0.1050, 0.0150, 0.3180, 0.4850,$$

$$0.3180, 0.0150, -0.1050, -0.0149, 0.0619,$$

$$0.0148, -0.0430, -0.0146, 0.0322\}$$

The values of the Hamming window with length $N + 1$ are

$$0.54 - 0.46\cos\left(\frac{2\pi n}{N}\right), \quad n = 0, 1, 2, \ldots, N$$

For $N = 18$, the samples of the window are

$$\{0.0800, 0.1077, 0.1876, 0.3100, 0.4601, 0.6199, 0.7700, 0.8924, 0.9723, 1$$

$$0.9723, 0.8924, 0.7700, 0.6199, 0.4601 0.3100, 0.1876, 0.1077, 0.0800\}$$

Multiplying the window values by the impulse response values, we get the 19 windowed values of $h_s'(n)$ of the lowpass filter

$$\{0.0026, -0.0016, -0.0081, 0.0046, 0.0285, -0.0092, -0.0809,$$

$$0.0134, 0.3091, 0.4850, 0.3091, 0.0134, -0.0809, -0.0092,$$

$$0.0285, 0.0046, -0.0081, -0.0016, 0.0026\}$$

The Impulse Response of the Highpass Filter
Let us design two lowpass filters, one with cutoff frequency $\omega_{ci} = \pi$ and another with ω_{ci}, the desired frequency. The impulse response of the highpass filter with the same cutoff frequency ω_{ci} is the difference between the two lowpass filters. That is,

$$h_{hp}(n) = h\pi_{lp}(n) - hd_{lp}(n)$$

Therefore, the impulse response of a highpass filter is given by

$$h_{hp}(n) = \frac{\sin\left(\pi\left(\frac{N}{2} - n\right)\right)}{\pi\left(\frac{N}{2} - n\right)} - \frac{\sin\left(\omega_{ci}\left(\frac{N}{2} - n\right)\right)}{\pi\left(\frac{N}{2} - n\right)}, \quad n = 0, 1, \ldots, N$$

$$= \delta\left(n - \frac{N}{2}\right) - \frac{\sin\left(\omega_{ci}\left(\frac{N}{2} - n\right)\right)}{\pi\left(\frac{N}{2} - n\right)}$$

If the cutoff frequency is π, then the frequency response is 1 from $-\pi$ to π with its inverse DTFT $\delta(n)$. That is,

$$h_{hp}(n) = \delta\left(n - \frac{N}{2}\right) - h_s'(n)$$

The shifted impulse response values, with a precision of four digits after the decimal point, of the highpass filter are

$$\{h_{hp}(n), n = 0, 1, \ldots, 18\} = \{-0.0026, 0.0016, 0.0081, -0.0046,$$
$$-0.0285, 0.0092, 0.0809, -0.0134,$$
$$-0.3091, 0.5150, -0.3091, -0.0134,$$
$$0.0809, 0.0092, -0.0285, -0.0046,$$
$$0.0081, 0.0016, -0.0026\}$$

The frequency response is

$$H\left(e^{j\omega}\right) = e^{-j9\omega} \sum_{n=0}^{9} c(n)\cos(\omega n), \quad c(n) = \begin{cases} h(\frac{N}{2}) & \text{for } n = 0 \\ 2h\left(\frac{N}{2} - n\right) & \text{for } n = 1, 2, \ldots, 9 \end{cases}$$

The filter coefficients $c(n)$, starting with zero, for this example are

$$\{0.5150, -0.6183, -0.0267, 0.1618, 0.0185,$$
$$-0.0570, -0.0092, 0.0161, 0.0032, -0.0052\}$$

The impulse response, and the magnitude and phase of the frequency response of the filter are shown in Fig. 6.12a, b, and d, respectively. The passband in (b) is shown in expanded linear scale in Fig. 6.12c. ∎

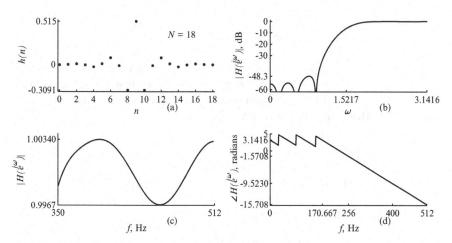

Fig. 6.12 (a) Causal filter impulse response; (b) the magnitude of the frequency response; (c) the passband in (b) shown in expanded linear scale; (d) the phase response

The phase response in the passband is linear. Remember that the highpass filter is a combination of two lowpass filters. The phase response of the lowpass filter has been shifted up by 4π radians. The initial phase is π radians. The 3 jumps also add 3π radians, making the shift 4π radians. The phase of the lowpass filter alone at the frequency $\omega = \pi/3$ radians is $-9\pi/3 = -3\pi$ radians. Therefore, the net phase of the filter at $\pi/3$ is $4\pi - 3\pi = \pi$ radians. From this value, we can compute the phase of the highpass filter at any frequency f in the passband using the formula

$$\theta = \pi - (9(f - \pi/3))\pi/512$$

For example, at $f = 256$, $f = 400$, and $f = 512$, we get

$$\theta = \pi - (9(0.5\pi - \pi/3)) = -0.5\pi$$

$$\theta = \pi - (9(400\pi/512 - \pi/3)) = -9.5230$$

$$\theta = \pi - (9(\pi - \pi/3)) = -15.7080$$

as shown in Fig. 6.12d.   ∎

6.1.7   Kaiser Window

Let the passband and stopband edge frequencies of a lowpass filter be ω_c and ω_s and the corresponding magnitudes of the attenuation be A_c and A_s in dB. Then, the order of the filter and the parameter b are computed using the following formulas.

$$\delta_c = \frac{10^{0.05A_c} - 1}{10^{0.05A_c} + 1}, \qquad \delta_s = 10^{-0.05A_s}$$

The minimum of δ_c and δ_s is the value of δ. Filters designed using windows result in $\delta_c = \delta_s$. This results in a smaller ripple in one of the bands if $\delta_c \neq \delta_s$. The attenuation is $A = -20\log_{10}\delta$. With only A_c or A_s given, δ corresponds to that specification. The parameter b is specified as

$$b = \begin{cases} 0 & A < 21 \\ 0.5842(A - 21)^{0.4} + 0.07886(A - 21) & 21 \leq A \leq 50 \\ 0.1102(A - 8.7) & A > 50 \end{cases}$$

The order of the filter, N, is given as

$$N \geq \begin{cases} 0.9222\frac{2\pi}{\omega_s - \omega_c} & A < 21 \\ \frac{(A-7.95)}{14.36}\frac{2\pi}{\omega_s - \omega_c} & \text{otherwise} \end{cases}$$

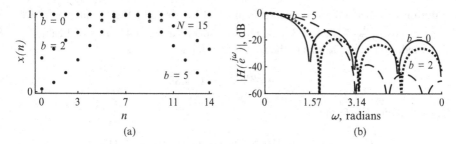

Fig. 6.13 (a) Typical Kaiser windows with $b = 0, 2$ and 5 and (b) the corresponding frequency responses

The Kaiser window, which is near optimal, is defined as

$$kai_w(n) = \begin{cases} \dfrac{I\left(b\sqrt{(1-(1-\frac{2n}{N})^2)}\right)}{I(b)} & \text{for } n = 0, 1, \ldots, N \\ 0 & \text{otherwise} \end{cases}$$

where

$$I(x) = 1 + \sum_{m=1}^{\infty} \left(\frac{(x/2)^m}{m!}\right)^2$$

For practical filter design, the summation is evaluated using a finite number of terms, typically 20, such that the accuracy is adequate. In addition to the number of coefficients ($N + 1$), the additional parameter b provides different window shapes. Figure 6.13a and b show some of the window shapes and the corresponding frequency responses of Kaiser windows. With $b = 0$, the window reduces to the rectangular window. Increasing the value of b increases the attenuation but the transition bandwidth becomes wider and vice versa, a tradeoff between the attenuation and the transition bandwidth.

Then, the impulse response is specified as

$$h(n) = kai_w(n)\frac{\sin\left(\omega_{ci}\left(\frac{N}{2} - n\right)\right)}{\pi\left(\frac{N}{2} - n\right)}, \quad n = 0, 1, \ldots, N, \quad \omega_{ci} = \frac{\omega_s + \omega_c}{2}$$

The formula for N is approximate and some trial and error may be required.

Example 6.3 (Type II Lowpass) The passband and stopband edge frequencies of a lowpass filter are specified, respectively, as $\omega_c = 0.3\pi$ radians and $\omega_s = 0.61\pi$ radians, respectively. The minimum attenuation required in the stopband is 41 dB. Design the lowpass filter using the Kaiser window. The sampling frequency is $f_s = 1024$ Hz.

Solution Frequency 0.3π corresponds to $\frac{0.3\pi}{2\pi}1024 = 153.6\,\text{Hz}$. Frequency 0.61π corresponds to $\frac{0.61\pi}{2\pi}1024 = 312.32\,\text{Hz}$. The specification of 41 dB is realizable by this window. The ripple size in the stopband is found as

$$\delta = 10^{-0.05(41)} = 0.0089$$

The parameter b is found as

$$b = 0.5842(A - 21)^{0.4} + 0.07886(A - 21) = 3.5135$$

Now, the order of the filter is found as

$$N = ((A - 7.95)(2\pi))/(14.36(\omega_s - \omega_c)) = 14.8486 \approx 15$$

The 16 Kaiser window coefficients, starting with index 0, are

$$kai_w(n) = \{0.1340, 0.2585, 0.4052, 0.5619, 0.7138, 0.8453, 0.9421, 0.9934,$$
$$0.9934, 0.9421, 0.8453, 0.7138, 0.5619, 0.4052, 0.2585, 0.1340\}$$

The cutoff frequency of the corresponding ideal filter is computed as

$$\omega_{ci} = \frac{\omega_s + \omega_c}{2} = \frac{0.3\pi + 0.61\pi}{2} = 0.455\pi$$

The shifted impulse response of the filter is given by

$$h(n) = \frac{\sin\left(\omega_{ci}\left(\frac{N}{2} - n\right)\right)}{\pi\left(\frac{N}{2} - n\right)} = \frac{\sin(0.455\pi(7.5 - n))}{\pi(7.5 - n)}, \quad n = 0, 1, \ldots, 15$$

The shifted impulse response values, with a precision of four digits after the decimal point, are

$$\{h_s(n), n = 0, 1, \ldots, 15\} = \{-0.0408, 0.0065, 0.0579, 0.0105, -0.0871,$$
$$-0.0533, 0.1783, 0.4172, 0.4172, 0.1783, -0.0533,$$
$$-0.0871, 0.0105, 0.0579, 0.0065, -0.0408\}$$

Multiplying with the window coefficients $kai_w(n)$, we get the shifted and windowed impulse response values

$$\{h_{sw}(n), n = 0, 1, \ldots, 15\} = \{-0.0055, 0.0017, 0.0234, 0.0059, -0.0622,$$
$$-0.0451, 0.1680, 0.4145, 0.4145, 0.1680, -0.0451,$$

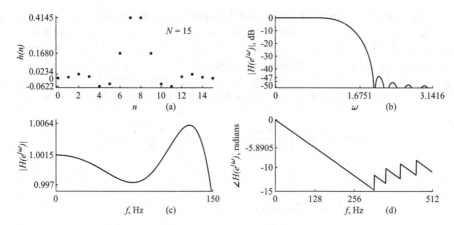

Fig. 6.14 (a) Causal filter impulse response; (b) the magnitude of the frequency response; (c) the passband in (b) shown in expanded linear scale; (d) the phase response

$$-0.0622, 0.0059, 0.0234, 0.0017, -0.0055\}$$

The impulse response, and the magnitude and phase of the frequency response of the filter are shown in Fig. 6.14a, b, and d, respectively. The passband in (b) is shown in expanded linear scale in Fig. 6.14c. ■

The frequency response is

$$H(e^{j\omega}) = e^{-j\omega\frac{N}{2}} \sum_{n=0}^{\frac{N-1}{2}} c(n) \cos(\omega(n + 0.5))$$

where

$$c(n) = 2h(\tfrac{N-1}{2} - n) \text{ for } n = 0, 1, \ldots, \tfrac{N-1}{2}$$

The filter coefficients $c(n)$, starting with index 0, are

$$c(n) = \{0.8290, \ 0.3359, \ -0.0901, \ -0.1244, \ 0.0118, \ 0.0469, \ 0.0034, \ -0.0109\}$$

The phase response in the passband is linear. For example, at the frequency $f = 128$ Hz, the phase is

$$(128)(-7.5)\frac{\pi}{512} = -5.8905 \text{ radians,}$$

as shown in Fig. 6.14d. The specifications have been met. ■

The Relation Between the Ripples and Attenuations in the Bands
In the design using Kaiser window, the ripples oscillate between $1 - \delta$ and $1 + \delta$ in the passband, where δ is the largest amplitude in either bands. In the stopband, the ripples oscillate between 0 and δ. The specification is usually given in terms of peak-to-peak ripple in the passband as A_c in dB. In the stopband, the specification is usually given in terms of attenuation as A_s in dB. Therefore, in the passband,

$$A_c = 20 \log_{10} \left(\frac{1 + \delta_c}{1 - \delta_c} \right) \quad \text{and} \quad \delta_c = \frac{10^{0.05 A_c} - 1}{10^{0.05 A_c} + 1}$$

In the stopband,

$$A_s = -20 \log_{10} \delta_s \quad \text{and} \quad \delta_s = 10^{-0.05 A_s}$$

The Impulse Response of the Bandpass Filter
The frequency response of the ideal bandpass filter is defined as

$$H_{BP}\left(e^{j\omega}\right) = \begin{cases} 1 \text{ for } \omega_{ci1} \leq |\omega| \leq \omega_{ci2} \\ 0 \text{ for } 0 \leq |\omega| < \omega_{ci1} \text{ and } \omega_{ci2} < |\omega| \leq \pi \end{cases}$$

The inverse of the DTFT of $H_{BP}(e^{j\omega})$ is the impulse response of the ideal bandpass filter $h_{BP}(n)$. Therefore, the shifted impulse response of a practical bandpass filter is given by

$$h_{BP}(n) = \frac{\sin\left(\omega_{ci2}\left(\frac{N}{2} - n\right)\right)}{\pi\left(\frac{N}{2} - n\right)} - \frac{\sin\left(\omega_{ci1}\left(\frac{N}{2} - n\right)\right)}{\pi\left(\frac{N}{2} - n\right)}, \quad n = 0, 1, \ldots, N$$

The specification of the sharper transition band of the two should be used in the design.

Example 6.4 (Type I Bandpass)   The lower passband and stopband edge frequencies of a bandpass filter are 0.3π and 0.1π, respectively. The upper passband and stopband edge frequencies of the bandpass filter are 0.5π and 0.75π, respectively. The minimum attenuation required in the stopband is $A_s = 45\,\text{dB}$. The maximum deviation acceptable in the passband is $A_c = 0.1\,\text{dB}$. Design the bandpass filter using the Kaiser window. The sampling frequency is $f_s = 512\,\text{Hz}$.

Solution With $A_c = 0.1\,\text{dB}$ and $A_s = 45\,\text{dB}$, compute

$$\delta_c = \frac{10^{0.05 A_c} - 1}{10^{0.05 A_c} + 1} = 0.0058,$$

$$\delta_s = 10^{-0.05 A_s} = 0.0056$$

Taking the minimum value of 0.0056, we get $A = -20 \log_{10} \delta = 45$. With $A = 45$, the window shape parameter b is computed as

$$b = 0.5842(A - 21)^{0.4} + 0.07886(A - 21) = 3.9754$$

With the sharper transition band 0.2π, the order of the filter, N, is given as

$$B = \frac{(A - 7.95)}{14.36} = 2.5801 \text{ and } N \geq B \frac{2\pi}{0.3\pi - 0.1\pi} = 25.8008 \approx 26$$

With the order of the filter N and the parameter b known, we get the coefficients of the window as

$$kai_w(n) = \{0.0904, 0.1514, 0.2241, 0.3066, 0.3969, 0.4918, 0.5881,$$
$$0.6820, 0.7696, 0.8474, 0.9118, 0.9600, 0.9899,$$
$$1, 0.9899, 0.9600, 0.9118, 0.8474, 0.7696, 0.6820, 0.5881,$$
$$0.4918, 0.3969, 0.3066, 0.2241, 0.1514, 0.0904\}$$

With one-half of the shorter transition band $0.5(0.3 - 0.1)\pi = 0.1\pi$, the cutoff frequencies of the ideal lowpass filters are computed as

$$\omega_{ci1} = 0.3\pi - 0.1\pi = 0.2\pi, \qquad \omega_{ci2} = 0.5\pi + 0.1\pi = 0.6\pi$$

$$h(n) = kai_w(n) \left(\frac{\sin(0.6\pi(13 - n))}{\pi(13 - n)} - \frac{\sin(0.2\pi(13 - n))}{\pi(13 - n)} \right)$$

The impulse response of the filter, before multiplying by the window, is given by

$$h(n) = \{-0.0377, -0.0408, 0.0105, -0.0000, -0.0128, 0.0612, 0.0700,$$
$$-0.0193, 0.0000, 0.0289, -0.1633, -0.2449, 0.1156, 0.4000,$$
$$0.1156, -0.2449, -0.1633, 0.0289, 0.0000, -0.0193, 0.0700,$$
$$0.0612, -0.0128, -0.0000, 0.0105, -0.0408, -0.0377\}$$

Multiplying the impulse response values with that of the Kaiser window values, we get the impulse response values, with a precision of four digits after the decimal point, are

$$\{h_{BP}(n), n = 0, 1, \ldots, 26\} = \{-0.0034, -0.0062, 0.0024, -0.0000, -0.0051,$$
$$0.0301, 0.0412, -0.0131, 0.000, 0.0245,$$
$$-0.1489, -0.2351, 0.1145, 0.4000, 0.1145,$$
$$-0.2351, -0.1489, 0.0245, 0.000, -0.0131,$$

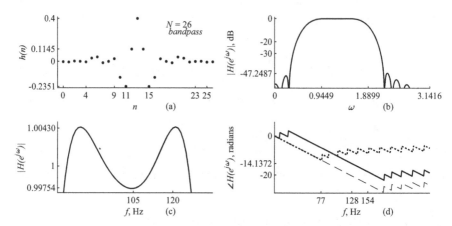

Fig. 6.15 (a) Causal filter impulse response; (b) the magnitude of the frequency response; (c) the passband in (b) shown in expanded linear scale; (d) the phase response (solid line)

$$0.0412, 0.0301, -0.0051 - 0.0000, 0.0024,$$

$$-0.0062, -0.0034\}$$

The impulse response and the magnitude and phase of the frequency response of the filter are shown in Fig. 6.15a, b, and d, respectively. The passband in (b) is shown in expanded linear scale in Fig. 6.15c. Figure 6.15d shows the phase response (solid line) of the bandpass filter along with those of its two constituent lowpass filters. The specifications have been met.

The passband of the bandpass filter is the stopband of the lowpass filter with the shorter cutoff frequency and its phase response is of a zigzag pattern. However, the phase response of the bandpass filter in the passband is linear. The reason is that the amplitude response of the lowpass filter in the stopband is a very small fraction to affect the linearity to a significant extent. ∎

The Impulse Response of the Bandstop Filter

The frequency response of the ideal bandstop filter is defined as

$$H_{BS}(e^{j\omega}) = \begin{cases} 0 \text{ for } \omega_{ci1} \leq |\omega| \leq \omega_{ci2} \\ 1 \text{ for } 0 \leq |\omega| < \omega_{ci1} \text{ and } \omega_{ci2} < |\omega| \leq \pi \end{cases}$$

The inverse of the DTFT of $H_{BS}(e^{j\omega})$ is the impulse response of the ideal bandstop filter $h_{BS}(n)$. The specification of the sharper transition band of the two should be used in the design. Therefore, the shifted impulse response of a practical bandstop filter, which is $1 - h_{BP}(n)$, is given by

$$h_{BS}\left(\frac{N}{2}\right) = 1 + \frac{(\omega_{ci1} - \omega_{ci2})}{\pi}$$

$$h_{BS}(n) = \left(\frac{\sin(\omega_{ci1}(\frac{N}{2} - n))}{\pi(\frac{N}{2} - n)} - \frac{\sin(\omega_{ci2}(\frac{N}{2} - n))}{\pi(\frac{N}{2} - n)}\right), \quad n = 0, 1, 2 \ldots, N, \quad n \neq \frac{N}{2}$$

Example 6.5 (Type I Bandstop) The lower passband and stopband edge frequencies of a bandstop filter are 0.1π and 0.3π, respectively. The upper passband and stopband edge frequencies of the bandpass filter are 0.75π and 0.5π, respectively. The minimum attenuation required in the stopband is $A_s = 45$ dB. The maximum deviation acceptable in the passband is $A_c = 0.1$ dB. Design the bandstop filter using the Hamming window. The sampling frequency is $f_s = 512$ Hz.

Solution With the sharper transition band 0.2π, the order of the filter, N, is given as

$$N \geq 3.3\frac{2\pi}{0.3\pi - 0.1\pi} = 33$$

Since a Type I filter is required to design a bandstop filter, we add 1 to get the filter order as $N = 34$. With $N = 34$, the Hamming window coefficients are

hamming$_w$(n)

$$= \{0.0800, 0.0878, 0.1111, 0.1489, 0.2001, 0.2628, 0.3350, 0.4141,$$

$$0.4976, 0.5824, 0.6659, 0.7450, 0.8172, 0.8799, 0.9311, 0.9689, 0.9922,$$

$$1, 0.9922, 0.9689, 0.9311, 0.8799, 0.8172, 0.7450, 0.6659, 0.5824,$$

$$0.4976, 0.4141, 0.3350, 0.2628, 0.2001, 0.1489, 0.1111, 0.0878, 0.0800\}$$

With one-half of the shorter transition band $0.5(0.3 - 0.1)\pi = 0.1\pi$, the cutoff frequencies of the ideal lowpass filters are computed as

$$\omega_{ci1} = 0.3\pi - 0.1\pi = 0.2\pi, \qquad \omega_{ci2} = 0.5\pi + 0.1\pi = 0.6\pi$$

$$h(NB2) = 1 + \frac{(0.2 - 0.6)\pi}{\pi} = 0.6$$

$$h(n) = \left(\frac{\sin(0.2\pi(17 - n))}{\pi(17 - n)} - \frac{\sin(0.6\pi(17 - n))}{\pi(17 - n)}\right) \text{ otherwise}$$

The impulse response of the, before multiplying by the window, filter is given by

$$h(n) = \{-0.0288, 0.0072, -0.0000, -0.0083, 0.0377, 0.0408, -0.0105, 0.0000,$$

$$0.0128, -0.0612, -0.0700, 0.0193, -0.0000, -0.0289, 0.1633, 0.2449,$$

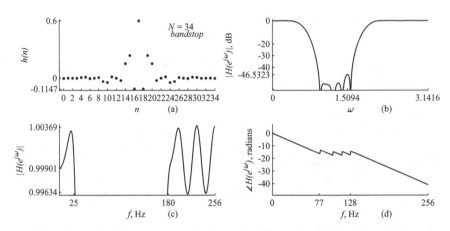

Fig. 6.16 (a) Causal filter impulse response; (b) the magnitude of the frequency response; (c) the passband in (b) shown in expanded linear scale; (d) the phase response

$$-0.1156, 0.6000, -0.1156, 0.2449, 0.1633, -0.0289, -0.0000, 0.0193,$$

$$-0.0700, -0.0612, 0.0128, 0.0000, -0.0105, 0.0408,$$

$$0.0377, -0.0083, -0.0000, 0.0072, -0.0288\}$$

Multiplying the impulse response values with that of the Hamming window values, we get the impulse response values, with a precision of four digits after the decimal point, as

$$\{h_{BS}(n), n = 0, 1, \ldots, 34\} = \{-0.0023, 0.0006, 0.0000, -0.0012, 0.0075,$$

$$0.0107, -0.0035, -0.0000, 0.0064, -0.0357,$$

$$-0.0466, 0.0144, 0.0000, -0.0254, 0.1520,$$

$$0.2373, -0.1147, 0.6000, -0.1147, 0.2373,$$

$$0.1520, -0.0254, 0.0000, 0.0144, -0.0466,$$

$$-0.0357, 0.0064, -0.0000, -0.0035, 0.0107,$$

$$0.0075, -0.0012, 0.0000, 0.0006, -0.0023\}$$

The impulse response and the magnitude and phase of the frequency response of the filter are shown in Fig. 6.16a, b, and d, respectively. The passband in (b) is shown in expanded linear scale in Fig. 6.16c. The specifications have been met. ■

Example 6.6 (Type III Hilbert Transformer) Find the impulse response of the 66th order fullband Type III Hilbert transformer filter using the Hamming window. The sampling frequency is $f_s = 512\,\text{Hz}$.

Solution From the impulse response of the ideal Hilbert transformer is given in an earlier chapter, the shifted version of the impulse response for a finite order N filter is obtained as

$$h(n) = \begin{cases} \dfrac{2\sin^2(\frac{\pi(NB2-n)}{2})}{\pi(NB2-n)} & \text{for } n \neq NB2, \quad 0 \leq n \leq N \\ 0 & \text{for } n = NB2 \end{cases}$$

Note that, in window based design, trial-and-error method is necessary to design the filter to the requirements. The 67 shifted and windowed impulse response values, with a precision of four digits after the decimal point, of the Hilbert filter are

$$\{h_{hilbert}(n), n = 0, 1, \ldots, 66\}$$

$$= \{-0.0015, 0, -0.0018, 0, -0.0025, 0, -0.0036, 0, -0.0053, 0,$$

$$-0.0076, 0, -0.0106, 0, -0.0145, 0, -0.0194, 0, -0.0257, 0,$$

$$-0.0338, 0, -0.0446, 0, -0.0595, 0, -0.0820, 0, -0.1208, 0,$$

$$-0.2083, 0, -0.6353, 0, 0.6353, 0, 0.2083, 0, 0.1208, 0,$$

$$0.0820, 0, 0.0595, 0, 0.0446, 0, 0.0338, 0, 0.0257, 0,$$

$$0.0194, 0, 0.0145, 0, 0.0106, 0, 0.0076, 0, 0.0053, 0,$$

$$0.0036, 0, 0.0025, 0, 0.0018, 0, 0.0015\}$$

The impulse response, and the magnitude and phase of the frequency response of the filter are shown in Fig. 6.17a, b, and d, respectively. The passband in (b) is shown in expanded linear scale in Fig. 6.17c. ∎

Example 6.7 (Type III Differentiating Filter) Find the impulse response of the 30th order Type III fullband differentiating filter using the window FIR filter design method. Use the Hamming window. The sampling frequency is $f_s = 512\,\text{Hz}$.

Solution From the impulse response of the ideal differentiating filter is given in an earlier chapter, the shifted version of the impulse response for a finite order N filter is obtained as

$$h(n) = \begin{cases} \dfrac{(-1)^{(n-15)}}{n-15} & \text{for } n \neq NB2, \quad 0 \leq n \leq N \\ 0 & \text{for } n = NB2 \end{cases}$$

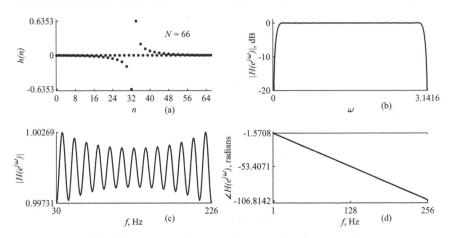

Fig. 6.17 (a) Causal filter impulse response; (b) the magnitude of the frequency response; (c) the passband in (b) shown in expanded linear scale; (d) the phase response

Note that, in window based design, trial-and-error method is necessary to design the filter to the requirements. The 31 shifted and widowed impulse response values, with a precision of four digits after the decimal point, of the differentiator are

$$\{h_{diff}(n), n = 0, 1, \ldots, 30\} =$$

$$\{0.0053, -0.0064, 0.0092, -0.0140, 0.0211, -0.0310, 0.0442,$$

$$-0.0615, 0.0840, -0.1137, 0.1540, -0.2120,$$

$$0.3040, -0.4801, 0.9899, 0, -0.9899,$$

$$0.4801, -0.3040, 0.2120, -0.1540, 0.1137, -0.0840,$$

$$0.0615, -0.0442, 0.0310, -0.0211,$$

$$0.0140, -0.0092, 0.0064, -0.0053\}$$

The impulse response, and the magnitude and phase of the frequency response of the filter are shown in Fig. 6.18a, b, and d, respectively. The passband in (b) is shown in expanded linear scale in Fig. 6.18c.   ∎

6.2   Summary of Characteristics of Linear-Phase FIR Filters

Table 6.1 shows the symmetry of the impulse response $h(n)$ of the four types of linear-phase FIR filters. The table also shows the filter coefficients $c(n)$ in terms of $h(n)$. Table 6.2 shows the frequency response $H(e^{j\omega})$ of the four types of linear-

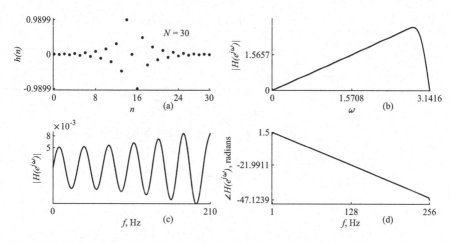

Fig. 6.18 (a) Causal filter impulse response; (b) the magnitude of the frequency response; (c) the passband in (b) shown in expanded linear scale; (d) the phase response

Table 6.1 Linear-Phase FIR filters. Type and symmetry

Type	Symmetry of the impulse response $h(n)$ and the filter coefficients $c(n)$
I	$h(n) = h(N-n)$, Order $N = 2, 4, 6, \ldots$,
	$c(n) = 2h(\frac{N}{2} - n)$, $n = 1, 2, \ldots, \frac{N}{2}$
	$c(0) = h(\frac{N}{2})$
II	$h(n) = h(N-n)$, Order $N = 1, 3, 5, \ldots$,
	$c(n) = 2h(\frac{N-1}{2} - n)$, $n = 0, 1, \ldots, \frac{N-1}{2}$
III	$h(n) = -h(N-n)$, Order $N = 2, 4, 6, \ldots$,
	$c(n) = 2h(\frac{N}{2} - n)$, $n = 1, 2, \ldots, \frac{N}{2}$
	$c(0) = h(\frac{N}{2}) = 0$
IV	$h(n) = -h(N-n)$, Order $N = 1, 3, 5, \ldots$,
	$c(n) = 2h(\frac{N-1}{2} - n)$, $n = 0, 1, \ldots, \frac{N-1}{2}$

phase FIR filters in terms of the filter coefficients $c(n)$. The frequency response is also the DFT of the sufficiently zero-padded impulse response $h(n)$ to get a denser spectrum and usually computed this way.

6.2.1   Location of Zeros

Zeros of a polynomial are the values of the variable for which the value of the polynomial becomes zero. For a real-valued symmetric impulse response, the possibility of the locations are: (1) single zero occurs only at $z = \pm 1$, (2) other zeros on the real axis occur in reciprocal pairs as, $z = r$ and $z = \frac{1}{r}$, (3) zeros on the unit-circle occur as complex-conjugate pairs, and (4) other zeros in the z-plane occur in quadruples as

Type	Frequency response in terms of filter coefficients
I	$H(e^{j\omega}) = e^{-j\omega\frac{N}{2}} \sum_{n=0}^{\frac{N}{2}} c(n)\cos(\omega n)$
II	$H(e^{j\omega}) = e^{-j\omega\frac{N}{2}} \sum_{n=0}^{\frac{N-1}{2}} c(n)\cos(\omega(n+0.5))$
III	$H(e^{j\omega}) = e^{-j\omega\frac{N}{2}} e^{j\frac{\pi}{2}} \sum_{n=0}^{\frac{N}{2}} c(n)\sin(\omega n)$
IV	$H(e^{j\omega}) = e^{-j\omega\frac{N}{2}} e^{j\frac{\pi}{2}} \sum_{n=0}^{\frac{N-1}{2}} c(n)\sin(\omega(n+0.5))$

Table 6.2 Linear-phase FIR filters. Type and frequency response

$$z = r\angle\theta, \ z = r\angle-\theta, \ z = \frac{1}{r}\angle\theta, \ z = \frac{1}{r}\angle-\theta$$

From the number of zeros located at $z = \pm 1$, the four types of filters can be identified.

Type I   0,2,4,..., at $z = \pm 1$.
Type II   1,3,5,..., at $z = -1$ and 0,2,4,..., at $z = 1$.
Type III   1,3,5,..., at $z = \pm 1$.
Type IV   1,3,5,..., at $z = 1$ and 0,2,4,..., at $z = -1$.

For Type I filter, the order is even. Therefore, the zeros must occur as pairs and quadruples implying even symmetry. With a zero at $z = -1$, the order of the Type II filter becomes odd implying even symmetry. With a zero at $z = 1$, the order of the Type IV filter becomes odd implying odd symmetry. With a zero at $z = 1$ and a zero at $z = -1$, the order of the Type III filter becomes even implying odd symmetry. Typical linear-phase FIR filter pole-zero plots are shown in Fig. 6.19 and the specific property for each type can be identified. Due to the constrained zeros of the frequency response of the four type of FIR filters, each one is more suitable to design certain types of filters. Table 6.3 shows, for each type, the most appropriate filters they can be used to design.

Type I FIR Filter
The z-transform transfer function of a second-order Type I FIR filter is given as Consider the transfer function with roots $\{0.5, 2\}$.

$$H(z) = \sum_{n=0}^{2} h(n)z^{-n} = 1 - 2.5h(1)z^{-1} + 1z^{-2}$$

$$= z^{-2}(1 - 2.5z^{1} + 1z^{2}) = z^{-2}\sum_{n=0}^{2} h(n)z^{n} = z^{-2}H(z^{-1})$$

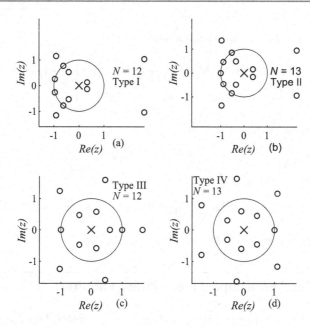

Fig. 6.19 Typical linear-phase FIR filter pole-zero plots. (**a**) Type I filter; (**b**) Type II filter; (**c**) Type III filter; (**d**) Type IV filter

Table 6.3 Endpoint zeros of the frequency response of FIR filters, phase offset, and candidate filters

Type	Endpoint zeros	Phase offset	Most appropriate for
I	Nowhere	$0, \pi$	All frequency-selective filters
II	$\omega = \pi$	$0, \pi$	LP and BP filters
III	$\omega = 0, \pi$	$\pm 0.5\pi$	Differentiators and Hilbert transformers
IV	$\omega = 0$	$\pm 0.5\pi$	Differentiators and Hilbert transformers

which implies that if $H(z)$ has a zero at $z = z_0$, then $z = (1/z_0)$ must also be its zero. The factors of the transfer function of linear-phase filters for a combination of 4 zeros are, in polar form,

$$H(z) = K\left(1 - r_1 e^{-j\theta_1} z^{-1}\right)\left(1 - r_1 e^{j\theta_1} z^{-1}\right)\left(1 - \frac{1}{r_1} e^{-j\theta_1} z^{-1}\right)\left(1 - \frac{1}{r_1} e^{j\theta_1} z^{-1}\right)$$

Type II FIR Filter
As this type has a zero at $z = -1$,

$$H(z) = \left(1 - 2.5h(1)z^{-1} + 1z^{-2}\right)(z + 1) = 1 - 1.5z^{-1} - 1.5z^{-2} + 1z^{-3}$$

The zero locations are $\{0.5, 2, -1\}$.

Type III FIR Filter
As this type has a zero at $z = -1$ and a zero at $z = 1$,

$$H(z) = \left(1 - 2.5h(1)z^{-1} + 1z^{-2}\right)(z+1)(z-1)$$
$$= 1 - 3.5z^{-1} + 0z^{-2} + 3.5z^{-3} - 1z^{-4}$$

The zero locations are $\{2, -1, 1, 0.5\}$.

Type IV FIR Filter
As this type has a zero at $z = 1$,

$$H(z) = \left(1 - 2.5h(1)z^{-1} + 1z^{-2}\right)(z-1) = 1 - 3.5z^{-1} + 3.5z^{-2} - 1z^{-3}$$

The zero locations are $\{0.5, 2, 1\}$.

6.3   Optimal Equiripple Linear-Phase FIR filters

In the design of this type of filters, the weighted difference between frequency responses of the desired and the designed filters is minimized iteratively. These filters are characterized by: (1) satisfying a given specification by the lowest order; (2) having equiripple in all the bands except in the transition bands. The equiripples in the bands could be of different sizes; (3) having specified cutoff band edge frequencies; and (4) requiring only one algorithm for various type of filters such as lowpass and highpass, etc. The design procedure for these filters is better presented through examples. For illustrative purposes, the order of the filters is kept small. In practice, the order is much higher.

6.3.1   Type I FIR Lowpass and Highpass

A Type I 8-th order optimal equiripple linear-phase lowpass filter with the passband and stopband edge frequencies, respectively, $\omega_p = 0.25\pi$ and $\omega_s = 0.4\pi$ radians is required. The size of the ripple δ_s in the stopband is to be 4 times that in the passband δ_p. The sampling frequency is $f_s = 512\,\text{Hz}$. Design the filter.

Solution The specifications of the filter are shown in Fig. 6.20. Frequencies 0.25π and 0.4π correspond to $\frac{0.25\pi}{2\pi}512 = 64\,\text{Hz}$ and $\frac{0.4\pi}{2\pi}512 = 102.4\,\text{Hz}$, respectively. The magnitude of the desired frequency response is given in terms of the amplitudes at the band edges. The ripple size in each band is specified by a weight vector. Transition band is ignored in the approximation procedure. The amplitude of the frequency response oscillates uniformly between the tolerance bound of each band. Each band is specified by its two edge points and the response is the line connecting

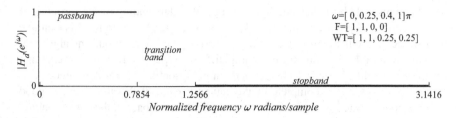

Fig. 6.20 Specifications of a Type I lowpass filter

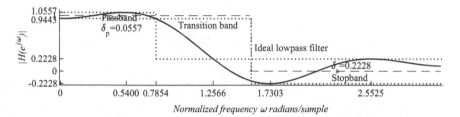

Fig. 6.21 Specifications of a Type I lowpass filter as related to its actual frequency response

the points. That is, the response is specified as any piecewise linear desired function with any required transition band.

Figure 6.21 shows the specifications of the example Type I lowpass filter as related to its actual frequency response. The passband frequency 64 Hz and the stopband frequency 102.4 are exact in the actual response as specified. In the passband, the ripple oscillates between 1.0557 and 0.9443 with the amplitude $\delta = 0.0557$. In the stopband, the ripple oscillates between 0.2228 and −0.2228 with the amplitude $\delta = 0.2228$, which is 4 times of that in the passband as specified. Ideal lowpass filter frequency response is also shown for comparison. Exact formulas to find the required order for a certain specification are not available. Trial-and-error method is good to find the required order. The frequencies where peaks of the ripples occur, called extremal (extreme point) frequencies, are shown in the figure. Since the amplitude response is real and even-symmetric, the response can be approximated by a cosine Fourier series (trigonometric polynomial) of order $L = N/2$, where N is the order of the filter. A trigonometric polynomial of order L can have up to $L-1$ interior peaks, where the slope of the response is zero. For a lowpass or highpass filter, the two interior band edges are always in the extremal set of frequencies, making the number to $L+1$. Further, one or both of the endpoint of the response may be extremal frequencies. Therefore, the number of extremal frequencies is $L+2$ or $L+3$. In the case of $L+3$, an endpoint frequency is omitted to form a $L+2$ extremal frequency set. Other filters may have more extremal frequencies. But, only $L+2$ frequencies are required for optimization. Finding these frequencies, starting with an initially guessed set, is the iterative algorithm. With the initial set, the frequency response is unlikely to be truly equiripple. In a few iterations, the set converges to the exact extremal frequencies. Sometimes, even with no programming mistakes, the design may not yield acceptable response. In this case, changing the filter specifications little bit is the solution. For a good understanding of the formulas of the algorithm and programming purposes, a trace of the algorithm for this example is provided. Trace is the information regarding program's execution. The algorithm remains essentially the same for designing different types of FIR filters, except in a few steps. The differences will be pointed out in presenting the various filters.

Now, we present the complete mathematical version of the algorithm. A good understanding of the algorithm can be obtained by looking back at the corresponding equations, when the trace is traversed. The amplitude of the frequency response of the Type I Nth order FIR filter is given by

$$H\left(e^{j\omega}\right) = \sum_{k=0}^{L} c(k) \cos(\omega k)$$

where $L = \frac{N}{2}$ and

$$c(n) = \begin{cases} h\left(\frac{N}{2}\right) & \text{for } n = 0 \\ 2h\left(\frac{N}{2} - n\right) & \text{for } n = 1, 2, \ldots, \frac{N}{2} \end{cases}$$

The $(N + 1)$ impulse response values, which are also the same as the filter coefficients, are denoted by

$$h(n), n = 0, 1, 2, \ldots, N$$

Let the desired and the actual amplitude response of the filter be $H_d(e^{j\omega})$ and $H(e^{j\omega})$, respectively. Then the weighted error function is defined as

$$err(\omega) = WT(\omega)\left(H_d\left(e^{j\omega}\right) - H\left(e^{j\omega}\right)\right) \tag{6.3}$$

where $WT(\omega)$ is a positive weighting function. The design, the filter coefficients, is to satisfy such that the maximum amplitude of the ripples of the error function is to be minimum. In the figure, there are $L + 2 = 4 + 2 = 6$ extremal frequencies over the passband and stopband with maximum deviation δ from the desired frequency response. This is the necessary and sufficient condition for $H(e^{j\omega})$ to be the unique and best weighted Chebyshev approximation to $H_d(e^{j\omega})$. The six extremal frequencies in Fig. 6.21 have maximum deviation with alternating signs. That is,

$$err(\omega_n) = -err(\omega_{n+1})$$

Then,

$$err(\omega_n) = WT(\omega_n)\left(H_d\left(e^{j\omega_n}\right) - H\left(e^{j\omega_n}\right)\right) = (-1)^n \delta, \ n = 0, 1, \ldots, (L + 1)$$

Dividing throughout by $WT(\omega_n)$ and having the desired response on one side, we get

$$H\left(e^{j\omega_n}\right) + \frac{(-1)^n \delta}{WT(\omega_n)} = H_d\left(e^{j\omega_n}\right), \ n = 0, 1, \ldots, (L+1)$$

Substituting the expression for $H(e^{j\omega_n})$, we get

$$\sum_{k=0}^{L} c(k) \cos(\omega_n k) + \frac{(-1)^n \delta}{WT(\omega_n)} = H_d\left(e^{j\omega_n}\right), \ n = 0, 1, \ldots, (L+1)$$

Expanding this equation, we get $L+2$ simultaneous equations

$$c(0) + c(1) \cos(\omega_0) + \cdots + c(L) \cos(L\omega_0) + \frac{\delta}{WT(\omega_0)} = H_d\left(e^{j\omega_0}\right)$$
$$c(0) + c(1) \cos(\omega_1) + \cdots + c(L) \cos(L\omega_1) - \frac{\delta}{WT(\omega_1)} = H_d\left(e^{j\omega_1}\right)$$
$$\vdots$$
$$c(0) + c(1) \cos(\omega_{L+1}) + \cdots + c(L) \cos(L\omega_{L+1}) + \frac{(-1)^{(L+1)}\delta}{WT(\omega_{L+1})} = H_d\left(e^{j\omega_{L+1}}\right)$$

While these equations can be solved to find the filter coefficients and hence the frequency response in each iteration, a computationally more efficient procedure is followed to solve the filter design problem. First, the value of δ is computed analytically from the equation

$$\delta = \frac{d_0 H_d\left(e^{j\omega_0}\right) + d_1 H_d\left(e^{j\omega_1}\right) + \cdots + d_{L+1} H_d\left(e^{j\omega_{L+1}}\right)}{\frac{d_0}{WT(\omega_0)} - \frac{d_1}{WT(\omega_1)} + \cdots + \frac{(-1)^{(L+1)} d_{L+1}}{WT(\omega_{L+1})}} \tag{6.4}$$

where ω_i are the extremal frequencies of the iteration and

$$d_k = \prod_{n=0, \ n\neq k}^{L+1} \frac{1}{\cos(\omega_k) - \cos(\omega_n)}, \ k = 0, 1, \ldots, (L+1) \tag{6.5}$$

Using the value of δ, we get the frequency response at the initially guessed $L+2$ extremal frequencies as

$$H\left(e^{j\omega_n}\right) = H_d\left(e^{j\omega_n}\right) - \frac{(-1)^n \delta}{WT(\omega_n)}, \ n = 0, 1, \ldots, L+1 \tag{6.6}$$

as these values are adequate to interpolate a polynomial of order L. Using these values, the frequency response is obtained by interpolation at a dense set of frequencies. In practice, response has to be interpolated at a minimum of $16L$ points. The formula for the interpolation is given as

$$H\left(e^{j\omega}\right) = \frac{\sum_{n=0}^{L} H\left(e^{j\omega_n}\right)\frac{b_n}{\cos(\omega)-\cos(\omega_n)}}{\sum_{n=0}^{L}\frac{b_n}{\cos(\omega)-\cos(\omega_n)}} \tag{6.7}$$

where $\omega \neq \omega_n$ and

$$b_k = \prod_{n=0,\ n\neq k}^{L} \frac{1}{\cos(\omega_k)-\cos(\omega_n)}, \quad k = 0, 1, \ldots, L \tag{6.8}$$

Typically, the algorithm converges in less than 10 iterations. That is, the extremal frequencies of two consecutive iterations are the same and $|err(\omega)| \leq \delta$ for all frequencies. The frequency response is interpolated at $L + 1$ equally spaced frequencies using the response at $L + 1$ extremal frequencies and the impulse response of the filter is computed using IDFT. From the impulse response $h(n)$, the coefficients can be computed using the formulas for each type of filters. For Type I filter, the second half of the impulse response is computed as

$$h\left(n+\frac{N}{2}\right) = \frac{1}{NC}\left(H\left(e^{j\frac{2\pi 0}{NC}}\right) + 2\sum_{k=1}^{L} H\left(e^{j\frac{2\pi k}{NC}}\right)\cos\left(\frac{2\pi kn}{NC}\right)\right), \quad n = 0, 1, \ldots, L \tag{6.9}$$

where $NC = N + 1$.

Seven examples of filter design are presented. The corresponding MATLAB programs are available online. The main purpose of these programs is to give a good understanding of the filter design algorithm. The programs fail for higher-order filters due to finite wordlength of the computers. To avoid this problem the higher-order filters have to be decomposed into lower-order filters, which is a common practice.

Setting Up the Grids
The design procedure is better presented by a specific example. The filter designed is to be Type I lowpass filter of order $N = 8$ with the edge frequencies

$$\{0, 0.7854, 1.2566, 3.1416\}$$

The amplitude of the desired frequency response $HI(e^{j\omega})$ in the passband and stopband are, respectively,

$$HI\left(e^{j\omega}\right) = H_d\left(e^{j\omega}\right) = \{1,\ 0\}$$

and the weights associated with the error are

$$WT = \{1,\ 0.25\}$$

The passband weight is given 4 times that of the stopband. That is, the size of the ripple in the stopband is 4 times in the passband. Grid density GD is 16, which is minimum for good results. Our task is to find the $L + 1$ unique filter coefficients. Therefore, the range of the frequency of the grid is one-half. That is, $\omega = 0$ to $\omega = \pi$ radians including the edge frequencies. To find the coefficients, we need the corresponding frequency response. The algorithm starts with the desired frequency response and approximates it with the best possible minimum equiripple error in the passband and stopband. As it is a numerical algorithm, all the continuous quantities, such as the frequency bands, have to be sampled.

The values of N, L, and GD are

$$N = 8, L = \frac{N}{2} = 4, GD = 16$$

The frequency increment is

$$\omega_{inc} = \frac{\pi}{((L+1)GD)} = 0.0393 \text{ radians}$$

Transition band is ignored. The passband and stopband widths are 0.25π and 0.6π, respectively. The total width of the two bands is 0.85π. The number of frequency samples in the passband is

$$\frac{0.25\pi}{\frac{\pi}{(L+1)GD}} = 20$$

The number of frequency samples in the stopband is

$$\frac{0.6\pi}{\frac{\pi}{(L+1)GD}} = 48$$

There are $NG = 20 + 48 = 68$ grid points with the condition that all the interior edge frequencies must be included. The desired amplitude frequency response $H_d(n)$ at the 68 indices of the grid frequencies ω_g are shown in Fig. 6.22a.

$$\omega_g = \{0, \ 0.0393, \ 0.0785, \ldots, 0.6676, \ 0.7069, \ 0.7854,$$

$$1.2566, \ 1.2959, \ 1.3352, \ldots, 3.0238, \ 3.0631, \ 3.1416\}$$

The desired frequency response values at the grid frequencies are

$$HI(n) = \{1, 1, \ldots, 1, 1, 1, \quad 0, 0, 0, \ldots, 0, 0, 0\}$$

The corresponding 68 values of the weighing sequence for the error err in the desired frequency response at the indices of the grid frequencies ω_g are shown in Fig. 6.22b.

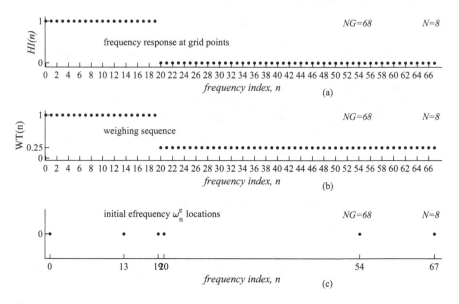

Fig. 6.22 Setting up the grids of a Type I lowpass filter

$$WT(n) = \{1, 1, 1, \ldots, 1, 1, 1, \quad 0.25, 0.25, 0.25, \ldots, 0.25, 0.25, 0.25\}$$

Initial extremal frequencies are located at increments of $NG/(L+1) = 68/(4+1) = 13.6$. Start with the first and increment with truncation up to the last frequency. The index of the first one is 0, that of the second one is $\lfloor 0 + 13.6 \rfloor = 13$ and that of the third one is $\lfloor 0 + 13.6 + 13.6 \rfloor = 27$

$$\omega_1^e = \{0, \ 13, \ 27, \ 40, \ 54, \ 67\}$$

However, we know that frequencies at the interior edge locations are extremal frequencies. Therefore, 27 and 40 are replaced to get

$$\omega_1^e = \{0, \ 13, \ 19, \ 20, \ 54, \ 67\}$$

This step may reduce the number of iterations. The indices of the initial extremal frequency points $\omega_1^e(n)$ on the grid are shown in Fig. 6.22c.

Iteration 1
Find δ

$$\delta = \frac{d_0 H_d\left(e^{j\omega_0}\right) + d_1 H_d\left(e^{j\omega_1}\right) + \cdots + d_{L+1} H_d\left(e^{j\omega_{L+1}}\right)}{\frac{d_0}{WT(\omega_0)} - \frac{d_1}{WT(\omega_1)} + \cdots + \frac{(-1)^{(L+1)} d_{L+1}}{WT(\omega_{L+1})}} \qquad (6.10)$$

where ω_i are the extremal frequencies of the iteration and

$$d_k = \prod_{n=0,\ n\neq k}^{L+1} \frac{1}{\cos(\omega_k) - \cos(\omega_n)}, \quad k = 0, 1, \ldots, (L+1) \tag{6.11}$$

The cosine of the initial set of extremal frequencies

$$\{0,\ 0.5105,\ 0.7854,\ 1.2566,\ 2.5918,\ 3.1416\}$$

are

$$\{1,\ 0.8725,\ 0.7071,\ 0.3090,\ -0.8526,\ -1\}$$

$$d_0 = \frac{1}{(1-0.8725)(1-0.7071)(1-0.3091)(1+0.8526)(1+1)} = 10.4587$$

d_1

$$= \frac{1}{(0.8725-1)(0.8725-0.7071)(0.8725-0.3091)(0.8725+0.8526)(0.8725+1)}$$
$$= -26.0524$$

The 6 values of d for the first iteration are

$$\{10.4587,\ -26.0524,\ 19.4755,\ -4.2428,\ 1.1719,\ -0.8109\}$$

$$\delta = \frac{10.4587(1)+(-26.0524)(1)+19.4755(1)+(-4.2428)(0)+1.1719(0)+(-0.8109)(0)}{10.4587/1-(-26.0524)/1+19.4755/1-(-4.2428)(4)+1.1719(4)-(-0.8109)(4)}$$
$$= 0.0480$$

$$b_0 = \frac{1}{(1-0.8725)(1-0.7071)(1-0.3091)(1+0.8526)} = 20.9174$$

$$b_1 = \frac{1}{(0.8725-1)(0.8725-0.7071)(0.8725-0.3091)(0.8725+0.8526)}$$
$$= -48.7829$$

The 5 values of b are

$$\{20.9174,\ -48.7829,\ 33.2467,\ -5.5539,\ 0.1727\}$$

These are one less and involves one less term in the denominator. With $\delta = 0.0480$, the 5 values of $H(e^{j\omega_n^e})$ are

$$\{0.9520, 1.0480, 0.9520, 0.1920, -0.1920\}$$

$$H(e^{j\omega_0^e}) = 1 - 0480 = 0.9520$$

$$H(e^{j\omega_1^e}) = 1 + 0480 = 1.0480$$

$$H(e^{j\omega_2^e}) = 1 - 0480 = 0.9520$$

$$H(e^{j\omega_3^e}) = 0 - 0480(4) = 0.1920$$

$$H(e^{j\omega_4^e}) = 0 + 0480(4) = -0.1920$$

We interpolate the unknown values at other grid points. The interpolation formula is

$$H(e^{j\omega}) = \frac{\sum_{n=0}^{L} H(e^{j\omega_n}) \frac{b_n}{\cos(\omega)-\cos(\omega_n)}}{\sum_{n=0}^{L} \frac{b_n}{\cos(\omega)-\cos(\omega_n)}} \qquad (6.12)$$

For example, let us interpolate the third point, 0.0785 radians, on the grid. $\cos(0.0785) = 0.9969$

$$t_1 = \frac{0.9520(20.9174)}{(0.9969 - 1)} + \frac{1.0480(-48.7829)}{(0.9969 - 0.8725)} + \frac{0.9520(33.2467)}{(0.9969 - 0.7071)}$$
$$+ \frac{(0.1920)(-5.5539)}{(0.9969 - 0.3091)} + \frac{(-0.1920)(0.1727)}{(0.9969 + 0.8526)} = -6.7270e + 03$$

$$t_2 = \frac{20.9174}{(0.9969 - 1)} + \frac{(-48.7829)}{(0.9969 - 0.8725)} + \frac{33.2467}{(0.9969 - 0.7071)}$$
$$+ \frac{(-5.5539)}{(0.9969 - 0.3091)} + \frac{0.1727}{(0.9969 + 0.8526)} = -7.0330e + 03$$

$$H(e^{j\omega_2}) = \frac{t1}{t2} = 0.9565$$

The first few of the 68 values of $H(e^{j\omega_n})$ are

$$H(e^{j\omega_n}) = \{0.9520, \; 0.9531, \; 0.9565, \; 0.9619, \; 0.9692, \; 0.9780,$$
$$0.9880, \; 0.9988, \; 1.0097, \; 1.0204, \; 1.0301, \; 1.0384, \; 1.0446, \; 1.0480, \; \dots$$

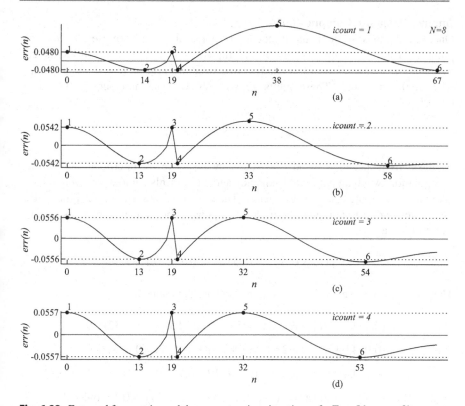

Fig. 6.23 Extremal frequencies and the error at various iterations of a Type I lowpass filter

The 68 values of error between the desired and the Chebyshev approximation,

$$H_d\left(e^{j\omega_n}\right) - H\left(e^{j\omega_n}\right)$$

are shown in Fig. 6.23a. The indices of the updated frequencies are

$$\omega_n^{e2} = \{0, \ 14, \ 19, \ 20, \ 38, \ 67\}$$

as can be seen in Fig. 6.23a.

All the interior points in the grid with zero slope are extremal frequencies. They touch or cross the dotted lines with ordinate at the ripple amplitude δ. Although there are points with greater error than δ, they are not alternating. The iterations continue with the new set of extremal frequencies. In each iteration, the error curve gets closer to the equiripple form. In a relatively small number iterations, the algorithm converges. That is, the extremal frequencies remain the same in two consecutive iterations. Figure 6.23b, c, and d show that the error gets reduced with more iterations. In this case, Fig. 6.23d indicates convergence, as the amplitudes at all the extremal frequencies are equal to δ.

Impulse Response Computation

Once the samples of the actual frequency response are found, we use $L + 1 = 5$-point IDFT to find the impulse response. After convergence, the uniformly sampled frequency response $H(e^{j\omega_n^\mu})$ is interpolated over the grid at $L + 1 = 5$ points between 0 to π radians. These points are, with $NC = 9$ and increment $2\pi/(NC) = 0.6981$,

$$\{0, 0.6981, 1.3963, 2.0944, 2.7925\}$$

At these frequencies, we find the frequency response by interpolation. During interpolation, when an extremal frequency and a uniformly sampled are very close, take the value at the extremal frequency. Then, we use the formula for Type I FIR filter given earlier to compute the impulse response. The frequency response at the five points are

$$\{0.9443, \ 1.0092, \ 0.0051, \ -0.0104, \ 0.1732\}$$

The impulse response $\{h(n), n = 4, 5, 6, 7, 8\}$ are computed as

$$h(0 + 4) = (0.9443 + 2(1.0092 + 0.0051 + (-0.0104) + 0.1732))/9 = 0.3665$$

$$h(1 + 4) = (0.9443 + 2(1.0092\cos(0.6981) + 0.0051\cos(0.6981(2))$$
$$+ (-0.0104)\cos(0.6981(3)) + 0.1732\cos(0.6981(4))))/9 = 0.2419$$

$$h(2 + 4) = (0.9443 + 2(1.0092\cos((2)0.6981) + 0.0051\cos((2)0.6981(2))$$
$$+ (-0.0104)\cos((2)0.6981(3))$$
$$+ 0.1732\cos((2)0.6981(4))))/9 = 0.1734$$

$$h(3 + 4) = (0.9443 + 2(1.0092\cos((3)0.6981) + 0.0051\cos((3)0.6981(2))$$
$$+ (-0.0104)\cos((3)0.6981(3))$$
$$+ 0.1732\cos((3)0.6981(4))))/9 = -0.0293$$

$$h(4 + 4) = (0.9443 + 2(1.0092\cos((4)0.6981) + 0.0051\cos((4)0.6981(2))$$
$$+ (-0.0104)\cos((4)0.6981(3))$$
$$+ 0.1732\cos((4)0.6981(4))))/9 = -0.0971$$

The noncausal impulse response is

$$h(n), n = \{-4, -3, -2, -1, 0, 1, 2, 3, 4\}$$

$$\{-0.0971, \ -0.0293, \ 0.1734, \ 0.2419, \ 0.3665, \ 0.2419, \ 0.1734,$$
$$-0.0293, \ -0.0971\}$$

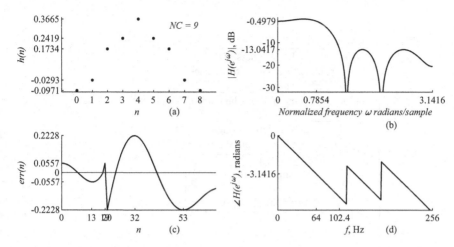

Fig. 6.24 (**a**) The causal impulse response (coefficients) of the filter; the frequency response in decibel; (**c**) the error versus grid points. The passband and stopband amplitudes of the ripples are 0.0557 and 0.2228, respectively; (**d**) the phase response

Figure 6.24a shows the causal impulse response (coefficients) of the filter. Figure 6.24b shows the frequency response in decibel. Figure 6.24c shows the error versus grid points. The passband and stopband amplitudes of the ripples are 0.0557 and 0.2228, respectively. Figure 6.24d shows the phase response. The phase response is linear in the passband and the phase at 64 Hz is

$$-\frac{N}{2}\omega = -4 \times \frac{64 \times \pi}{256} = -\pi \text{ radians}$$

The specifications have been met.

Flowchart of the Optimal Equiripple Linear-Phase FIR Filter Design Algorithm

The flowchart of the optimal FIR filter design algorithm is shown in Fig. 6.25. These are the steps we went through in designing the example filter. The algorithm remains the same for other types of filters, except in setting up the frequency grid, initial setting up and updating the extremal frequencies and the computation of the impulse response. These differences will be taken care of as shown in the following examples.

Type I Highpass

A Type I 8-th order optimal linear-phase highpass filter with the passband and stopband edge frequencies, respectively, $\omega_p = 0.4\pi$ and $\omega_s = 0.25\pi$ radians is required. The size of the ripple δ_s in the stopband is to be 4 times that in the passband δ_p. The sampling frequency is $f_s = 512$ Hz. Design the filter.

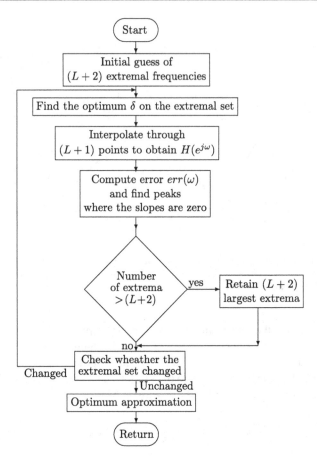

Fig. 6.25 Flowchart of the optimal equiripple linear-phase FIR filter design algorithm

Solution The specifications of the filter are shown in Fig. 6.26. Frequencies 0.25π and 0.4π correspond to $\frac{0.25\pi}{2\pi}512 = 64$ Hz and $\frac{0.4\pi}{2\pi}512 = 102.4$ Hz, respectively. The desired frequency response is defined in terms of band edges and their amplitudes. The ripple size in each band is specified by a weight. Transition band is ignored in the approximation procedure. The amplitude of the frequency response oscillates uniformly between the tolerance bound of each band. Each band is specified by its two edge points and the response is the line connecting the points. That is, response is specified as any piecewise linear desired function with any required transition band.

Figure 6.27 shows the specifications of the example Type I highpass filter as related to its actual frequency response. The passband frequency 102.4 Hz and the stopband frequency 64 are exact in the actual response as specified. In the passband, the ripple oscillates between 1.0819 and 0.9181 with the amplitude 0.0819. In the

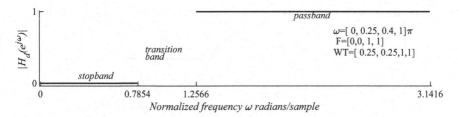

Fig. 6.26 Specifications of a Type I highpass filter

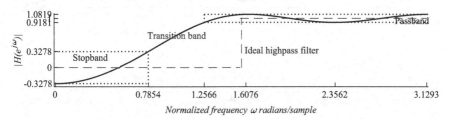

Fig. 6.27 Specifications of a Type I highpass filter as related to its frequency response

stopband, the ripple oscillates between 0.3278 and −0.3278 with the amplitude 0.3278, which is 4 times of that in the passband as specified. Ideal highpass filter frequency response is also shown for comparison.

The algorithm remains the same as for lowpass filter except that appropriate edge frequencies, amplitudes, and weights have to be entered. Extremal frequencies at various iterations of a Type I highpass filter are shown in Fig. 6.28. In this case, Fig. 6.28d indicates convergence, as the amplitudes at all the extremal frequencies are equal to δ. Figure 6.29a shows the causal impulse response (coefficients) of the filter; Fig. 6.29b shows the frequency response in decibel. Figure 6.29c shows the error versus grid points. The passband and stopband amplitudes of the ripple are 0.0819 and 0.3278, respectively. Figure 6.29d shows the phase response.

The phase response is linear in the passband. The phase of the highpass filter at any frequency f in the passband can be computed using the formula

$$\theta = \pi - (4(f - \pi/4))\pi/256$$

For example, at $f = 128$ and $f = 256$, we get

$$\theta = \pi - (4(128 - 64))\pi/256 = 0$$

$$\theta = \pi - (4(256 - 64))\pi/256 = -2\pi$$

The specifications have been met.

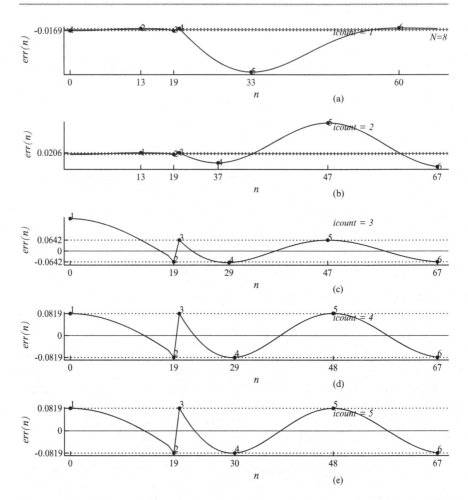

Fig. 6.28 Extremal frequencies at various iterations of a Type I highpass filter

6.3.2 Type II Lowpass

The major change required to apply the optimal filter design algorithm for the other types, II, III, and IV of FIR filters is that the frequency response has to be expressed in an equivalent form. Further, some changes are required in updating the extremal frequencies in the iterations of the algorithm. Once the algorithm converges, the filter coefficients are obtained using the respective formulas defining the relationship between the coefficients and the frequency response.

The impulse response of Type II filter is shifted by one-half of a sampling interval from the origin and even-symmetric, $h(n) = h(N - n)$ with $N + 1$ terms, where N is the order of the filter and odd. Therefore, the number of coefficients NC is even. The frequency response of the Type II FIR filter is given by

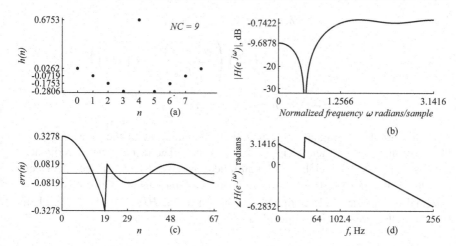

Fig. 6.29 (a) The causal impulse response (coefficients) of the filter; (b) the frequency response in decibel; (c) the error versus grid points. The passband and stopband amplitudes of the ripples are 0.0819 and 0.3278, respectively; (d) the phase response

$$H\left(e^{j\omega}\right) = e^{-j\omega\frac{N}{2}} \sum_{n=0}^{\frac{N-1}{2}} c(n) \cos(\omega(n+0.5))$$

where $c(n) = 2h(\frac{N-1}{2} - n)$, $n = 0, 1, \ldots, \frac{N-1}{2}$.

The frequency response of the Type II FIR filter, without the linear-phase term, can be equivalently expressed as

$$H\left(e^{j\omega}\right) = \cos\left(\frac{\omega}{2}\right) \sum_{k=0}^{\frac{N-1}{2}} c'(k) \cos(\omega k) = \cos\left(\frac{\omega}{2}\right) H'\left(e^{j\omega}\right)$$

where

$$H'\left(e^{j\omega}\right) = \sum_{k=0}^{L} c'(k) \cos(\omega k) \qquad \text{and} \qquad L = \frac{N-1}{2}$$

After the algorithm converges, we get $H'(e^{j\omega})$, which has to be multiplied by $\cos(\frac{\omega}{2})$ to get $H(e^{j\omega})$. We can get the coefficients using $H(e^{j\omega})$. Therefore, the linear relation between the coefficients $c'(k)$ and $c(k)$ is not required for the filter design. As Type II filters has a zero at $\omega = \pi$, there is no peak with value δ at the end of the frequency grid. The weighted error function is written as

$$err(\omega_n) = WT(\omega_n)\left(H_d\left(e^{j\omega_n}\right) - \cos\left(\frac{\omega_n}{2}\right)H'\left(e^{j\omega_n}\right)\right)$$

$$= (-1)^n \delta, \quad n = 0, 1, \ldots, (L+1)$$

$$= WT(\omega_n) \cos\left(\frac{\omega_n}{2}\right) \left(\frac{H_d\left(e^{j\omega_n}\right)}{\cos\left(\frac{\omega_n}{2}\right)} - H'\left(e^{j\omega_n}\right)\right)$$

$$= (-1)^n \delta, \quad n = 0, 1, \ldots, (L+1)$$

Now, the problem formulation is essentially the same as in the case of Type I filter with the response and weight terms modified. The response term is divided by $\cos(\frac{\omega_n}{2})$ and the weight term multiplied by $\cos(\frac{\omega_n}{2})$. After the algorithm converges, the frequency response $H'(e^{j\omega_n})$ is interpolated at the final set of extremal frequencies and multiplied by $\cos(\frac{\omega_n}{2})$ to get $H(e^{j\omega_n})$. The second half of the impulse response is computed as

$$h\left(n + \frac{N+1}{2}\right) = \frac{1}{NC}\left(H\left(e^{j\frac{2\pi 0 T_s}{NC}}\right) + 2\sum_{k=1}^{L} H\left(e^{j\frac{2\pi k T_s}{NC}}\right) \cos\left(\frac{2\pi k(n+0.5)}{NC}\right)\right),$$

$$n = 0, 1, \ldots, L \tag{6.13}$$

where $NC = N + 1$ and $T_s = 0.5$.

A Type II 9-th order optimal linear-phase lowpass filter with the passband and stopband edge frequencies are 0.25π and 0.45π radians, respectively, is required. The size of the ripple in the stopband is to be 2 times that of the passband. The sampling frequency is $f_s = 512$ Hz. Design the filter.

Solution Frequencies 0.25π and 0.45π correspond to $\frac{0.25\pi}{2\pi}512 = 64$ Hz and $\frac{0.45\pi}{2\pi}512 = 115.2$ Hz, respectively.

$$N = 9, L = (N-1)/2 = 4, GD = 16, inc = \pi/((L+1)GD) = 0.0393$$

Edge frequencies are

$$\{0, 0.7854, 1.4137, 3.1416\}$$

Frequency grid points in 2 bands must include all the edge frequencies. Some points are

$$\{0, 0.0393, 0.0785, 0.1178, 0.1571, \ldots, 2.9845, 3.0238, 3.0631, 3.1023\}$$

The number of grid points is $NG = 64$. The extremal frequencies at various iterations of the Type II lowpass filter are shown in Fig. 6.30. The error curve slowly approaches true equiripple form. The specifications have been met. Figure 6.31a shows the causal impulse response (coefficients) of the filter. Figure 6.31b shows the frequency response in decibel. Figure 6.31c shows the error versus grid points. The

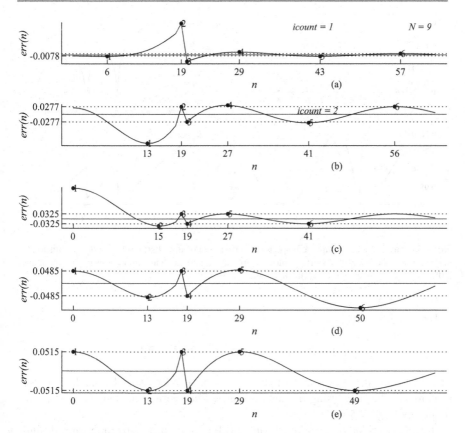

Fig. 6.30 Extremal frequencies at various iterations of the Type II lowpass filter

passband and stopband amplitudes of the ripple are 0.0515 and 0.1030, respectively. Figure 6.31d shows the phase response. The phase response is linear in the passband and the phase at 64 Hz is

$$-\frac{N}{2}\omega = -4.5 \times \frac{64 \times \pi}{256} = -3.5343 \ \text{radians}$$

6.3.3   Type I Bandstop

The six edge frequencies of a 22nd order linear-phase Type I bandstop filter in radians are

$$\{0, \ 0.3\pi, \ 0.4\pi, \ 0.6\pi, \ 0.7\pi, \ \pi\}$$

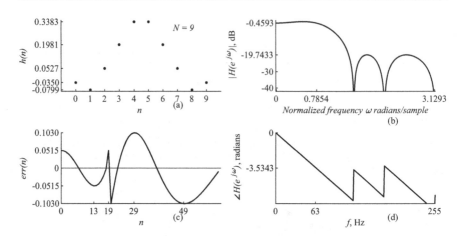

Fig. 6.31 (**a**) The causal impulse response (coefficients) of the filter; (**b**) the frequency response in decibel; (**c**) the error versus grid points. The passband and stopband amplitudes of the ripple are 0.0515 and 0.1030, respectively; (**d**) the phase response

In Hz,

$$\{0, \ 76.8, \ 102.4, \ 153.6, \ 179.2, \ 256\}$$

The ripple ratio in the stopband is to be 4 times that in the passband. Find the impulse response of the filter using the optimum FIR filter design method. The sampling frequency is $f_s = 512\,\text{Hz}$.

The major change in the algorithm for Type I is to initialize the filter design parameters appropriately. The extremal frequencies at various iterations of the Type I bandstop filter are shown in Fig. 6.32.

The causal impulse response values are

$$h(n) = \{0, -0.0316, 0, 0.0063, -0, 0.0561, -0, -0.1578, -0, 0.2548, 0, 0.7051,$$

$$0, 0.2548, -0, -0.1578, -0, 0.0561, -0, 0.0063, 0, -0.0316, -0\}$$

Figure 6.33a shows the causal impulse response (coefficients) of the filter. Figure 6.33b shows the frequency response in decibel. Figure 6.33c shows the error versus grid points. The passband and stopband amplitudes of the ripple are 0.0391 and 0.1566, respectively. Figure 6.33d shows the phase response. The phase response is linear in the passband and the phase at 102.4 Hz is

$$-\frac{N}{2}\omega = -11 \times \frac{102.4 \times \pi}{256} = -13.8230 \ \text{radians}$$

The specifications have been met.

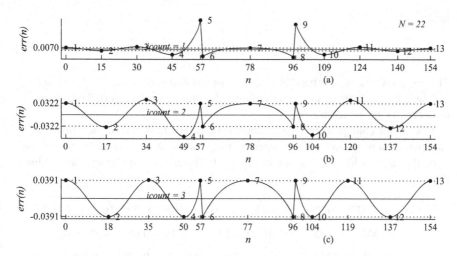

Fig. 6.32 Extremal frequencies at various iterations of the Type I bandstop filter

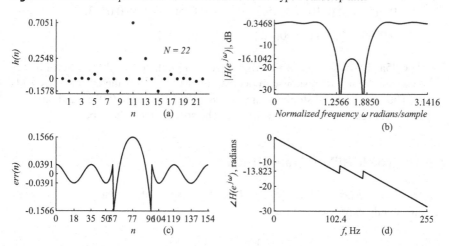

Fig. 6.33 (a) The causal impulse response (coefficients) of the filter; (b) the frequency response in decibel; (c) shows the error versus grid points. The passband and stopband amplitudes of the ripple are 0.0391 and 0.1566, respectively; (d) the phase response

6.3.4   Type II Bandpass

The six edge frequencies of a 33th order linear-phase Type II bandpass filter in radians are

$$\{0, \ 0.1, \ 0.2, \ 0.8, \ 0.9, \ 1\}\pi$$

In Hz,

$$\{0, \ 25.6, \ 51.2, \ 204.8, \ 230.4, \ 256\}$$

The ripple ratio in the stopband is to be 4 times that in the passband. Find the impulse response of the filter using the optimum FIR filter design method. The sampling frequency is $f_s = 512\,\text{Hz}$.

One major change in the algorithm for Type II is to initialize the filter design parameters appropriately. The extremal frequencies at various iterations of the Type II bandpass filter are shown in Fig. 6.34. The causal impulse response values are

$$h(n) = \{-0.0027, -0.0087, 0.0094, -0.0045, 0.0199, 0.0113, 0.0199, 0.0350,$$

$$-0.0047, 0.0445, -0.0510, 0.0114, -0.0896, -0.0848, -0.0607,$$

$$-0.2957, 0.4767, 0.4767, -0.2957, -0.0607, -0.0848, -0.0896,$$

$$0.0114, -0.0510, 0.0445, -0.0047, 0.0350, 0.0199, 0.0113,$$

$$0.0199 - 0.0045, 0.0094, -0.0087, -0.0027\}$$

Figure 6.35a shows the causal impulse response (coefficients) of the filter. Figure (b) shows the frequency response in decibel. Figure (c) shows the error versus grid points. The passband and stopband amplitudes of the ripple are 0.0128 and 0.0513, respectively. Figure (d) the phase response. The specifications have been met.

6.3.5 Type III Hilbert Transformer

The frequency response of the Type III FIR filter, without the linear-phase term, can be equivalently expressed as

$$H\!\left(e^{j\omega}\right) = \sin(\omega) \sum_{k=0}^{\frac{N}{2}-1} c'(k)\cos(\omega k) = \sin(\omega) H'\!\left(e^{j\omega}\right)$$

where

$$H'\!\left(e^{j\omega}\right) = \sum_{k=0}^{\frac{N}{2}-1} c'(k)\cos(\omega k)$$

and $L = \frac{N}{2} - 1$. The magnitude of the frequency response of a practical Hilbert transformer is given as

$$\left|H\!\left(e^{j\omega}\right)\right| = 1 \ \text{for} \ \omega_{c1} \le \omega \le \omega_{c2}$$

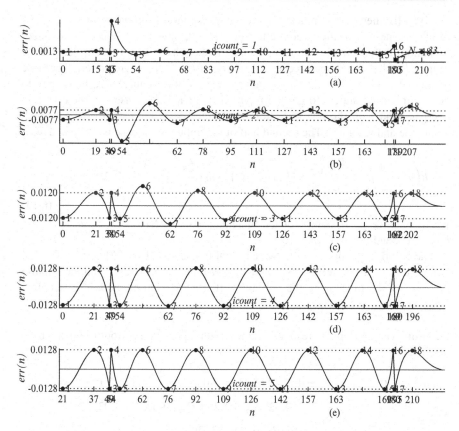

Fig. 6.34 Extremal frequencies at various iterations of the Type II bandpass filter

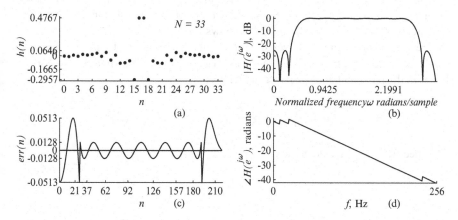

Fig. 6.35 (a) The causal impulse response (coefficients) of the filter; (b) the frequency response in decibel; (c) shows the error versus grid points. The passband and stopband amplitudes of the ripple are 0.0128 and 0.0513, respectively; (d) the phase response

For Type III filters, the symmetrical frequency response specification $\omega_{c2} = \pi - \omega_{c1}$ results in one-half of the filter coefficients zero. It, obviously, reduces the number of multiplications in the implementation. Although Hilbert transformer can be designed with even number of coefficients also, Type III filters are preferred because of this advantage. After the algorithm converges, the frequency response has to be multiplied by $\sin(\omega)$ to get the required frequency response. As there is only one passband with the rest transition band, filter design is carried out with both amplitude and weight 1. The second half of the impulse response is computed as

$$h\left(n + \frac{N}{2}\right) = \frac{2}{NC} \sum_{k=1}^{L+1} \mathrm{Im}\left(H\left(e^{j\frac{2\pi k}{NC}}\right)\right) \sin\left(\frac{2\pi kn}{NC}\right), \quad n = 1, 2, \ldots, L+1$$

(6.14)

where $\mathrm{Im}(H(e^{j\frac{2\pi k}{NC}}))$ stands for the imaginary part of $H(e^{j\frac{2\pi k}{NC}})$, $H(e^{j\frac{2\pi 0}{NC}}) = 0$, $h(\frac{N}{2}) = 0$, and $L = (N/2) - 1$.

The cutoff frequencies of the two edges of the passband of a 20th order Type III Hilbert transformer filter are 0.1π and 0.9π radians, respectively. Find the impulse response of the filter using the optimum FIR filter design method. The sampling frequency is $f_s = 512\,\mathrm{Hz}$.

Solution The 21 impulse response values, with a precision of four digits after the decimal point, are

$$\{h(n), n = 0, 1, \ldots, 20\} = \{0, -0.0273, 0, -0.0479, 0, -0.0932, 0, -0.1902, 0,$$

$$-0.6290, 0, 0.6290, 0, 0.1902, 0, 0.0932, 0,$$

$$0.0479, 0, 0.0273, 0\}$$

The extremal frequencies at various iterations of the Type III Hilbert transformer are shown in Fig. 6.36. Figure 6.37a shows the causal impulse response (coefficients) of the filter. Figure 6.37b shows the frequency response in decibel. Figure 6.37c shows the error versus grid points. The passband amplitude of the ripple is 0.0227. Figure 6.37d shows the phase response. The phase response is linear in the passband and the phase at 26 Hz is

$$\frac{\pi}{2} - \frac{N}{2}\omega = \frac{\pi}{2} - 10 \times \frac{26 \times \pi}{256} = -1.6199 \text{ radians}$$

The specifications have been met.

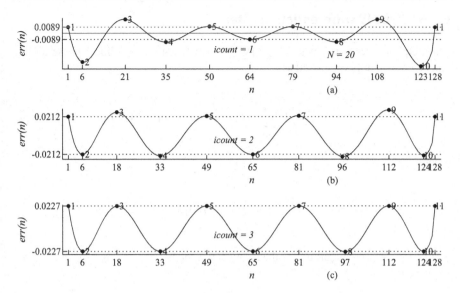

Fig. 6.36 Extremal frequencies at various iterations of a Type III Hilbert transformer

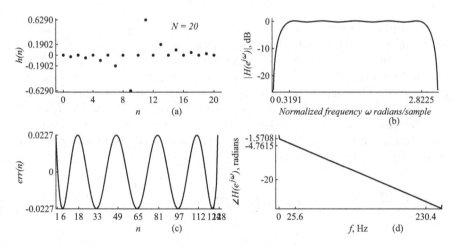

Fig. 6.37 (a) The causal impulse response (coefficients) of the filter; (b) the frequency response in decibel; (c) shows the error versus grid points. The passband amplitude of the ripple is 0.0227; (d) the phase response

6.3.6 Type IV Differentiating Filter

The frequency response of the Type IV FIR Filter, without the linear-phase term, can be equivalently expressed as

$$H(e^{j\omega}) = \sin\left(\frac{\omega}{2}\right) \sum_{k=0}^{\frac{N-1}{2}} c'(k) \cos(\omega k) = \sin\left(\frac{\omega}{2}\right) H'(e^{j\omega})$$

where

$$H'(e^{j\omega}) = \sum_{k=0}^{\frac{N-1}{2}} c'(k) \cos(\omega k)$$

and $L = \frac{N-1}{2}$. The magnitude of the frequency response of a practical differentiator is given as

$$|H(e^{j\omega})| = \begin{cases} \omega & \text{for } 0 \le \omega \le \omega_c \\ 0 & \text{for } \omega_s \le \omega \le \pi \end{cases}$$

Type III FIR filter frequency response is zero at $\omega = \pi$. Therefore, we cannot design a differentiator over the full range and its performance is inferior to that of Type IV filter even otherwise. If the phase response must be proportional to an integer, then the choice is Type III filter. Otherwise, Type IV filters are preferred for designing differentiators. The gain is set as $k\omega$ in the passband, where k is the slope of the response. As the gain is linearly increasing with the frequency, the weight function must decrease correspondingly to keep the relative error the same. The factor $\sin(\frac{\omega}{2})$ has to be taken into account in the design. The second half of the impulse response is computed as

$$h\left(n + \frac{N+1}{2}\right) = -\frac{1}{NC} \left(\text{Im}\left(H\left(e^{j\frac{2\pi \frac{NC}{2} T_s}{NC}}\right)\right)(-1)^n + 2 \sum_{k=1}^{L} \text{Im}\left(H\left(e^{j\frac{2\pi k T_s}{NC}}\right)\right) \right.$$

$$\left. \times \sin\left(\frac{2\pi k(n+0.5)}{NC}\right)\right) \tag{6.15}$$

where $\text{Im}(H(e^{j\frac{2\pi k T_s}{NC}}))$ stands for the imaginary part of $H(e^{j\frac{2\pi k T_s}{NC}})$, $H(e^{j\frac{2\pi 0 T_s}{NC}}) = 0$, $n = 0, 1, \ldots, L$, and $L = (N-1)/2$. The frequency response of Type III and IV filters is pure imaginary. In the algorithm implementation, the interpolated values are $\text{Im}(H(e^{j\frac{2\pi k T_s}{NC}}))$.

The cutoff frequencies of the passband and stopband of a 21th order Type IV differentiating filter are 0.5π and 0.6π radians, respectively. The magnitude of the frequency response is to be 1 at the passband edge frequency. The ripple ratio in the stopband is to be 4 times that in the passband. Find the impulse response of the filter using the optimum FIR filter design method. The sampling frequency is $f_s = 512$ Hz.

Solution The slope is $1/(0.5\pi)$ and the frequency response in the passband is $(2/\pi)\omega$. The 22 impulse response values, with a precision of four digits after the decimal point, are

$$h(n) = \{-0.0153, 0.0028, 0.0445, -0.0468, -0.0117, 0.0522, 0.0103,$$

$$-0.0984, 0.0185, 0.2465, 0.1659, -0.1659, -0.2465, -0.0185,$$

$$0.0984, -0.0103, -0.0522, 0.0117, 0.0468, -0.0445, -0.0028, 0.0153\}$$

The extremal frequencies at various iterations of the Type IV differentiating filter are shown in Fig. 6.38. Figure 6.39a shows the causal impulse response (coefficients) of the filter. Figure 6.39b shows the frequency response and the peak stopband amplitude of the ripples 0.112. Figure 6.39c shows the peak passband amplitude (normalized) of the ripples 0.0280. Figure 6.39d shows the phase

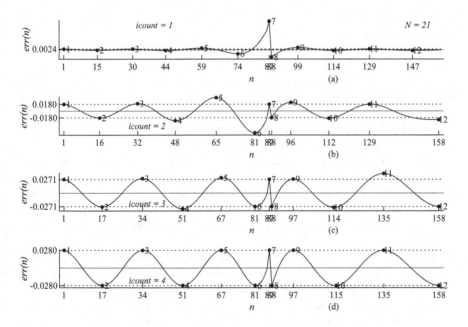

Fig. 6.38 Extremal frequencies at various iterations of a Type IV differentiating filter

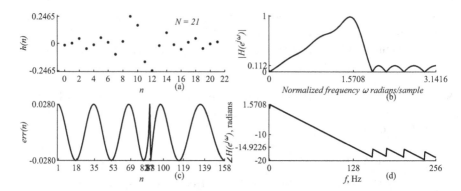

Fig. 6.39 (a) The causal impulse response (coefficients) of the filter; (b) the frequency response and the peak stopband amplitude of the ripples 0.112; (c) shows the error versus grid points and the peak passband amplitude of the ripple (normalized) 0.112, respectively; (d) the phase response

response. The phase response is linear in the passband and the phase at 128 Hz is

$$\frac{\pi}{2} - \frac{N}{2}\omega = \frac{\pi}{2} - 10.5 \times \frac{128 \times \pi}{256} = -14.9226 \text{ radians}$$

The specifications have been met.

6.4   Summary

- Filter means that removes something that passes through it.
- FIR filters are characterized by finite impulse response.
- The output of FIR filters is an exclusive function of past and present input samples only, without feedback. That means no inherent stability problem.
- The frequency response of analog filters with resistors, inductors, and capacitors are approximately simulated by the coefficients in the case of digital filters. The coefficients of the filters weigh the input samples appropriately to provide the filter action.
- The infinite extent impulse response of the ideal FIR filter that meets the specification is first found by taking the inverse DTFT of the frequency response. The impulse response is made realizable by shifting and truncation of the ideal impulse response. The truncation is carried out by multiplying the impulse response by a window of suitable length and characteristics.
- In the time domain, the causal FIR filter is characterized by a difference equation that is a linear combination of past and present input samples $x(n)$ multiplied by the coefficients $h(n)$.

- The order N of a FIR filter is the filter length minus one ($N = M - 1$) or the longest delay of the input terms in number of samples.
- As impulse is composed of all frequency components, its transform is both the transfer function and the frequency response with respect to the transform used.
- The frequency components of the input signal in the passband are delayed when it passes through practical filters. If the phase response of the filter in the passband is linear (of the form $-n_0\omega$), then the frequency components of the output signal are a delayed version of the input frequency components in the passband. Such a phase delay results in no phase distortion.
- The group delay gives the delay of the input frequency components in the passband are subjected to in passing through a filter. If the phase response of the filter is linear, then the shape of the signal formed by the frequency components remains the same.
- Linear-phase FIR filters are of four types with different characteristics.
- In the design of optimum equiripple FIR filters, the weighted difference between frequency responses of the desired and the designed filters is minimized iteratively.
- These filters are characterized by: (1) satisfying a given specification by the lowest order; (2) having equiripple in all the bands except in the transition bands. The equiripples in the bands could be of different sizes; (3) having specified cutoff band edge frequencies; and (4) requiring only one algorithm for various type of filters such as lowpass and highpass, etc.

Exercises

* **6.1** The passband and stopband edge frequencies of a lowpass filter are specified, respectively, as $\omega_c = 0.3\pi$ radians and $\omega_s = 0.4\pi$ radians, respectively. The minimum attenuation required in the stopband is 18 dB. Design the lowpass filter using the rectangular window. The sampling frequency is $f_s = 1024$ Hz.

  6.2 The passband and stopband edge frequencies of a highpass filter are specified, respectively, as $\omega_c = 0.7\pi$ radians and $\omega_s = 0.4\pi$ radians, respectively. The minimum attenuation required in the stopband is 40 dB. Design the highpass filter using the Hamming window. The sampling frequency is $f_s = 1024$ Hz.

* **6.3** The passband and stopband edge frequencies of a lowpass filter are specified, respectively, as $\omega_c = 0.3\pi$ radians and $\omega_s = 0.54\pi$ radians, respectively. The minimum attenuation required in the stopband is 41 dB. Design the lowpass filter using the Kaiser window. The sampling frequency is $f_s = 1024$ Hz.

  6.4 The lower passband and stopband edge frequencies of a bandpass filter are 0.3π and 0.1π, respectively. The upper passband and stopband edge frequencies of the bandpass filter are 0.5π and 0.7π, respectively. The minimum attenuation required in the stopband is $A_s = 48$ dB. The maximum deviation acceptable in the passband is $A_c = 0.1$ dB. Design the bandpass filter using the Kaiser window. The sampling frequency is $f_s = 512$ Hz.

* **6.5** The lower passband and stopband edge frequencies of a bandstop filter are 0.1π and 0.4π, respectively. The upper passband and stopband edge frequencies of the bandstop filter are 0.8π and 0.5π, respectively. The minimum attenuation required in the stopband is $A_s = 35$ dB. The maximum deviation acceptable in the passband is $A_c = 0.2$ dB. Design the bandstop filter using the Hamming window. The sampling frequency is $f_s = 512$ Hz.

6.6 Find the impulse response of the 22nd order Type III Hilbert transformer using the window FIR filter design method. Use the Hamming window. The sampling frequency is $f_s = 512$ Hz.

* **6.7** Find the impulse response of the 20th order Type III differentiating filter using the window FIR filter design method. Use the Hamming window. The sampling frequency is $f_s = 512$ Hz.

6.8 A Type I 22nd order optimal equiripple linear-phase lowpass filter with the passband and stopband edge frequencies, respectively, $\omega_p = 0.25\pi$ and $\omega_s = 0.4\pi$ radians is required. The size of the ripple δ_s in the stopband is to be 3 times that in the passband δ_p. The sampling frequency is $f_s = 512$ Hz. Design the filter.

* **6.9** A Type I 20th order optimal linear-phase highpass filter with the passband and stopband edge frequencies, respectively, $\omega_p = 0.4\pi$ and $\omega_s = 0.25\pi$ radians is required. The size of the ripple δ_s in the stopband is to be 4 times that in the passband δ_p. The sampling frequency is $f_s = 512$ Hz. Design the filter.

6.10 A Type II 19th order optimal linear-phase lowpass filter with the passband and stopband edge frequencies are 0.34π and 0.39π radians, respectively, is required. The size of the ripple in the stopband is to be $3/2$ times that of the passband. The sampling frequency is $f_s = 512$ Hz. Design the filter.

* **6.11** The six edge frequencies of a 18th order linear-phase Type I bandstop filter in radians are

$$\{0, \ 0.1\pi, \ 0.2\pi, \ 0.8\pi, \ 0.9\pi, \ \pi\}$$

The ripple ratio in the stopband is to be 3 times that in the passband. Find the impulse response of the filter using the optimum FIR filter design method. The sampling frequency is $f_s = 512$ Hz.

6.12 The six edge frequencies of a 23rd order linear-phase Type II bandpass filter in radians are

$$\{0, \ 0.1\pi, \ 0.2\pi, \ 0.8\pi, \ 0.9\pi, \ \pi\}$$

The ripple ratio in the stopband is to be 4 times that in the passband. Find the impulse response of the filter using the optimum FIR filter design method. The sampling frequency is $f_s = 512$ Hz.

* 6.13 The cutoff frequencies of the two edges of the passband of a 20th order Type III Hilbert transformer filter are 0.2π and 0.8π radians, respectively. Find the impulse response of the filter using the optimum FIR filter design method. The sampling frequency is $f_s = 512\,\text{Hz}$.

6.14 The cutoff frequencies of the passband and stopband of a 19th order Type IV differentiating filter filter are 0.6π and 0.7π radians, respectively. The magnitude of the frequency response is to be 1 at the passband edge frequency. The ripple ratio in the stopband is to be 4 times that in the passband. Find the impulse response of the filter using the optimum FIR filter design method. The sampling frequency is $f_s = 512\,\text{Hz}$.

Infinite Impulse Response Filters

7

The spectrum of a signal is an ordered list of the values (usually complex) of its constituent frequency components. Electrical and electronic filters suppress certain parts of the spectrum of a signal, while passing the rest. In electrical and electronic engineering, the fundamental circuit elements are resistors, inductors, and capacitors. A filter can be constructed if the volt-ampere characteristics of some of our basic components is a function of frequency. The volt-ampere characteristic of the resistance is independent of frequency, while those of the other two are linearly dependent on frequency. That is, the reactance of the inductors is directly proportional to the frequency. The reactance of the capacitors is inversely proportional to the frequency. Most probably, the first example presented in introducing electric filters is the resistor–capacitor filter, shown in Fig. 7.1. It is a circuit with a resistor, capacitor, and the voltage signal to be filtered connected in series. Let the output of the filter be the voltage across the capacitor. The reactance of the capacitor decreases from ∞ to zero as the frequency varies from zero to ∞. Therefore, considering the circuit as a voltage divider, the amplitude of the low frequency components will be high at the output and those of the high frequency components will be low. The circuit is an analog lowpass filter. This is a first-order filter. With more inductors and capacitors, higher-order filters can be built. Relatively much later in time, digital devices came into existence and overshadowed the analog devices. Nowadays, digital filters are widely used in practice. While analog filters are not much used due to the obsolescence of analog devices, the design formulas of well-established families of analog filters, such as the Butterworth filter, can still be used. That is, design an analog filter to the required specifications and then use suitable transformation formulas, from the Laplace domain to the z-transform domain, to get the required transfer function of the digital filter. The major advantages of FIR filters are that they are always stable and easy to design linear-phase filters. The major disadvantage is that they require more number of coefficients for the same specification. However, due to the availability of fast DFT algorithms, this

Fig. 7.1 A series RC filter circuit

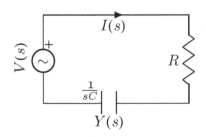

problem is considerably minimized. The IIR filters require much smaller number of coefficients but prone to instability due to feedback. In practice, both are important and have their applications.

7.1   Analog Filters

We present, in this section, tables of the transfer functions of two popular families of filters, the Butterworth and Chebyshev filters. These filters, in practice, approximate the response of the ideal filter to a required extent.

7.1.1   Butterworth Filters

Consider the Laplace domain representation of the resistor–capacitor lowpass filter circuit shown in Fig. 7.1. The reactance, $\frac{1}{sC}$, of the capacitor with $s = j\omega$ varies inversely proportional with the frequency. Therefore, the circuit can be considered as a voltage divider, in which the voltage of the high frequency components of the input signal is low across the capacitor and those of the low frequency components is high. Therefore, the circuit is an analog lowpass filter.

Let us derive the transfer function of the circuit. Let the Laplace transform of the input and output be $V(s)$ and $Y(s)$, respectively. The impedance of the circuit is $R+\frac{1}{sC}$. The current $I(s)$ in the circuit is the input voltage divided by the impedance,

$$\frac{X(s)}{R + \frac{1}{sC}}$$

The output voltage, $Y(s)$, across the capacitor is the current multiplied by the capacitive reactance.

$$Y(s) = \left(\frac{X(s)}{R + \frac{1}{sC}} \right) \left(\frac{1}{sC} \right)$$

Therefore, the transfer function, the transform of the output divided by the transform of the input, is

$$H(s) = \frac{Y(s)}{X(s)} = \frac{1}{1 + sRC}$$

To get the frequency response, $H(j\omega)$, of the filter, we replace s by $j\omega$.

$$H(j\omega) = \frac{1}{1 + j\omega RC}$$

Let the cutoff frequency, ω_c, of the filter be equal to $1/RC = 1$ rad/s. Then,

$$H(j\omega) = \frac{1}{1 + j\frac{\omega}{\omega_c}} = \frac{1}{1 + j\omega} \quad \text{and} \quad |H(j\omega)| = \frac{1}{\sqrt{1 + \omega^2}}$$

This filter circuit is of first-order $N = 1$, since there is only one storage device (the capacitor). It is a lowpass Butterworth filter, since the frequency variable ω appears in the denominator of the transfer function. Therefore, the magnitude response will be decreasing with increasing frequency. For higher-order Butterworth filters, the magnitude of the frequency response, with $\omega_c = 1$, is

$$|H(j\omega)| = \frac{1}{\sqrt{1 + \omega^{(2N)}}}$$

where N is the order of the filter. Obviously, the attenuation of the filter increases with increasing order. The transfer functions of Butterworth filter of higher order are available with $\omega_c = 1$, called the normalized filter. For arbitrary cutoff frequencies and other type of filters, formulas for transformation are available. The factored form of the transfer functions of normalized lowpass Butterworth filters of order up to six are shown in Table 7.1.

The squared magnitude of the frequency response is equal to the product of the transfer function $H(j\omega)$ with its conjugate. That is,

$$H(s)H(-s) = \frac{1}{1 + j\omega^N}\frac{1}{1 - j\omega^N} = \frac{1}{1 + \omega^{(2N)}} = \frac{1}{1 + \left(\frac{s}{j}\right)^{(2N)}}$$

Table 7.1 The transfer functions $H(s)$, in factored form, of normalized lowpass Butterworth filters of order up to six

Order	$H(s)$
1	$\frac{1}{(s+1)}$
2	$\frac{1}{(s^2+1.4142 1s+1)}$
3	$\frac{1}{(s+1)(s^2+s+1)}$
4	$\frac{1}{(s^2+0.76537s+1)(s^2+1.84776s+1)}$
5	$\frac{1}{(s+1)(s^2+0.61803s+1)(s^2+1.61803s+1)}$
6	$\frac{1}{(s^2+0.51764s+1)(s^2+1.41421s+1)(s^2+1.93185s+1)}$

with $s = j\omega$ or $s/j = \omega$. The roots of $1 + \left(\frac{s}{j}\right)^{(2N)}$ on the left-half of the s-plane are the poles of $H(j\omega)$, since the filter must be stable. Solving the equation $1 + \left(\frac{s}{j}\right)^{(2N)} = 0$, we get

$$s^{2N} = (-1)(j)^{2N} = e^{j\pi(2n-1)}(e^{j\frac{\pi}{2}})^{(2N)} = e^{j\pi(2n-1+N)}$$

where n is an integer. Note that $e^{j\pi(2n-1)} = -1$ for any integer n and $e^{j\frac{\pi}{2}} = j$.

$$p_n = e^{\frac{j\pi}{2N}(2n+N-1)}, \ n = 1, 2, \dots, N$$

The transfer function is given by

$$H(s) = \frac{1}{\prod_{n=1}^{N}(s - p_n)}$$

The pole locations of the filter for $N = 2, 3, 5$ are shown in Fig. 7.2. The poles of the filter are equally spaced around the left-half of the unit-circle. There is a pole on the real axis for N odd.

The magnitude of the frequency response $|H(j\omega)|$, in decibels, of normalized first- and second-order Butterworth lowpass filters is shown in Fig. 7.3a. Passbands are shown, in an expanded scale, in Fig. 7.3b. The frequency response is an even function of ω. Therefore, the response for the positive half of the frequency range only is shown. The gain is monotonically decreasing in both the passband and the stopband. The asymptotic falloff rate in the transition band is $-6N$ dB per octave (as the frequency is doubled) or $-20N$ dB per decade (as the frequency becomes ten

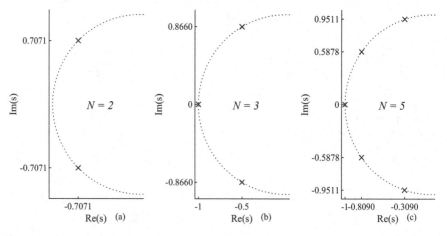

Fig. 7.2 Pole locations of (**a**) second-, (**b**) third-, and (**c**) fifth-order normalized lowpass Butterworth analog filters

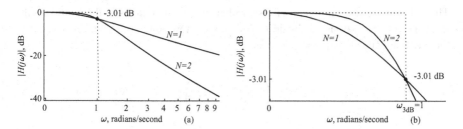

Fig. 7.3 (a) The magnitude of the frequency response of the first- and second-order normalized lowpass Butterworth analog filters; (b) passbands in an expanded scale

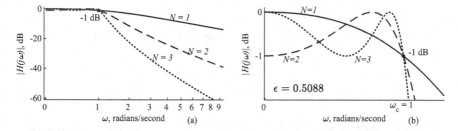

Fig. 7.4 (a) The magnitude of the frequency response of the first-, second-, and third-order normalized lowpass Chebyshev analog filters with 1 dB passband ripple; (b) the passbands in an expanded scale

times) approximately. Normalized filters of any order have the $-3\,\text{dB}$ $(-10\log_{10}(2)$ to be more precise) or $\frac{1}{\sqrt{2}}$ response point at the same frequency, $\omega_{3dB} = 1$ rad/s. Higher-order filters approximate the ideal response, shown by the dotted line, closer than lower-order filters.

7.1.2 Chebyshev Filters

The magnitude of the frequency response of the first-, second-, and third-order normalized lowpass Chebyshev analog filters with 1 dB passband ripple is shown in Fig. 7.4a. The passbands, in an expanded scale, are shown in Fig. 7.4b.

This type of filters has only poles. There is another type of filters with monotonic response in the passband and equiripple response in the stopband. As shown in the figure, the magnitude gain at $\omega = 0$ is zero for an odd order filter and it is $\frac{1}{\sqrt{1+\epsilon^2}}$ for an even order filter, where ϵ is a parameter of the filter related to the ripple in the passband. The size of the passband ripple δ_p is

$$\delta_p = 1 - \frac{1}{\sqrt{1+\epsilon^2}}$$

The poles of the filter lie on an ellipse, rather than on a circle as in the Butterworth filter. The pole locations can be derived with the knowledge of ϵ, filter order N, and the passband cutoff frequency ω_p. Let

$$\alpha = \left(\frac{1 + \sqrt{1 + \epsilon^2}}{\epsilon} \right)^{\frac{1}{N}}$$

Then, the length of the major and minor axes of the ellipse are, respectively,

$$a = \omega_p \frac{\alpha^2 + 1}{2\alpha} \text{ and } b = \omega_p \frac{\alpha^2 - 1}{2\alpha}$$

The angles at which the poles lie are the same as in the Butterworth filter and are defined by

$$\theta_n = \frac{\pi}{2} + \frac{(2n + 1)\pi}{2N}, \quad n = 0, 1, 2, \ldots, N - 1$$

Now, the $2N$ coordinates of the locations of the N poles of the filter are defined as

$$x_n = b \cos(\theta_n) \text{ and } y_n = a \sin(\theta_n), \quad n = 0, 1, 2, \ldots, N - 1$$

Figures 7.5c–a show the pole locations of the first-, second-, and third-order normalized lowpass Chebyshev analog filters with 1 dB passband ripple.

The gain is monotonically decreasing outside the passband. The passband is an equiripple response with all the ripples of equal magnitude. The ripples are compressed towards the cutoff frequency. Beyond the cutoff frequency, the asymptotic falloff rate is inversely proportional to the magnitude of the passband ripple. The tradeoff between these two parameters can be used to suit the requirements. At the cutoff frequency, $\omega_c = 1$ rad/s, filters of all order have the same response. A higher-

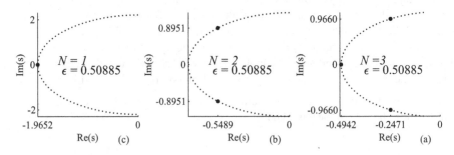

Fig. 7.5 (c–a) The pole locations of the first-, second-, and third-order normalized lowpass Chebyshev analog filters with 1 dB passband ripple

order filter has the same passband ripple magnitude but with a steeper response beyond the cutoff frequency. At $\omega = 0$, the ripple has the maximum value for N odd and it has the minimum value for N even. Excluding the passband edge, the number of minimums and maximums of the ripple is equal to the order of the filter, N.

For the same order and passband attenuation, due to equiripple in the passband, Chebyshev filter has a steeper transition band response compared with a Butterworth filter. That is, for a given specification, the required order is lower for the Chebyshev filter than that of the Butterworth filter. The asymptotic falloff rate beyond the cutoff frequency is $(-20N \log_{10} \omega - 6(N-1) - 20 \log_{10} \epsilon)$ dB approximately, where N is the order and ϵ is related to the passband loss as $\epsilon^2 = (10^{0.1A_c} - 1)$.

Chebyshev filter with passband ripple is an all-pole filter with a zero only at $\omega = \infty$. For a given passband and stopband attenuation A_p and A_s, the order of the filter N is given by

$$N \geq \frac{\cosh^{-1}\left(\sqrt{\frac{10^{0.1A_s}-1}{10^{0.1A_p}-1}}\right)}{\cosh^{-1}(\frac{\omega_s}{\omega_p})} \tag{7.1}$$

The normalized transfer function is given by

$$H(s) = \frac{g}{\prod_{n=1}^{N}(s - p_n)}$$

where the constant g, for N odd, is given by

$$g = \prod_{n=1}^{N}(-p_n) \tag{7.2}$$

For N even, in order to ensure that the filter gain is not more than unity in the passband, the gain is decreased. Therefore, the constant g is given as

$$g = 10^{-0.05A_c} \prod_{n=1}^{N}(-p_n) \tag{7.3}$$

There is a pole on the real axis for N odd.

The factored form of the transfer functions of normalized lowpass Chebyshev filters of order up to six with ripples of various sizes is shown in Tables 7.2, 7.3, 7.4, and 7.5.

Table 7.2 The factored form of the transfer functions of normalized lowpass Chebyshev filters of order up to six with 0.5 dB ripple ($\epsilon = 0.34931$)

$H(s)$
$\dfrac{2.86278}{(s + 2.86278)}$
$\dfrac{1.43139}{(s^2 + 1.42562s + 1.5162)}$
$\dfrac{0.71569}{(s + 0.62646)(s^2 + 0.62646s + 1.14245)}$
$\dfrac{0.35785}{(s^2 + 0.35071s + 1.06352)(s^2 + 0.84668s + 0.35641)}$
$\dfrac{0.17892}{(s + 0.36232)(s^2 + 0.22393s + 1.03578)(s^2 + 0.58625s + 0.47677)}$
$\dfrac{0.08946}{(s^2+0.1553s+1.02302)(s^2+0.42429s+0.59001)(s^2+0.57959s+0.157)}$

Table 7.3 The factored form of the transfer functions of normalized lowpass Chebyshev filters of order up to six with 1.0 dB ripple ($\epsilon = 0.50885$)

$H(s)$
$\dfrac{1.96523}{(s + 1.96523)}$
$\dfrac{0.98261}{(s^2 + 1.09773s + 1.10251)}$
$\dfrac{0.49131}{(s + 0.49417)(s^2 + 0.49417s + 0.9942)}$
$\dfrac{0.24565}{(s^2 + 0.67374s + 0.2794)(s^2 + 0.27907s + 0.98650)}$
$\dfrac{0.12283}{(s + 0.28949)(s^2 + 0.46841s + 0.4293)(s^2 + 0.17892s + 0.98831)}$
$\dfrac{0.06141}{(s^2+0.12436s+0.99073)(s^2+0.33976s+0.55772)(s^2+0.46413s+0.12471)}$

7.2   Frequency Transformations

We derived the transfer functions of the normalized analog lowpass Butterworth and Chebyshev filters. Using transformations, we can get the transfer functions of analog lowpass, highpass, bandpass, and bandstop filters with arbitrary passband cutoff frequencies from that of the lowpass filter by simple transformations, shown in Table 7.6. Let ω_l is the passband cutoff frequency of the prototype lowpass filter and ω_d be the desired passband cutoff frequency of the required lowpass or highpass filter. For bandpass filters, we have lower and upper edge frequencies, ω_{bl} and ω_{bu} of the required filter. For bandstop filters, we have lower and upper edge frequencies, ω_{sl} and ω_{su} of the required filter.

Table 7.4 The factored form of the transfer functions of normalized lowpass Chebyshev filters of order up to six with 2.0 dB ripple ($\epsilon = 0.76478$)

$H(s)$
$\dfrac{1.30756}{(s + 1.30756)}$
$\dfrac{0.65378}{(s^2 + 0.80382s + 0.82306)}$
$\dfrac{0.32689}{(s + 0.36891)(s^2 + 0.36891s + 0.8861)}$
$\dfrac{0.16345}{(s^2 + 0.20977s + 0.92868)(s^2 + 0.50644s + 0.22157)}$
$\dfrac{0.08172}{(s + 0.21831)(s^2 + 0.13492s + 0.95217)(s^2 + 0.35323s + 0.39315)}$
$\dfrac{0.04086}{(s^2 + 0.09395s + 0.96595)(s^2 + 0.25667s + 0.53294)(s^2 + 0.35061s + 0.09993)}$

Table 7.5 The factored form of the transfer functions of normalized lowpass Chebyshev filters of order up to six with 3.01029996 dB ripple ($\epsilon = 1$)

$H(s)$
$\dfrac{1}{(s + 1)}$
$\dfrac{0.5}{(s^2 + 0.64359s + 0.70711)}$
$\dfrac{0.25}{(s + 0.29804)(s^2 + 0.29804s + 0.83883)}$
$\dfrac{0.125}{(s^2 + 0.17001s + 0.9029)(s^2 + 0.41044s + 0.19579)}$
$\dfrac{0.0625}{(s + 0.17719)(s^2 + 0.10951s + 0.9359)(s^2 + 0.2867s + 0.37689)}$
$\dfrac{0.03125}{(s^2 + 0.07631s + 0.95475)(s^2 + 0.20849s + 0.52173)(s^2 + 0.2848s + 0.08872)}$

7.2.1 Lowpass to Lowpass

This transformation can be considered as scaling of the frequency axis. In the desired filter, the attenuation at ω_d must be the same as that at ω_l. Consider the transfer function of a lowpass filter

$$H(s) = \frac{0.8}{s + 0.8}$$

Table 7.6 Frequency transformations for analog filters

Type	Transformation	Edge frequencies of the desired filter
Lowpass to lowpass	$s \rightarrow s\frac{\omega_l}{\omega_d}$	ω_d
Lowpass to highpass	$s \rightarrow \frac{\omega_l \omega_d}{s}$	ω_d
Lowpass to bandpass	$s \rightarrow \omega_l \frac{(s^2 + \omega_{bl}\omega_{bu})}{s(\omega_{bu} - \omega_{bl})}$	ω_{bl}, ω_{bu}
Lowpass to bandstop	$s \rightarrow \omega_l \frac{s(\omega_{su} - \omega_{sl})}{(s^2 + \omega_{sl}\omega_{su})}$	ω_{sl}, ω_{su}

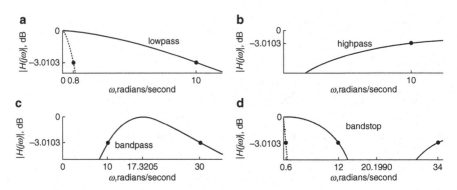

Fig. 7.6 Analog filter frequency transformations; (**a**) lowpass to lowpass; (**b**) lowpass to highpass; (**c**) lowpass to bandpass; (**d**) lowpass to bandstop

with the 3 dB point at $\omega = 0.8$, as shown in Fig. 7.6a by the dotted line. The 3 dB point at $\omega = 10$ will occur, if s is replaced by $0.8s/10$. The desired transfer function is

$$H_{dl}(s) = \frac{0.8}{(0.8s/10) + 0.8} = \frac{10}{s + 10}$$

The 3 dB point occurs at $s = 10$ rad/s, as shown in Fig. 7.6a. This is just frequency scale expansion by a factor of 12.5. That is, what occurs at $\omega = 0.8$ before transformation occurs at $\omega = 10$ after transformation.

7.2.2 Lowpass to Highpass

In the highpass filter, the passband is centered at $\omega = \infty$ and the stopband is centered at $\omega = 0$, just opposite to that in the case of lowpass filter. Therefore, the transformation for lowpass to highpass is just the inverse of that for lowpass to lowpass transformation. That is, s is replaced by $\omega_l \omega_d / s$. Consider the transfer function of a lowpass filter

$$H(s) = \frac{0.8}{s + 0.8}$$

with the 3 dB point at $\omega = 0.8$, as shown in Fig. 7.6a by the dotted line. The 3 dB point at $\omega = 10$ will occur in a highpass filter, if s is replaced by $((0.8)(10))/s$. The desired transfer function is

$$H_{dh}(s) = \frac{0.8}{((0.8)(10))/s) + 0.8} = \frac{s}{s + 10}$$

The 3 dB point occurs at $s = 10$ rad/s after transformation, as shown in Fig. 7.6b.

7.2.3 Lowpass to Bandpass

The transformation for lowpass to bandpass is obtained replacing s by

$$\left(\omega_l \frac{s^2 + \omega_{bl}\omega_{bu}}{s(\omega_{bu} - \omega_{bl})} \right)$$

where ω_{bl} and ω_{bu} are the lower and upper band edge frequencies, respectively, of the bandpass filter. At the frequency, which is the center frequency of the passband of the bandpass filter,

$$\pm\omega_0 = \pm\sqrt{\omega_{bl}\omega_{bu}}$$

the response is maximum. That is, the passband center frequency of the lowpass filter is shifted to $\pm\omega_0$. At frequencies $\omega = 0$ or $\omega = \infty$, of the stopband center frequency, the response of the lowpass filter at $\omega = \infty$ is produced. The net effect of the transformation is that it leaves a lowpass function centered at ω_0 and another centered at $-\omega_0$, $-\infty \le \omega \le \infty$, which is a bandpass filter.

Consider the transfer function of a lowpass filter

$$H(s) = \frac{0.8}{s + 0.8}$$

with the 3 dB point at $\omega = 0.8$. Let the band edge frequencies of a bandpass filter with 3 dB points are $\omega_{bl} = 10$ and $\omega_{bu} = 30$. Replacing s by $0.8(s^2+10(30))/(20s)$, we get the desired transfer function as

$$H_{db}(s) = \frac{0.8}{(0.8(s^2 + 10(30))/(20s)) + 0.8} = \frac{20s}{s^2 + 20s + 300}$$

The magnitude response of the bandpass transfer function with 3 dB points at 10 and 30 rad/s after transformation is shown in Fig. 7.6c.

7.2.4 Lowpass to Bandstop

The transformation, shown in Table 7.6, is essentially the inverse of that of the bandpass. Consider the transfer function of a lowpass filter

$$H(s) = \frac{0.6}{s + 0.6}$$

with the 3 dB point $\omega = 0.6$. A bandstop filter with 3 dB points at $\omega = 6$ and $\omega = 34$ is obtained by replacing s by $0.6(28s)/(s^2 + 6(34))$

$$H_{ds}(s) = \frac{0.6}{(0.6(28s)/(s^2 + 6(34))) + 0.6} = \frac{s^2 + 204}{s^2 + 28s + 204}$$

The magnitude response of the bandstop transfer function with 3 dB points at 6 and 34 rad/s after transformation is shown in Fig. 7.6d.

7.3 The Bilinear Transformation

Now that we got the required analog transfer function, it has to be transformed to the z domain. That is, we have to derive a suitable transformation for replacing the variable s of the analog filter transfer function by the digital variable z. While there are some methods for the transformation, the transformation more often used in practice is called the bilinear (linear with respect to each of two variables) transformation. It is based on transforming a differential equation into a difference equation.

Consider the transfer function of the first-order Butterworth analog filter

$$H(s) = \frac{Y(s)}{X(s)} = \frac{1}{s + 1}$$

The corresponding differential equation is

$$\frac{dy(t)}{dt} + y(t) = x(t)$$

While the derivative term can be approximated by a finite difference, in the bilinear transformation, the differential equation is integrated and the integral is approximated by the trapezoidal formula for numerical integration. In numerical integration, the area to be integrated is divided into subintervals and an approximation function is used to find the area enclosed in each subinterval. One of the approximation functions often used is based on the trapezoid. Let the sampling interval be T_s. Then, the trapezoidal rule is

$$y(n) = y(n-1) + \frac{T_s}{2}(x(n) + x(n-1))$$

The z-transform of this equation, in the transfer function form, is

$$H(z) = \frac{Y(z)}{X(z)} = \frac{T_s}{2} \frac{(z+1)}{(z-1)}$$

Since $1/s$ is integration in the s-domain, we have to make the substitution

$$s = \frac{2}{T_s} \frac{(z-1)}{(z+1)} \tag{7.4}$$

in the transfer function of the filter to get an equivalent discrete transfer function $H(z)$. Applying this transformation, called the bilinear transformation, to the transfer function of the Butterworth filter, we get

$$H(z) = \frac{Y(z)}{X(z)} = \frac{T_s(z+1)}{(T_s+2)z + (T_s-2)} \tag{7.5}$$

7.3.1 Frequency Warping

The frequency range in the s-plane is infinite. That is, $-\infty \leq \omega_a \leq \infty$. The effective frequency range in the z-plane is finite and it is periodic. That is, $-\pi < \omega_d T_s \leq \pi$. We have to find the relationship between ω_a and ω_d in the filter design.

Let $s = (\sigma + j\omega_a)$ and $z = re^{j\omega_d T_s}$. Substituting for s and z in the bilinear transformation formula, we get

$$s = \frac{2}{T_s} \frac{(z-1)}{(z+1)} = \frac{2}{T_s} \frac{(re^{j\omega_d T_s} - 1)}{(re^{j\omega_d T_s} + 1)}$$

Equating the real and imaginary parts of both sides, we get

$$\sigma = \frac{2}{T_s} \left(\frac{r^2 - 1}{1 + r^2 + 2r\cos(\omega_d T_s)} \right)$$

and

$$\omega_a = \frac{2}{T_s} \left(\frac{2r\sin(\omega_d T_s)}{1 + r^2 + 2r\cos(\omega_d T_s)} \right)$$

The real part $\sigma < 0$ if $r < 1$ and $\sigma > 0$ if $r > 1$. That is, the left-half of the s plane maps into the inside of the unit-circle and the right-half maps into the outside. If $r = 1$, $\sigma = 0$, the $j\omega$ axis maps on the unit-circle. With $r = 1$, we get

$$\omega_a = \frac{2}{T_s}\left(\frac{\sin(\omega_d T_s)}{1+\cos(\omega_d T_s)}\right) = \frac{2}{T_s}\tan\left(\frac{T_s}{2}\omega_d\right)$$

using trigonometric half-angle formula. The relationship between the frequency variables ω_d and ω_a is given as

$$\omega_a = \frac{2}{T_s}\tan\left(\frac{T_s}{2}\omega_d\right) \quad \text{and} \quad \omega_d = \frac{2}{T_s}\tan^{-1}\left(\frac{T_s}{2}\omega_a\right) \tag{7.6}$$

Figure 7.7 shows the relationship with $T_s = 1$ second. As an infinite range is mapped to a finite range, the relation between ω_d and ω_a is highly nonlinear, except for a short range.

The frequency responses of the analog and discrete ($T_s = 0.1$) transfer functions of the Butterworth filter are shown in Fig. 7.8. The shapes of the responses of the analog and digital transfer functions are similar and they are identical for some range. At some point, the discrete response shows higher attenuation due to the warping effect. For example, the attenuations of the analog and digital transfer functions at $\omega = 15\,\text{rad/s}$ are, respectively, $-23.5411\,\text{dB}$ and $-25.4176\,\text{dB}$ in Fig. 7.8a. If the discrete transfer function should have the same attenuation at a particular frequency $\omega_d = 15$ radians, then the corresponding analog frequency must be, with $T_s = 0.1$,

$$\omega_a = (2/0.1)\tan((0.1/2)15) = 18.6319$$

The frequency response after prewarping is shown in Fig. 7.8b.

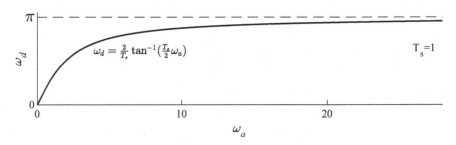

Fig. 7.7 Mapping between ω_a and ω_d in the bilinear transformation

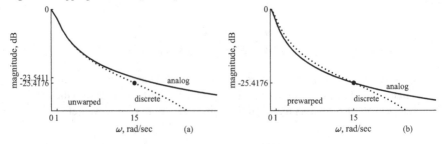

Fig. 7.8 The frequency response; (**a**) without prewarping; (**b**) with prewarping

7.3.2 Application of the Bilinear Transformation

Filters of any order are usually, for analytical or implementation advantages, decomposed into a set of first- and second-order sections. The formulas relating the analog and digital second-order transfer functions can be derived. Let the first- and second-order analog transfer functions be

$$H(s) = \frac{c_1 s + c_0}{d_1 s + d_0} \quad \text{and} \quad H(s) = \frac{c_2 s^2 + c_1 s + c_0}{d_2 s^2 + d_1 s + d_0}$$

The transformation is

$$s = \frac{2}{T_s} \frac{(z-1)}{(z+1)} = k \frac{(z-1)}{(z+1)}$$

Then, the corresponding digital second-order transfer function is

$$H(z) = \frac{\frac{(c_0 + c_1 k + c_2 k^2)}{D} z^2 + \frac{2(c_0 - c_2 k^2)}{D} z + \frac{(c_0 - c_1 k + c_2 k^2)}{D}}{z^2 + \frac{2(d_0 - d_2 k^2)}{D} z + \frac{(d_0 - d_1 k + d_2 k^2)}{D}},$$

where $D = (d_0 + d_1 k + d_2 k^2)$.

For the first-order analog transfer function, the corresponding digital transfer function is

$$H(z) = \frac{\frac{(c_0 + c_1 k)}{D} z + \frac{(c_0 - c_1 k)}{D}}{z + \frac{(d_0 - d_1 k)}{D}}$$

where $D = (d_0 + d_1 k)$

7.4 IIR Filter Design

The specifications of the digital filter is usually given in the digital domain. The frequencies are converted to the corresponding ones in the analog domain after prewarping. The required order of the analog filter of the desired type (such as Butterworth, Chebyshev, etc.) to meet the specifications is found. Then, the poles and zeros of the normalized lowpass filter of the required order are found. The filter is denormalized to the required type (lowpass, highpass, etc.) with the specified edge frequencies. The required digital filter is obtained from the transfer function of the denormalized filter using the bilinear transformation. Following examples exemplify the procedure.

7.4.1   Lowpass Filters

Design a lowpass digital filter with Butterworth response. The passband and stopband edge frequencies of the digital filter are $f_{dc} = 51$ Hz and $f_{ds} = 81$ Hz, respectively. The maximum passband attenuation is $A_c = 3$ dB and the minimum stopband attenuation is $A_s = 22$ dB. The sampling frequency is $f_s = 512$ Hz.

Step 1 Prewarp the frequencies.

In the analog frequency domain, the prewarped edge frequencies, which are slightly higher than the specified ones, correspond to

$$\omega_{ac} = (2 f_s) \tan \left(\frac{\pi f_{dc}}{f_s} \right) = 2(512) \tan \left(\frac{\pi 51}{512} \right) = 331.3290 \text{ rad/s}$$

$$\omega_{as} = (2 f_s) \tan \left(\frac{\pi f_{ds}}{f_s} \right) = 2(512) \tan \left(\frac{\pi 81}{512} \right) = 555.4444 \text{ rad/s}$$

For example, $f_{dc} = 51$ Hz corresponds to 52.7326 Hz. Now, we find the transfer function of the analog lowpass filter with $\omega_c = \omega_{ac}$ and $\omega_s = \omega_{as}$.

Step 2 The required order N of the filter has to be determined.

The order of the required filter is found as

$$N \geq \frac{\log_{10} \left(\frac{10^{0.1 A_s} - 1}{10^{0.1 A_c} - 1} \right)}{2 \log_{10} (\frac{\omega_s}{\omega_c})} = \frac{\log_{10} \left(\frac{10^{0.1(22)} - 1}{10^{0.1(3)} - 1} \right)}{2 \log_{10} \left(\frac{555.4444}{331.3290} \right)} = \frac{2.1993}{0.4488} = 4.9008$$

To make the filter order an integer, we find the nearest integer greater than or equal to 4.9008 to be $N = 5$.

Step 3 Find the factored form of the fifth-order normalized analog lowpass Butterworth filter transfer function.

The transfer function is given in Table 7.1 as

$$H(s) = \frac{1}{(s + 1)(s^2 + 0.61803s + 1)(s^2 + 1.61803s + 1)}$$

Of course, the formula defining the poles can also be used.

Step 4 Find the analog frequency transformation from the normalized frequency to the desired frequency.

The lowpass to lowpass analog frequency transformation is $s = \omega_l s / \omega_d$, where ω_l and ω_d are, respectively, the reference frequency of the normalized filter and the desired frequency. The normalized reference frequency ω_l has to be computed. As the estimated filter order N is usually a real value, it is rounded to the nearest higher integer. Due to this, the filter performance becomes better at the edge frequencies. Now, we have three choices. We can specify that the filter performs better at both the edge frequencies or one of the edge frequencies and meeting the given specification at the other.

We can specify that the specification of the filter is to be met exactly at the cutoff frequency. The cutoff frequency of the normalized filter, with the attenuation to be exact as specified, is computed as

$$\omega_l = (10^{0.1A_c} - 1)^{1/(2N)} = (10^{0.1(3)} - 1)^{1/(2(5))} = 0.999525218401061$$

With the desired prewarped cutoff frequency $\omega_{ac} = 331.3290$, the frequency transformation is given by

$$s = \frac{\omega_l s}{331.3290} = \frac{s}{331.4864018688932}$$

Step 5 Find the factored form of the fifth-order lowpass digital filter transfer function.

This transformation replacing s by $\frac{s}{331.4864018688932}$ can be combined with the bilinear transformation as

$$s = \frac{2f_s}{331.4896}\frac{(z-1)}{(z+1)} = 3.0891\frac{(z-1)}{(z+1)} = k\frac{(z-1)}{(z+1)}$$

Then, using the formulas given earlier, for each second-order section

$$H(z) = \frac{a_2 z^2 + a_1 z + a_0}{z^2 + b_1 z + b_0},$$

we get, with $k = 3.0891$,

$$a_2 = \frac{1}{D}, \ a_1 = \frac{2}{D}, \ a_0 = \frac{1}{D}, b_1 = \frac{2(1-k^2)}{D}, \ b_0 = \frac{(1-d_1k+k^2)}{D},$$

where $D = (1 + d_1 k + k^2)$. The values for d_1 for the two second-order sections are 0.61803 and 1.61803. For the first-order section, the transformation yields

$$\frac{\frac{1}{(1+k)}(z+1)}{z + \frac{(1-k)}{(1+k)}}$$

The transfer function of the lowpass digital filter $H(z)$ is obtained as

$$H(z) = \frac{0.0013(z^2 + 2z + 1)(z^2 + 2z + 1)(z+1)}{(z^2 - 1.3721z + 0.6933)(z^2 - 1.0994z + 0.3568)(z - 0.5109)}$$

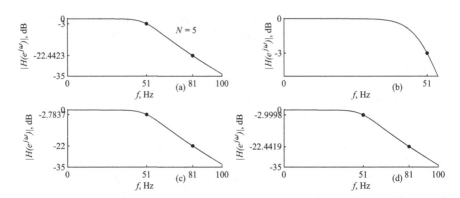

Fig. 7.9 The magnitude of the frequency response of the Butterworth lowpass digital filter; (**a**) with better stopband response; (**b**) passband response expanded; (**c**) with better passband response; (**d**) with better passband and stopband responses

Each of the three factors has a $1/D$ value and the constant in the numerator, 0.0013 is the product of these three values. We take the 512-point DFT of the numerator and denominator and divide point-by-point to get the frequency response, the magnitude of which is shown in Fig. 7.9a, b shows the passband alone. As designed, while the attenuation at the cutoff frequency is exactly equal as specified, the stopband attenuation is better than the specification.

Similarly, better passband response can be obtained using the reference frequency of the normalized filter as

$$\omega_l = (10^{0.1A_s} - 1)^{1/(2N)} = (10^{0.1(22)} - 1)^{1/(2(5))} = 1.6585$$

The analog frequency transformation is given by

$$s = 1.6585s/555.4444 = s/334.9003$$

and the response is shown in Fig. 7.9c. The transfer function of the digital filter $H(z)$ with better passband response is

$$H(z) = \frac{0.0013(z^2 + 2z + 1)(z^2 + 2z + 1)(z + 1)}{(z^2 - 1.3644z + 0.6912)(z^2 - 1.0916z + 0.3531)(z - 0.5071)}$$

Using $N = 4.9008$, we get the reference frequency as

$$\omega_l = (10^{0.1A_c} - 1)^{1/(2N)} = (10^{0.1(3)} - 1)^{1/(2(4.9008))} = 0.999515614715903$$

The frequency transformation is replacing s by $s/331.4895869027143$. The magnitude of the frequency response of the filter, shown in Fig. 7.9d is better at both the edge frequencies.

7.4.2 Highpass Filters

Design a highpass digital filter with Chebyshev response. The passband and stopband edge frequencies of the digital filter are $f_{dc} = 60\,\text{Hz}$ and $f_{ds} = 40\,\text{Hz}$, respectively. The maximum passband attenuation is $A_c = 1\,\text{dB}$ and the minimum stopband attenuation is $A_s = 31\,\text{dB}$. The sampling frequency is $f_s = 512\,\text{Hz}$.

Step 1 Prewarp the frequencies
  In the analog frequency domain, the edge frequencies correspond to

$$\omega_{ac} = (2f_s)\tan\left(\frac{\pi f_{dc}}{f_s}\right) = 2(512)\tan\left(\frac{\pi 60}{512}\right) = 395.0004 \text{ rad/s}$$

$$\omega_{as} = (2f_s)\tan\left(\frac{\pi f_{ds}}{f_s}\right) = 2(512)\tan\left(\frac{\pi 40}{512}\right) = 256.4986 \text{ rad/s}$$

Now, we design an analog lowpass filter with $\omega_c = \omega_{as}$ and $\omega_s = \omega_{ac}$.

Step 2 The required order N of the filter has to be determined.
  The order of the required filter is found as

$$N \geq \frac{\cosh^{-1}\left(\sqrt{\frac{10^{0.1A_s}-1}{10^{0.1A_c}-1}}\right)}{\cosh^{-1}\left(\frac{\omega_s}{\omega_c}\right)} = \frac{\cosh^{-1}\left(\sqrt{\frac{10^{0.1(31)}-1}{10^{0.1(1)}-1}}\right)}{\cosh^{-1}\left(\frac{395.0004}{256.4986}\right)} = \frac{4.9373}{0.9973} = 4.9504$$

To make the filter order an integer, we find the nearest integer greater than or equal to 4.9504 to be $N = 5$.

Step 3 Find the factored form of the fifth-order normalized analog lowpass Chebyshev filter transfer function.
  The transfer function is given in Table 7.3 as

$$H(s) = \frac{0.12283}{(s + 0.28949)(s^2 + 0.46841s + 0.4293)(s^2 + 0.17892s + 0.98831)}$$

Of course, the formula defining the poles can also be used.

Step 4 Find the analog frequency transformation from the normalized frequency to the desired frequency.

The lowpass to highpass analog frequency transformation is $s = (\omega_l\omega_d)/s$, where ω_l and ω_d are, respectively, the reference frequency of the normalized filter and the desired frequency. With $\omega_l = 1$ and $\omega_d = 395.0004$, the transformation is $s = (395.0004/s)$.

Step 5 Find the factored form of the fifth-order highpass digital filter function. This transformation

$$s = (395.0004/s)$$

can be combined with the bilinear transformation as

$$s = \frac{395.0004}{2f_s}\frac{(z+1)}{(z-1)} = 0.3857\frac{(z+1)}{(z-1)} = k\frac{(z+1)}{(z-1)}$$

Then, using the formulas given earlier, for each second-order section

$$H(z) = \frac{a_2z^2 + a_1z + a_0}{z^2 + b_1z + b_0},$$

we get, with $k = 0.3857$,

$$a_2 = \frac{1}{D},\ a_1 = -\frac{2}{D},\ a_0 = \frac{1}{D}, b_1 = \frac{-2(d_0 - k^2)}{D},\ b_0 = \frac{(d_0 - d_1k + k^2)}{D},$$

where $D = (d_0 + d_1k + k^2)$. The values for d_1 for the two second-order sections are 0.46841 and 0.17892. The values for d_0 for the two second-order sections are 0.4293 and 0.98831.

For the first-order section, the transformation yields

$$\frac{1}{k\frac{(z+1)}{(z-1)} + 0.28949}$$

The transfer function of the Chebyshev highpass digital filter $H(z)$ is obtained as

$$H(z) = \frac{0.1988(z^2 - 2z + 1)(z^2 - 2z + 1)(z - 1)}{(z^2 - 0.7393z + 0.5237)(z^2 - 1.3921z + 0.8856)(z + 0.1425)}$$

The magnitude of the frequency response of the Chebyshev digital filter is shown in Fig. 7.10a. Figure 7.10b shows the passband alone.

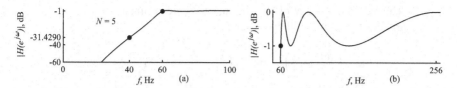

Fig. 7.10 (a) The magnitude of the frequency response of the Chebyshev highpass filter; (b) passband response alone

7.4.3 Design of Bandpass Filters

Design a lowpass digital filter with Butterworth response. The cutoff and stopband edge frequencies of the filter are $f_{dc1} = 800$, $f_{dc2} = 1700$ Hz, and $f_{ds1} = 300$, $f_{ds2} = 2700$ Hz, respectively. The maximum attenuation in the passband is $A_c = 1$ dB. The minimum attenuation in the stopband attenuation is $A_s = 25$ dB. Design the bandpass digital filter using the analog Butterworth filter. The sampling frequency is $f_s = 8192$ Hz.

Step 1 Prewarp the frequencies

In the analog frequency domain, the edge frequencies correspond to

$$\omega_{ac1} = (2f_s) \tan\left(\frac{\pi f_{dc1}}{f_s}\right) = 2(8192) \tan\left(\frac{\pi\, 800}{8192}\right) = 5190.4 \text{ rad/s}$$

$$\omega_{ac2} = (2f_s) \tan\left(\frac{\pi f_{dc2}}{f_s}\right) = 2(8192) \tan\left(\frac{\pi\, 1700}{8192}\right) = 12{,}505 \text{ rad/s}$$

$$\omega_{as1} = (2f_s) \tan\left(\frac{\pi f_{ds1}}{f_s}\right) = 2(8192) \tan\left(\frac{\pi\, 300}{8192}\right) = 1893.3 \text{ rad/s}$$

$$\omega_{as2} = (2f_s) \tan\left(\frac{\pi f_{ds2}}{f_s}\right) = 2(8192) \tan\left(\frac{\pi\, 2700}{8192}\right) = 27{,}623 \text{ rad/s}$$

Bandpass frequency transformation is derived assuming that $(\omega_{ac1}\omega_{ac2}) = (\omega_{as1}\omega_{as2})$. Assume that the passband edges are fixed. If $(\omega_{ac1}\omega_{ac2}) \neq (\omega_{as1}\omega_{as2})$, ω_{as1} has to be increased or ω_{as2} has to be decreased. The frequency ω_{as1} is increased to $(\omega_{ac1}\omega_{ac2})/\omega_{as2} = 2349.8$, since $(\omega_{ac1}\omega_{ac2}) > (\omega_{as1}\omega_{as2})$. Then, the cutoff frequency of the normalized lowpass filter becomes

$$\omega_c = \omega_{ac2} - \omega_{ac1} = 7315$$

The stopband edge frequency becomes

$$\omega_s = \omega_{as2} - \omega_{as1} = 25{,}273$$

Step 2 The required order N of the filter has to be determined.

The order of the required filter is found as

$$N \geq \frac{\log_{10}\left(\frac{10^{0.1A_s}-1}{10^{0.1A_c}-1}\right)}{2\log_{10}\left(\frac{\omega_s}{\omega_c}\right)} = \frac{\log_{10}\left(\frac{10^{0.1(25)}-1}{10^{0.1(1)}-1}\right)}{2\log_{10}\left(\frac{25,273}{7315}\right)} = \frac{3.0854}{1.0769} = 2.8652$$

To make the filter order an integer, we find the nearest integer greater than or equal to 2.8652 to be $N = 3$.

Step 3 Find the factored form of the third-order normalized analog lowpass Butterworth filter transfer function.

The transfer function is given in Table 7.1 as

$$H(s) = \frac{1}{(s+1)(s^2+s+1)}$$

Of course, the formula defining the poles can also be used.

Step 4 The center frequency of the bandpass filter is

$$Wo = \sqrt{(\omega_{ac1}\omega_{ac2})} = \sqrt{(\omega_{as1}\omega_{as2})} = 8056.6$$

The factored form of the sixth-order denormalized Butterworth analog bandpass filter transfer function, in terms of s_n, where $s_n = \frac{s}{Wo}$.
The lowpass to bandpass frequency transformation is

$$s = \omega_l \frac{s^2 + \omega_{d1}\omega_{d2}}{s(\omega_{d2} - \omega_{d1})}$$

where ω_l is the reference frequency of the normalized lowpass filter and ω_{d1} and ω_{d2} are the corresponding desired frequencies of the bandpass filter. Let us specify that the specification of the lowpass filter is to be met exactly at the passband edge frequency. In this case, the passband edge frequency of the normalized lowpass filter is computed as

$$\omega_l = (10^{0.1A_c} - 1)^{1/(2N)} = (10^{0.1(1)} - 1)^{1/(2(3))} = 0.7984$$

The passband bandwidth Bw is

$$\frac{\omega_{ac2} - \omega_{as1}}{\omega_l} = 7315/0.7984 = 9162.6$$

For this design, the frequency transformation becomes

$$s = \frac{s^2 + \omega_{ac1}\omega_{ac2}}{9162.6s} = \frac{s^2 + Wo^2}{9162.6s} = \frac{s^2 + (8056.6)^2}{9162.6s}$$

$$= \frac{(s/8056.6)^2 + 1}{1.1373(s/8056.6)} = \frac{s_n^2 + 1}{1.1373s_n}$$

This transformation is applied to each first- and second-order denominator terms of the normalized transfer function. The order of the filter doubles to $N = 6$. The polynomial

$$(s^2 + as + b) \quad \text{becomes} \quad (s_n^2 + a_1 s_n + b_1)(s_n^2 + a_2 s_n + b_2),$$

where $s_n = s/Wo$. The constants a_1, a_2, b_1, and b_2 are computed as follows. Let $x = +\sqrt{b}$, $y = x/a$, $z = (Bw/Wo)x$, and $p = 1 + \frac{4}{z^2}$. Then,

$$m = \frac{y}{\sqrt{2}}\sqrt{p + \sqrt{p^2 - \frac{4}{(yz)^2}}}$$

Let $q = zm/y$. Then,

$$n_1 = 0.5(q + \sqrt{q^2 - 4}), \qquad n_2 = 0.5(q - \sqrt{q^2 - 4})$$

$$b_1 = n_1^2, \quad b_2 = n_2^2, \quad a_1 = n_1/m, \quad a_2 = n_2/m$$

The numerator is

$$(1.1373s_n)^N = 1.4710s_n^3,$$

where 1.1373 is the constant in the denominator of the transformation formula. The transfer function $H(s_n)$ is

$$H(s_n) = \frac{1.4710s_n^3}{(s_n^2 + 1.1373s_n + 1)(s_n^2 + 0.3107s_n + 0.3758)(s_n^2 + 0.8266s_n + 2.6608)}$$

Step 5 Find the factored form of the sixth order bandpass digital filter.

The transformation $s_n = s_n/Wo = s_n/8056.6$ can be combined with the bilinear transformation as

$$s_n = \frac{2f_s}{8056.6}\frac{(z-1)}{(z+1)} = 2.0336\frac{(z-1)}{(z+1)} = k\frac{(z-1)}{(z+1)}$$

Then, for each second-order section, we get

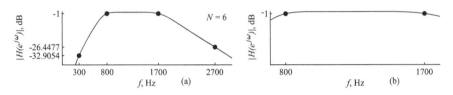

Fig. 7.11 (a) The magnitude of the frequency response of the Butterworth bandpass filter; (b) passband response alone

$$a_2 = \frac{k}{D}, \quad a_1 = 0, \quad a_0 = \frac{-k}{D}, \quad b_1 = \frac{2(d_0 - k^2)}{D}, \quad b_0 = \frac{(d_0 - d_1 k + k^2)}{D},$$

where $D = (d_0 + d_1 k + k^2)$. The transfer function of the digital filter $H(z)$ is obtained as

$$H(z) = \frac{0.0381(z^2 - 1)(z^2 - 1)(z^2 - 1)}{(z^2 - 0.8420z + 0.3790)(z^2 - 1.4620z + 0.7543)(z^2 - 0.3479z + 0.6034)}$$

The constant is the product of the k/D terms of each of the three terms multiplied by 1.471. The magnitude of the frequency response of the filter is shown in Fig. 7.11a. Figure 7.11b shows the passband alone.

The transfer function of the digital filter $H(z)$ with better passband response is

$$H(z) = \frac{0.0429(z^2 - 1)(z^2 - 2z + 1)(z^2 + 2z + 1)}{(z^2 - 0.8272z + 0.3549)(z^2 - 1.4727z + 0.7499)(z^2 - 0.3030z + 0.5877)}$$

7.4.4 Design of Bandstop Filters

The cutoff and stopband edge frequencies of a bandstop filter are $f_{dc1} = 300$, $f_{dc2} = 2700$ Hz and $f_{ds1} = 800$, $f_{ds2} = 1700$ Hz, respectively. The maximum passband attenuation is specified as $A_c = 2$ dB. The minimum stopband attenuation is specified as $A_s = 22$ dB. Design the bandstop digital filter using the analog Chebyshev filter. The sampling frequency is $f_s = 8192$ Hz.

Step 1 Prewarp the frequencies and find the specification of the prototype lowpass filter.

We prewarp the frequencies to get

$$\omega_{ac1} = (2f_s) \tan\left(\frac{\pi f_{dc1}}{f_s}\right) = 2(8192) \tan\left(\frac{\pi 300}{8192}\right) = 1893.3 \text{ rad/s}$$

$$\omega_{ac2} = (2f_s) \tan\left(\frac{\pi f_{dc2}}{f_s}\right) = 2(8192) \tan\left(\frac{\pi 2700}{8192}\right) = 27{,}623 \text{ rad/s}$$

$$\omega_{as1} = (2f_s)\tan\left(\frac{\pi f_{ds1}}{f_s}\right) = 2(8192)\tan\left(\frac{\pi 800}{8192}\right) = 5190.4 \text{ rad/s}$$

$$\omega_{as2} = (2f_s)\tan\left(\frac{\pi f_{ds2}}{f_s}\right) = 2(8192)\tan\left(\frac{\pi 1700}{8192}\right) = 12{,}505 \text{ rad/s}$$

The bandstop frequency transformation is derived assuming that $(\omega_{ac1}\omega_{ac2}) = (\omega_{as1}\omega_{as2})$. If $(\omega_{ac1}\omega_{ac2}) \neq (\omega_{as1}\omega_{as2})$, with fixed passband edges, ω_{as1} has to be decreased or ω_{as2} has to be increased. The frequency ω_{as1} is decreased to $(\omega_{ac1}\omega_{ac2})/\omega_{as2} = 4182$ since $(\omega_{ac1}\omega_{ac2}) < (\omega_{as1}\omega_{as2})$. Now, we design a lowpass filter with cutoff frequency $\omega_c = \omega_{ac2} - \omega_{ac1} = 8323.4$ and stopband edge frequency $\omega_s = \omega_{as2} - \omega_{as1} = 25{,}729$.

Step 2 Find the required filter order N.

The order of the required filter is found as

$$N \geq \frac{\cosh^{-1}\left(\sqrt{\frac{10^{0.1A_s}-1}{10^{0.1A_c}-1}}\right)}{\cosh^{-1}(\frac{\omega_s}{\omega_c})} = \frac{\cosh^{-1}\left(\sqrt{\frac{10^{0.1(22)}-1}{10^{0.1(2)}-1}}\right)}{\cosh^{-1}(\frac{25{,}729}{8323.4})} = \frac{3.4901}{1.7944} = 1.9449$$

As the filter order must be an integer, by rounding 1.9449 up to the nearest integer, we get $N = 2$.

Step 3 Find the factored form of the second-order normalized analog lowpass Chebyshev filter transfer function.

The transfer function is given in Table 7.4 as

$$H(s) = \frac{0.65378}{(s^2 + 0.80382s + 0.82306)}$$

Of course, the formula defining the poles can also be used.

Step 4 Find the factored form of the fourth-order denormalized Chebyshev analog bandstop filter transfer function in terms of s_n, where $s_n = \frac{s}{Wo}$ and $Wo = \sqrt{\omega_{c1}\omega_{c2}} = 7231.7$ is the center frequency of the bandstop filter. The frequency transformation for lowpass to bandstop is, with $\omega_l = 1$,

$$s = \frac{s(\omega_{c2} - \omega_{c1})}{s^2 + \omega_{c1}\omega_{c2}}$$

For this design, the frequency transformation becomes

$$s = \frac{25729s}{s^2 + \omega_{c1}\omega_{c2}} = \frac{Bws}{s^2 + Wo^2} = \frac{25729s}{s^2 + (7231.7)^2} = \frac{3.5578(s/7231.7)}{(s/7231.7)^2 + 1}$$

The polynomial

$$(s^2 + as + b) \quad \text{becomes} \quad (s_n^2 + a_1 s_n + b_1)(s_n^2 + a_2 s_n + b_2),$$

where $s_n = s/Wo$. The constants a_1, a_2, b_1, and b_2 are computed as follows. Let $x = +\sqrt{b}$, $y = x/a$, $z = Bw/(Wo\,x)$, and $p = 1 + \frac{4}{z^2}$. Then,

$$m = \frac{y}{\sqrt{2}} \sqrt{p + \sqrt{p^2 - \frac{4}{(yz)^2}}}$$

Let $q = zm/y$. Then,

$$n_1 = 0.5(q + \sqrt{q^2 - 4}) \qquad n_2 = 0.5(q - \sqrt{q^2 - 4})$$

$$b_1 = n_1^2, \quad b_2 = n_2^2, \quad a_1 = n_1/m, \quad a_2 = n_2/m$$

The numerator is $0.65378(1/b)(s_n^2 + 1)^2$. The transfer function $H(s_n)$ is

$$\frac{0.7943(s_n^2 + 1)^2}{(s_n^2 + 3.2780 s_n + 16.6748)(s_n^2 + 0.1966 s_n + 0.06)}$$

Step 5   Find the factored form of the fourth-order digital bandstop filter.

The transformation $s_n = s/Wo = s/7231.7$ can be combined with the bilinear transformation as

$$s = \frac{2f_s}{7231.7}\frac{(z-1)}{(z+1)} = 2.2656\frac{(z-1)}{(z+1)} = k\frac{(z-1)}{(z+1)}$$

Then, for each second-order section, we get

$$a_2 = a_0 = \frac{1+k^2}{D}, \quad a_1 = \frac{2(1-k^2)}{D}, \quad b_1 = \frac{2(d_0 - k^2)}{D}, \quad b_0 = \frac{(d_0 - d_1 k + k^2)}{D},$$

where $D = (d_0 + d_1 k + k^2)$. The transfer function of the digital filter $H(z)$ is obtained as

$$\frac{0.1813(z^2 - 1.3478z + 1)(z^2 - 1.3478z + 1)}{(z^2 + 0.7896z + 0.4919)(z^2 - 1.7995z + 0.8420)}$$

The constant 0.1813 in the numerator is the product of the $(1 + k^2)/D$ of the two terms and the constant 0.7943. The magnitude of the frequency response of the filter is shown in Fig. 7.12a. Figure 7.12b shows the passband alone.

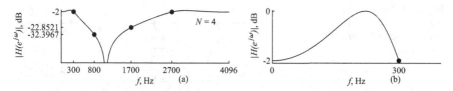

Fig. 7.12 (a) The magnitude of the frequency response of the Chebyshev bandstop filter; (b) passband response alone

Comparison of FIR and IIR Filters

- The transfer function $H(z)$ of a FIR filter is a polynomial in z. The transfer function $H(z)$ of an IIR filter is a ratio of polynomials in z.
- Reproducing the response of continuous-time filters by FIR filters is less accurate. IIR filters are more accurate in reproducing the response of continuous-time filters.
- FIR filters are inherently stable. Stability of IIR filters has to be ensured by proper design.
- FIR filters can produce linear-phase response with or without additional constant phase readily. It is difficult to design IIR linear-phase filters.
- Iterative procedures are commonly used to determine the filter coefficients of FIR filters. Closed-form formulas are available to determine the filter coefficients of IIR filters.
- FIR filters require more number of filter coefficients compared with IIR filters for the same specification.

7.5 Implementation of Digital Filters

It is straightforward to implement or realize digital filters from their difference equation or transfer function, called direct form. However, as the finite wordlength of digital systems produces coefficient quantization and round-off error in multiplication, it is difficult to design and implement higher-order digital filters in direct form. Therefore, it becomes necessary to decompose higher-order filters into several lower-order filters and implement them in cascade or parallel form. Further, specialized structures, those reduce coefficient sensitivity, are also available. The computational complexity, memory requirements, and finite wordlength effects are to be considered in selecting a structure for the implementation of a filter in software or hardware or firmware.

7.5.1 Type-I FIR Filter Implementation

Consider the direct-form realization of a Type I linear-phase FIR filter with 7 coefficients,

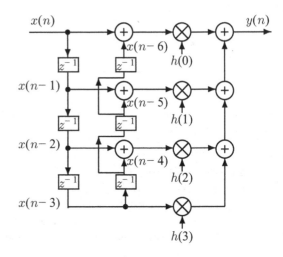

Fig. 7.13 Direct-form realization of a Type I linear-phase FIR filter with 7 coefficients

$$y(n) = h(0)x(n) + h(1)x(n-1) + h(2)x(n-2) + h(3)x(n-3)$$
$$+ h(2)x(n-4) + h(1)x(n-5) + h(0)x(n-6)$$
$$= h(0)(x(n) + x(n-6)) + h(1)(x(n-1) + x(n-5))$$
$$+ h(2)(x(n-2) + x(n-4)) + h(3)x(n-3)$$

shown in Fig. 7.13. As the impulse response is even-symmetric, only one multiplication is required for each pair of symmetric coefficients. Therefore, only 4 multipliers are sufficient. An implementation without consideration of the symmetry would have required 7 multipliers. Similar reduction in the number of multipliers is also possible in the case of Type III linear-phase FIR filters with odd-symmetric impulse response.

7.5.2   Type-II FIR Filter Implementation

Consider the direct-form realization of a Type II linear-phase FIR filter with 8 coefficients,

$$y(n) = h(0)x(n) + h(1)x(n-1) + h(2)x(n-2) + h(3)x(n-3)$$
$$+ h(3)x(n-4) + h(2)x(n-5) + h(1)x(n-6) + h(0)x(n-7)$$
$$= h(0)(x(n) + x(n-7)) + h(1)(x(n-1) + x(n-6))$$
$$+ h(2)(x(n-2) + x(n-5)) + h(3)(x(n-3) + x(n-4))$$

shown in Fig. 7.14. As the impulse response is even-symmetric, only one multiplication is required for each pair of symmetric coefficients. Therefore, only 4 multipliers

Fig. 7.14 Direct-form
realization of a Type II
linear-phase FIR filter with 8
coefficients

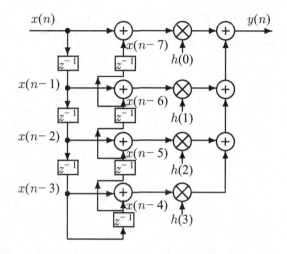

are sufficient. Similar reduction in the number of multipliers is also possible in the case of Type IV linear-phase FIR filters with odd-symmetric impulse response.

7.5.3 Second-Order IIR Filter Implementation

Consider the direct-form realization of a second-order IIR filter,

$$y(n) = b(0)x(n) + b(1)x(n-1) + b(2)x(n-2) - a(1)y(n-1) - a(2)y(n-2)$$

shown in Fig. 7.15. The corresponding transfer function $H(z)$ is

$$H(z) = \frac{Y(z)}{X(z)} = \frac{b(0) + b(1)z^{-1} + b(2)z^{-2}}{1 + a(1)z^{-1} + a(2)z^{-2}}$$

In cascade, the subsystems can be interchanged producing the same output. The transfer function can be decomposed into two systems in cascade, one characterized by the numerator polynomial of the $H(z)$ and the other characterized by the denominator polynomial. That is,

$$H(z) = \frac{1}{1 + a(1)z^{-1} + a(2)z^{-2}}(b(0) + b(1)z^{-1} + b(2)z^{-2}) = H_1(z)H_2(z)$$

The input $X(z)$ is applied to the first system producing the intermediate output $R(z)$. Now, $R(z)$ is the input to the second system producing the required output $Y(z)$. That is,

$$R(z) = H_1(z)X(z) \quad \text{and} \quad Y(z) = H_2(z)R(z)$$

Fig. 7.15 Direct-form canonic realization of a second-order IIR filter

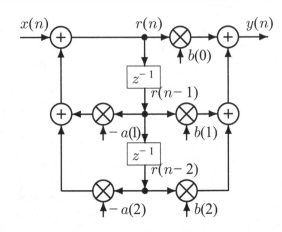

In the time domain, the implementation is characterized by

$$r(n) = x(n) - a(1)r(n-1) - a(2)r(n-2) \quad \text{and}$$

$$y(n) = b(0)r(n) + b(1)r(n-1) + b(2)r(n-2)$$

This realization requires only 2 delay elements, the minimum for a second-order filter, called canonic realization.

7.5.4 Implementation of the Transposed Form

Flow-graph reversal of a flow-graph of an algorithm often results in an alternative algorithm. For example, the flow-graphs of the algorithms for the fast computation of DFT have this property. A transposed form of a filter structure is obtained by: (1) interchanging the input and output points; (2) replacing the junction points by adders and vice versa; and (3) reversing the directions of all the signal flow paths. The transposed form of the flow-graph of the filter in Fig. 7.15 is shown in Fig. 7.16. The following difference equations describe this filter structure:

$$y(n) = b(0)x(n) + r_1(n-1)$$

$$r_1(n) = b(1)x(n) - a(1)y(n) + r_2(n-1)$$

$$r_2(n) = b(2)x(n) - a(2)y(n)$$

Fig. 7.16 The transposed direct-form realization of a second-order IIR filter

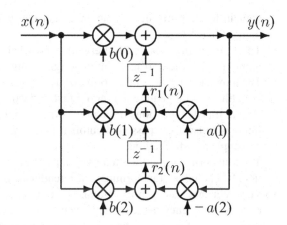

7.5.5 Cascade and Parallel Implementation

The direct-form realization of IIR and FIR filters are very sensitive to coefficient quantization due to the tendency of the poles and zeros to occur in clusters. Further, finite wordlength effects may make the design of higher-order filters difficult. For practical implementations, the most commonly used method is to decompose the required filter into a set of first- and second-order sections connected in cascade or parallel and then realize. The transfer function is decomposed, in the cascade form, into a product of first- and second-order transfer functions.

$$H(z) = H_1(z)H_2(z)\cdots H_N(z)$$

The transfer function is decomposed, in the parallel form, into a sum of first- and second-order transfer functions and then realize.

$$H(z) = k + H_1(z) + H_2(z)+, \cdots, +H_N(z)$$

where k is a constant. In each filter section, the maximum number of poles and zeros is limited to two. The clustering of poles and zeros is much reduced.

7.6 Summary

- Electrical and electronic filters suppress certain parts of the spectrum of a signal, while passing the rest.
- While analog filters are not much used due to the obsolescence of analog devices, the design formulas of well-established families of analog filters, such as the Butterworth filter, can still be used. That is, design an analog filter to the required specifications and then use suitable transformation formulas, from the Laplace

domain to the z-transform domain, to get the required transfer function of the digital filter.

- The transfer functions of Butterworth and Chebyshev analog filters are commonly used as the starting point in designing digital filters by transformation.

- The transfer functions of the normalized analog lowpass Butterworth and Chebyshev filters of the required order are first derived. Using transformations, we can get the transfer functions of analog lowpass, highpass, bandpass, and bandstop filters with arbitrary passband cutoff frequencies from that of the lowpass filter by simple transformations.

- The transformation more often used in practice is called the bilinear transformation. It is based on transforming a differential equation into a difference equation.

- The frequency range in the s-plane is infinite. That is, $-\infty \leq \omega_a \leq \infty$. The effective frequency range in the z-plane is finite and it is periodic. That is, $-\pi < \omega_d T_s \leq \pi$. We have to find the relationship between ω_a and ω_d in the filter design.

- Filters of any order are usually, for analytical or implementation advantages, decomposed into a set of first- and second-order sections. The formulas relating the analog and digital second-order transfer functions can be derived.

- The specifications of the digital filter is usually given in the digital domain. The frequencies are converted to the corresponding ones in the analog domain. The required order of the analog filter of the desired type (such as Butterworth, Chebyshev, etc.) to meet the specifications is found. Then, the poles and zeros of the normalized lowpass filter of the required order are found. The filter is denormalized to the required type (lowpass, highpass, etc.) with the specified edge frequencies. The required digital filter is obtained from the transfer function of the denormalized filter using the bilinear transformation.

- As the finite wordlength of digital systems produces coefficient quantization and round-off error in multiplication, it is difficult to design and implement higher-order digital filters in direct form. Therefore, it becomes necessary to decompose higher-order filters into several lower-order filters and implement them in cascade or parallel form.

Exercises

7.1 Design a lowpass digital filter with Butterworth response. The passband and stopband edge frequencies of the digital filter are $f_{dc} = 41\,\text{Hz}$ and $f_{ds} = 81\,\text{Hz}$, respectively. The maximum passband attenuation is $A_c = 3\,\text{dB}$ and the minimum stopband attenuation is $A_s = 10\,\text{dB}$. Let te stopband attenuation be better than the specification. The sampling frequency is $f_s = 512\,\text{Hz}$.

* 7.2 Design a highpass digital filter with Chebyshev response. The passband and stopband edge frequencies of the digital filter are $f_{dc} = 60\,\text{Hz}$ and $f_{ds} = 30\,\text{Hz}$, respectively. The maximum passband attenuation is $A_c = 1\,\text{dB}$ and the minimum stopband attenuation is $A_s = 11\,\text{dB}$. The sampling frequency is $f_s = 512\,\text{Hz}$.

7.3 Design a bandpass digital filter with Butterworth response. The cutoff and stopband edge frequencies of the filter are $f_{dc1} = 800$, $f_{dc2} = 1700$ Hz and $f_{ds1} = 300$, $f_{ds2} = 2700$ Hz, respectively. The maximum attenuation in the passband is $A_c = 3$ dB. The minimum attenuation in the stopband attenuation is $A_s = 11$ dB. Let the stopband attenuation be better than the specification. Design the bandpass digital filter using the analog Butterworth filter. The sampling frequency is $f_s = 8192$ Hz.

* 7.4 Design a bandstop digital filter with Chebyshev response. The cutoff and stopband edge frequencies of a bandstop filter are $f_{dc1} = 300$, $f_{dc2} = 2700$ Hz and $f_{ds1} = 800$, $f_{ds2} = 1700$ Hz, respectively. The maximum passband attenuation is specified as $A_c = 3$ dB. The minimum stopband attenuation is specified as $A_s = 10$ dB. Design the bandstop digital filter using the analog Chebyshev filter. The sampling frequency is $f_s = 8192$ Hz.

7.5 Design a lowpass digital filter with Butterworth response. The passband and stopband edge frequencies of the digital filter are $f_{dc} = 51$ Hz and $f_{ds} = 81$ Hz, respectively. The maximum passband attenuation is $A_c = 3$ dB and the minimum stopband attenuation is $A_s = 17$ dB. Let the stopband attenuation be better than the specification. The sampling frequency is $f_s = 512$ Hz.

* 7.6 Design a highpass digital filter with Chebyshev response. The passband and stopband edge frequencies of the digital filter are $f_{dc} = 60$ Hz and $f_{ds} = 40$ Hz, respectively. The maximum passband attenuation is $A_c = 2$ dB and the minimum stopband attenuation is $A_s = 31$ dB. The sampling frequency is $f_s = 512$ Hz.

7.7 Design a bandpass digital filter with Butterworth response. The cutoff and stopband edge frequencies of the filter are $f_{dc1} = 800$, $f_{dc2} = 1700$ Hz and $f_{ds1} = 300$, $f_{ds2} = 2700$ Hz, respectively. The maximum attenuation in the passband is $A_c = 2$ dB. The minimum attenuation in the stopband attenuation is $A_s = 15$ dB. Let the stopband attenuation be better than the specification. Design the bandpass digital filter using the analog Butterworth filter. The sampling frequency is $f_s = 8192$ Hz.

* 7.8 The cutoff and stopband edge frequencies of a bandstop filter are $f_{dc1} = 600$, $f_{dc2} = 5400$ Hz and $f_{ds1} = 1600$, $f_{ds2} = 3400$ Hz, respectively. The maximum passband attenuation is specified as $A_c = 2$ dB. The minimum stopband attenuation is specified as $A_s = 15$ dB. Design the bandstop digital filter using the analog Chebyshev filter. The sampling frequency is $f_s = 16,384$ Hz.

Multirate Digital Signal Processing

8

Although we are more used to digital systems with a single sampling rate from input to output, there are applications in which it is advantageous to use different sampling rates for different tasks. For example, reconstruction of signals is easier with a higher sampling rate. For processing operation, such as convolution, just more than two samples is adequate, resulting in faster execution. A system that uses different sampling rates is called a multirate system. Some of the applications of multirate systems are narrowband filters, sampling rate converters, multiplexing and demultiplexing of signals, analog-to-digital and digital-to-analog converters, and wavelet transforms. In addition to adders, multipliers, and delay units used in implementing single sampling rate digital systems, two more basic components, the upsampler and downsampler, are required in the implementation of multirate systems.

8.1 Decimation

Let the frequency components of a signal is in the range $0 \leq \omega \leq \pi$. Then, the decimation operation retains only the lower frequency components of the input signal in the range $|\omega| \leq \pi/M$, where M is an integer, called the decimation factor. If the frequency range is reduced, then the sampling rate can be reduced. While a highpass filter removes the high frequency components, a downsampler reduces the sampling rate.

8.1.1 DownSampler

The block diagram of a downsampler is shown in Fig. 8.1. The output of the downsampler retains every Mth sample, starting from the index $n = 0$, and discarding the rest. That is,

© The Author(s), under exclusive license to Springer Nature Switzerland AG 2021
D. Sundararajan, *Digital Signal Processing*,
https://doi.org/10.1007/978-3-030-62368-5_8

Fig. 8.1 Block diagram of a downsampler

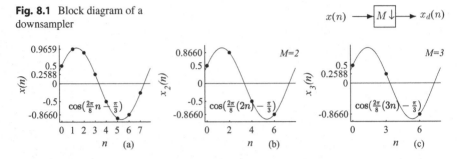

Fig. 8.2 (a) Input signal $x(n)$; (b) downsampled signal x_d with down-sampling factor $M = 2$; (c) downsampled signal x_d with down-sampling factor $M = 3$

$$x_d = x(nM)$$

Figure 8.2a–c show, respectively, the input signal $x(n)$, downsampled signal x_d with down-sampling factors $M = 2$ and $M = 3$.

DownSampler in the Frequency Domain

Let us derive the relationship of the input and output of a downsampler in the frequency domain with $M = 2$. Let $x(n) \leftrightarrow X(e^{j\omega})$, where $x(n)$ is real-valued with a periodic conjugate-symmetric spectrum. We have to express the DTFT of x_d in terms of that of $x(n)$. If we shift $X(e^{j\omega})$ by π radians to get $X(e^{j(\omega-\pi)})$, then its inverse DTFT is $(-1)^n x(n)$. Therefore,

$$\frac{x(n) + (-1)^n x(n)}{2} \leftrightarrow \frac{X(e^{j\omega}) + X(e^{j(\omega-\pi)})}{2}$$

This spectrum is periodic of period π. Consider the transform pairs

$$x(n) = \{x(0) = 3, x(1) = 1, x(2) = 2, x(3) = 4\} \leftrightarrow$$

$$X(e^{j\omega}) = 3 + 1e^{-j\omega} + 2e^{-j2\omega} + 4e^{-j3\omega}$$

$$(-1)^n x(n) = \{x(0) = 3, x(1) = -1, x(2) = 2, x(3) = -4\} \leftrightarrow$$

$$X(e^{j(\omega-\pi)}) = 3 - 1e^{-j\omega} + 2e^{-j2\omega} - 4e^{-j3\omega}$$

By adding the signals and spectra, we get

$$\frac{1}{2}\left(x(n) + (-1)^n x(n)\right) = \{x(0) = 3, x(1) = 0, x(2) = 2, x(3) = 0\}$$

$$\leftrightarrow \frac{1}{2}\left(X(e^{j\omega}) + X(e^{j(\omega-\pi)})\right) = 3 + 0 + 2e^{-j2\omega} + 0$$

The samples of the spectrum are

$$\{3 + 2e^{-j2\omega}, \; \omega = 0, \pi/2, \pi, 3\pi/2\} = \{5, 1, 5, 1\}$$

The DTFT of the down sampled signal $\{3, 2\}$ with $M = 2$, $5 + e^{-j\omega}$, is the compressed version of $3 + 0 + 2e^{-j2\omega} + 0$. Therefore,

$$x_d(n) = x(2n) = \{3, 2\} \leftrightarrow X_d(e^{j\omega}) = \frac{1}{2}\left(X(e^{j\frac{\omega}{2}}) + X(e^{j(\frac{\omega}{2}-\pi)})\right)$$

$$= 5 + e^{-j\omega}, \quad 0 < \omega < 2\pi \tag{8.1}$$

To make the visualization of the downsampling process easier, a pictorial example is provided.

Example 8.1 Find the DTFT $X(e^{j\omega})$ of $x(n)$. The nonzero samples of $x(n)$ are

$$x(n) = \{x(-2) = 2, x(-1) = 3, x(0) = 1, x(1) = 3, x(2) = 2\}$$

Express the DTFT $X_d(e^{j\omega})$ of $x(2n)$ in terms of $X(e^{j\omega})$.

Solution Figure 8.3 shows the details of the process of driving the DTFT of $x(2n)$. The signal $x(n)$ is shown in Fig. 8.3a. The DTFT of $x(n)$ is

$$X(e^{j\omega}) = 2e^{j2\omega} + 3e^{j\omega} + 1 + 3e^{-j\omega} + 2e^{-j2\omega} = 1 + 6\cos(\omega) + 4\cos(2\omega)$$

is shown in Fig. 8.3b. The signal $(-1)^n x(n)$

$$\{x(-2) = 2, x(-1) = -3, x(0) = 1, x(1) = -3, x(2) = 2\}$$

is shown in Fig. 8.3c. Its DTFT, which is the shifted (by π radians) version of that in Fig. 8.3b,

$$X(e^{j(\omega-\pi)}) = 1 + 6\cos(\omega - \pi)) + 4\cos(2(\omega - \pi)) = 1 - 6\cos(\omega) + 4\cos(2\omega)$$

is shown in Fig. 8.3d. The signal $(x(n) + (-1)^n x(n))/2$

$$\{x(-2) = 2, x(-1) = 0, x(0) = 1, x(1) = 0, x(2) = 2\}$$

is shown in Fig. 8.3e. Its DTFT

$$\frac{1}{2}\left(X(e^{j\omega}) + X(e^{j(\omega-\pi)})\right) = (1 + 6\cos(\omega) + 4\cos(\omega)) + (1 - 6\cos(\omega)$$

$$+ 4\cos(\omega))/2 = 1 + 4\cos(2\omega)$$

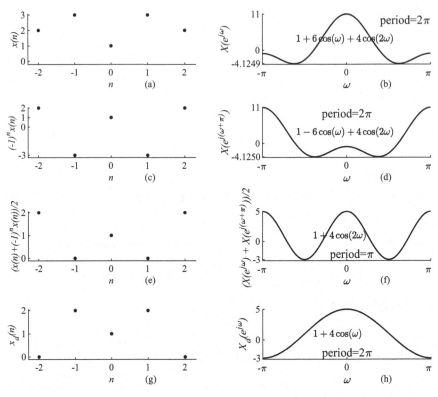

Fig. 8.3 (a) Signal $x(n)$; (b) its DTFT $X(e^{j\omega})$; (c) signal $(-1)^n x(n)$; (d) its DTFT $X(e^{j(\omega-\pi)})$; (e) signal $(x(n)+(-1)^n x(n))/2$; (f) its DTFT $(X(e^{j\omega})+X(e^{j(\omega-\pi)}))/2$; (g) downsampled signal $x_d(n)=x(2n)$; (h) its DTFT $X_d(e^{j\omega})=(X(e^{j\frac{\omega}{2}})+X(e^{j(\frac{\omega}{2}-\pi)}))/2$

is shown in Fig. 8.3f. The spectrum is periodic with period π. The downsampled signal $x_d(n)=x(2n)$

$$\{x(-2)=0, x(-1)=2, x(0)=1, x(1)=2, x(2)=0\}$$

is shown in Fig. 8.3g. Its DTFT, which is the expanded (by a factor of 2) version of that shown in Fig. 8.3f,

$$X_d(e^{j\omega})=1+4\cos(2(\omega/2))=1+4\cos(\omega)$$

is shown in Fig. 8.3h.   ∎

This expression we derived for downsampling with $M=2$ can be generalized for any integer M as

$$X_d(e^{j\omega}) = \frac{1}{M} \sum_{k=0}^{M-1} X(e^{j(\frac{\omega-2\pi k}{M})})$$

The derivation is as follows. The DTFT of the downsampled signal is

$$X_d(e^{j\omega}) = \sum_{n=-\infty}^{\infty} x(Mn)e^{-j\omega n}$$

replacing n by m/M, we get

$$X_d(e^{j\omega}) = \sum_{m=-\infty}^{\infty} x(m)e^{-j\omega \frac{m}{M}}$$

The expression

$$\frac{1}{M} \sum_{k=0}^{M-1} e^{j2\pi k \frac{m}{M}}$$

evaluates to 1 for m integer multiples of M and zero otherwise. Multiplying this expression with the definition for $X_d(e^{j\omega})$, we get

$$X_d(e^{j\omega}) = \frac{1}{M} \sum_{k=0}^{M-1} \sum_{m=-\infty}^{\infty} x(m)e^{-jm\left(\frac{\omega-2\pi k}{M}\right)} = \frac{1}{M} \sum_{k=0}^{M-1} X(e^{j\left(\frac{\omega-2\pi k}{M}\right)})$$

Downsampling compresses the given signal in the time domain and expands its spectrum in the frequency domain. To change the sampling rate by a rational fraction, a cascade of an upsampler and a downsampler has to be used, with the condition that the factors must be relatively prime (the factors do not have a common divisor other than unity). As down-sampling reduces the sampling rate, the high frequency components of the signal will alias as low frequency components. To make the signal usable, the signal has to be filtered appropriately before downsampling.

8.1.2 Decimator

With an adequate sampling frequency f_s, which yields at the least more than two samples of the highest constituent frequency component of the signal, the signal is properly represented by a set of discrete samples with sampling interval $T_s = 1/f_s$. Downsampling the signal by a factor of M reduces the sampling rate to f_s/M. Frequency components with frequencies from zero to $f_s/(2M)$ can only be properly represented. Therefore, the high frequency components have to be filtered out,

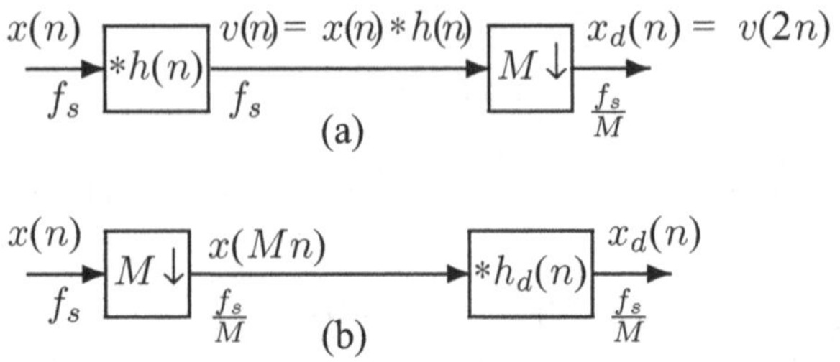

Fig. 8.4 Decimation of a signal $x(n)$ by a factor of M with (**a**) downsampling succeeding lowpass filtering; (**b**) downsampling preceding lowpass filtering

which are assumed of no interest. A signal that is filtered first by a filter with cutoff frequency $f_s/(2M)$ and then downsampled by a factor of M is called a decimated signal by a factor of M. The decimation operation is shown in Fig. 8.4a. The input signal $x(n)$ is filtered with a filter with cutoff frequency $f_s/(2M)$ and impulse response $h(n)$ by convolving $x(n)$ and $h(n)$, $v(n) = x(n)*h(n)$. The downsampling of $v(n)$ by a factor of M yields the decimated signal $x_d(n)$.

We present an example of the decimation process of a periodic signal using its DFT. For aperiodic signals, a larger set of samples is required. Remember that signals are usually aperiodic and continuous. They need infinite samples for exact representation in both the time and frequency domains. However, they are always processed using the DFT/IDFT in practice. This is so, since any practical signal can be adequately represented using finite number of samples in both the domains. The number of samples required depends on the frequency content of the signal. The design of filters for decimation, handling of the discontinuities of the signals at the edges, and the effect of the phase response of the filters are to be taken care of in the decimation and interpolation process.

Example 8.2 Let the samples of one period of the periodic input signal, starting with index zero, be

$$x(n) = \{\hat{2}, 1, 3, -4, -1, 2, 4, 3, 1, -3, 2, 4\}$$

Find the decimated signal with $M = 6$ and $M = 5$ using its DFT.

Solution The DFT spectrum $X(k)$ of the signal $x(n)$ is shown in Fig. 8.5a. The spectral values $X(0)$, $X(1)$, and $X(11)$ only are shown in Fig. 8.5b. The rest of the spectral components has been filtered out. The corresponding DTFT spectral width is 11 radians. Therefore, spectral width between each component is $\pi/6 = 0.5236$ radians. With the decimation factors $M = 5$ and $M = 6$, the cutoff frequencies are, respectively, $\pi/5 = 0.6283$ and $\pi/6 = 0.5236$ radians. Therefore, with $M = 5$, we are left with the DC component and the component with index 1 only. With $M = 6$, we are left with the DC component only.

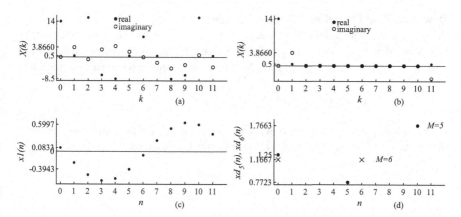

Fig. 8.5 (**a**) DFT spectrum $X(k)$ of the signal $x(n)$; (**b**) the spectral values $X(0)$, $X(1)$, and $X(11)$; (**c**) the time-domain samples corresponding to $X(1)$ and $X(11)$; (**d**) the values of the decimated signal with $M = 5$ and $M = 6$

The inverse DFT of $X(0)$ alone is the DC signal with all the 12 time-domain samples the same, $14/12 = 1.1667$. Therefore, downsampling the signal by 6 yields the decimated signal shown in Fig. 8.5d with crosses. That is,

$$\{x_6(0) = 1.1667, x_6(1) = 1.1667\}$$

Figure 8.5c shows the time-domain samples corresponding to $X(1)$ and $X(11)$. The real part of the coefficients is very small compared with the conjugated imaginary parts and the sign of the imaginary part of $X(1)$ is positive. That is, the time-domain signal is almost an inverted sine wave. Mathematically, the IDFT of the coefficients is the real time-domain signal

$$\frac{(0.5000 + j3.8660)e^{(j\frac{2\pi}{12}n)} + (0.5000 - j3.8660)e^{(-j\frac{2\pi}{12}n)}}{12}$$

$$= 0.6497 \cos\left(\frac{2\pi}{12}n + 1.4422\right)$$

With $M = 5$, the signals passed by the lowpass filter are DC and the frequency component with frequency index 1. Therefore, the sum of the samples of these two signals, downsampled by the factor $M = 5$, is the decimated signal shown in Fig. 8.5d by dots. That is,

$$\{x_5(0) = 1.25, x_5(1) = 0.7723, x_5(2) = 1.7663\}$$

Decimation does not destroy the retained signal components, if they are represented by sufficient number of samples to represent them.

Downsampling Preceding Lowpass Filtering

Placing the filter preceding the downsampler is inefficient as half of the filter output is discarded with $M = 2$. By placing the filter succeeding the downsampler, the filter works at half the sampling rate. As long as the impulse response is of the form

$$h(n) = \{h(0), 0, h(1), 0, \ldots, h(N-1), 0\} \leftrightarrow H(e^{j\omega})$$

the interchange of the filter and downsampler does not affect the output with making the operation efficient. The transform pair for the downsampled version of $h(n)$ is

$$h_d(n) = \{h(0), h(1), \ldots, h(N-1) \leftrightarrow H_d(e^{j\omega})$$

and

$$H_d(e^{j2\omega}) = H(e^{j\omega})$$

That is, $h(n)$ is the upsampled version of $h_d(n)$. Two implementations of the decimator are shown in Fig. 8.4a, b. In Fig. 8.4a, the output after the first stage is

$$X(e^{j\omega})H_d(e^{j2\omega}) = X(e^{j\omega})H(e^{j\omega}) \tag{8.2}$$

After downsampling, we get

$$Y(e^{j\omega}) = \frac{1}{2}\left(H_d(e^{j\frac{2\omega}{2}})X(e^{j\frac{\omega}{2}}) + H_d(e^{j2(\frac{\omega}{2}-\pi)})X(e^{j(\frac{\omega}{2}-\pi)})\right) \tag{8.3}$$

$$= \frac{1}{2}\left(H_d(e^{j\omega})X(e^{j\frac{\omega}{2}}) + H_d(e^{j\omega})X(e^{j(\frac{\omega}{2}-\pi)})\right) \tag{8.4}$$

In Fig. 8.4b, $x(n)$ is downsampled to get

$$\frac{\left(X(e^{j\frac{\omega}{2}}) + X(e^{j(\frac{\omega}{2}-\pi)})\right)}{2}$$

which is then convoluted with $H_d(e^{j\omega})$ to get the decimated as

$$Y(e^{j\omega}) = H_d(e^{j\omega})\frac{\left(X(e^{j\frac{\omega}{2}}) + X(e^{j(\frac{\omega}{2}-\pi)})\right)}{2} \tag{8.5}$$

The outputs of both the implementations are the same, while the implementation in Fig. 8.4b is more efficient, as the filter works at a lower sampling rate.

The following example illustrates only the process. The output is not correct, since the impulse response of the filter is arbitrarily chosen and also the discontinuities are not taken care of. Further, only proper part of the output has to be chosen.

Example 8.3 Let the input $x(n)$ and the impulse response of the filter $h(n)$ be

$$x(n) = \{3, 1, 2\} \quad \text{and} \quad h(n) = \{-2, 0, 3, 0, 1, 0\}$$

Verify that the outputs of the two decimator implementations shown in Fig. 8.4a, b are equivalent.

Solution

$$h_d(n) = \{-2, 3, 1\}$$

The convolution output of $x(n)$ and $h(n)$ in Fig. 8.4a is

$$x(n) * h(n) = \{3, 1, 2\} * \{-2, 0, 3, 0, 1, 0\} = \{-6, -2, 5, 3, 9, 1, 2, 0\}$$

Downsampling the convolution output, we get the decimator output as

$$\{-6, 5, 9, 2\}$$

In Fig. 8.4b the decimated input is $x_d(n) = \{3, 2\}$. The convolution of $x_d(n)$ with $h_d(n) = \{-2, 3, 1\}$ yields

$$x_d(n) * h_d(n) = \{3, 2\} * \{-2, 3, 1\} = \{-6, 5, 9, 2\}$$

∎

Both the outputs are the same.

8.2 Interpolation

The sampling rate of a signal $x(n)$ with sampling rate f_s can be increased by a factor of L by inserting $L - 1$ zeros, called upsampling, between successive values of the signal. The upsampled signal $x_u(n)$ is defined in terms of $x(n)$ and L as

$$x_u(n) = \begin{cases} x(\frac{n}{L}), & \text{for } n = 0, \pm L, \pm 2M, \ldots, \\ 0, & \text{otherwise} \end{cases}$$

The sampling rate of $x_u(n)$ becomes Lf_s, L times that of $x(n)$. The effect of zero-insertion is that the spectrum of $x_u(n)$ is L-fold periodic replication of that of $x(n)$. The additional frequencies, called image frequencies, created by replication are required to construct the upsampled signal. These unwanted components have to be filtered out by a lowpass filter with cutoff frequency $f_s/(2L)$ in order to get the interpolated signal $x_i(n)$. Therefore, the interpolation operation requires upsampling and lowpass filtering.

Example 8.4 Let the input signal be

$$x(n) = \cos\left(\frac{2\pi}{8}\right)n$$

Find the interpolated signal with $L = 2$ using its DFT.

Solution The signal $x(n)$ is shown in Fig. 8.6a. Its samples are

$$x(n) = \{1, 0.7071, 0, -0.7071, -1, -0.7071, 0, 0.7071\}$$

Its upsampled version by $L = 2$ (crosses) and the interpolated version (dots) are shown in Fig. 8.6b.

$$x_{u2}(n) = \{1, 0, 0.7071, 0, 0, 0, -0.7071, 0, -1, 0, -0.7071, 0, 0, 0, 0.7071, 0\}$$

The DFT spectrum $X(k)$ of the signal $x(n)$ is shown in Fig. 8.6c.

$$X(k) = \{0, 4, 0, 0, 0, 0, 0, 4\}$$

The DFT spectrum $X_{u2}(k)$ of the signal $x_{u2}(n)$ is shown in Fig. 8.6d.

$$X_2(k) = \{0, 4, 0, 0, 0, 0, 0, 0, 4, 0, 4, 0, 0, 0, 0, 0, 4\}$$

The spectral values $X_{u2}(7)$ and $X_{u2}(9)$ are the image frequencies required to construct the upsampled signal in conjunction with $X_{u2}(1)$ and $X_{u2}(15)$. Therefore, image frequency components are filtered out by a lowpass filter with cutoff frequency 4 and gain 2. The IDFT of this DFT spectrum yields the interpolated signal. The upsampled signal x_{u2}, with zero-valued samples, in between successive

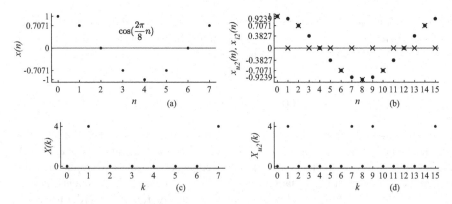

Fig. 8.6 (a) Signal $x(n)$; (b) signal $x_{u2}(n)$ and the output $x_{i2}(n)$; (c) DFT spectrum $X(k)$ of the signal $x(n)$; (d) DFT spectrum $X_{u2}(k)$ of the signal $x_{u2}(n)$

samples of $x(n)$ by crosses and the interpolated signal x_{i2} by dots are shown in Fig. 8.6b. The sampled values of x_{i2} are

$$x_{i2}(n) = \{1, 0.9239, 0.7071, 0.3827, 0, -0.3827, -0.7071, -0.9239,$$

$$-1, -0.9239, -0.7071, -0.3827, 0, 0.3827, 0.7071, 0.9239\}$$

Methods of Interpolation

Two methods of interpolation of a signal $x(n)$ by a factor of L are shown in Fig. 8.7. In Fig. 8.7a, upsampling is preceding lowpass filtering, which is inefficient as the filter has to work at a higher sampling rate than necessary. The first stage output is the upsampled input signal, $x_u(n)$, by a factor of L. Then, $x_u(n)$ is convoluted with the impulse response of the filter to get the interpolated signal $y(n)$.

The convolution of the upsampled input $x_u(n)$ with the filter impulse response $h(n)$ is given by

$$y(n) = \sum_{k=0,1,2,\ldots} x_u(k)h(n-k)$$

Since the odd-indexed values of the upsampled sequence $x_u(n)$ are zero, the even-indexed output is produced convolving the even-indexed values of the input with the even-indexed values of $h(n)$. The odd-indexed output is produced convolving the even-indexed values of the input with the odd-indexed values of $h(n)$. That is,

$$y_e(n) = \sum_{k=0,1,2,\ldots} x(k)h_e(n-k) \quad \text{and} \quad y_o(n) = \sum_{k=0,1,2,\ldots} x(k)h_o(n-k)$$

where $x(k)$ is the downsampled version of $x_u(k)$. The coefficients are right-shifted by one sample interval after computation of each output value. That is, when the input $x(n)$ is convolved individually with the even- and odd-indexed coefficients and the outputs are merged appropriately, the output is the same. With even number of shifts, even-indexed values of $h(n)$ match up with the nonzero values of $x_u(n)$. With odd number of shifts, odd-indexed values of $h(n)$ match up with the nonzero values of $x_u(n)$.

Fig. 8.7 Interpolation of a signal $x(n)$ by a factor of L with (**a**) upsampling preceding lowpass filtering; (**b**) upsampling succeeding lowpass filtering

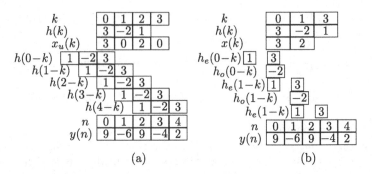

Fig. 8.8 (a) Convolution of an upsampled sequence $x_u(n)$ with $h(n)$; (b) convolution of $x(n)$ with $h_e(n)$ and $h_o(n)$ with the same output

The following example illustrates only the process. The output is not correct, since the filter impulse response is arbitrarily chosen and also the discontinuities are not taken care of. Further, only proper part of the output has to be chosen. The convolution of the upsampled input

$$\{x_u(0) = 3, x_u(1) = 0, x_u(2) = 2, x_u(3) = 0\}$$

with the impulse response $\{h(0) = 3, h(1) = -2, h(2) = 1\}$ is shown in Fig. 8.8a. Convolution of the downsampled input

$$\{x(0) = 3, x(1) = 2\}$$

with the even- and odd-indexed components of $h(n)$, $h_e(n) = \{3, 1\}$ and $h_o(n) = -2$, is shown in Fig. 8.8b. Convolution of $x(n)$ with $h_e(n)$ produces the even-indexed output and with $h_o(n)$ produces the odd-indexed output.

8.2.1   Upsampling in the Frequency Domain

Let $x(n) \leftrightarrow X(e^{j\omega})$. The signal $x(n)$ is made bumpy due to the insertion of zeros in upsampling it. Therefore, the upsampled signal requires additional high frequency components to reconstruct it. These high frequency components, called image frequencies, are created by the replication of the original spectrum. Let the upsampling factor be $L = 2$. The spectrum of the upsampled signal is obtained replacing ω by 2ω, which is spectral compression of the original spectrum. That is, the spectral contents over two periods are placed over the range 0 to 2π.

$$x_u(n) = \begin{cases} x(\frac{n}{2}), & \text{for } n = 0, \pm 2, \pm 4, \ldots \\ 0, & \text{otherwise} \end{cases}$$

$$\leftrightarrow X_u(e^{j\omega}) = X(e^{j2\omega}), \quad 0 < \omega < 2\pi$$

since only every second value of $x_u(n)$ is nonzero. In general, with the upsampling factor L,

$$X_u(e^{j\omega}) = X(e^{jL\omega}), \quad 0 < \omega < 2\pi$$

For example, let

$$\{x(0) = -1, x(1) = 3\} \leftrightarrow -1 + 3e^{-j\omega}$$

Then,

$$\{x(0) = -1, x(1) = 0, x(2) = 3, x(3) = 0\} \leftrightarrow -1 + 3e^{-j2\omega}$$

The spectrum $-1 + 3e^{-j2\omega}$ is periodic with period π, as both the terms are periodic with that period. Therefore, two copies of the original spectrum are compressed into one period. Upsampling increases the frequency components of the spectrum L-fold, while downsampling decreases the frequency components of the spectrum M times.

8.2.2 Filtering Followed by Upsampling

In Fig. 8.7a, input signal $x(n)$ is first upsampled and then the resulting image frequencies are eliminated by filtering to get the interpolated version of $x(n)$. The filter works at twice the sampling frequency of $x(n)$ unnecessarily. As in the case of downsampling, upsampling and filtering can be interchanged without affecting the output, if the impulse response of the filter is of the form

$$h(n) = \{h(0), 0, h(1), 0, \ldots, h(N-1), 0\}$$

The downsampled version of $h(n)$ is

$$h_d(n) = \{h(0), h(1), \ldots, h(N-1)\}$$

and

$$H_d(e^{j2\omega}) = H(e^{j\omega})$$

In Fig. 8.7b, the input signal $x(n)$ is convolved with the impulse response of the filter $h(n)$ to get the filtered version of $x(n)$. The filter works at the sampling rate of $x(n)$. The filtered signal is upsampled to get the interpolated signal more efficiently. In Fig. 8.7a, interpolation operation is carried out as

$$Y(e^{j\omega}) = X(e^{j2\omega})H_d(e^{j2\omega}) = X(e^{j2\omega})H(e^{j\omega}) \tag{8.6}$$

In Fig. 8.7b, the signal is filtered first resulting

$$X(e^{j\omega})H_d(e^{j\omega}) \tag{8.7}$$

the upsampling of which yields the same output.

$$Y(e^{j\omega}) = X(e^{j2\omega})H_d(e^{j2\omega}) = X(e^{j2\omega})H(e^{j\omega}) \tag{8.8}$$

Example 8.5 Let

$$x(n) = \{3, 1, 2, 4\} \quad \text{and} \quad h(n) = \{3, 0, 1\}$$

Verify that the two versions of interpolator implementations, shown in Fig. 8.7a, b, produce the same interpolated output signal.

Solution Then, $x_u(n) * h(n)$ yields

$$\{3, 0, 1, 0, 2, 0, 4\} * \{3, 0, 1\} = \{9, 0, 6, 0, 7, 0, 14, 0, 4\}$$

With

$$h_d(n) = \{3, 1\},$$

$x(n) * h(n)$ yields

$$\{3, 1, 2, 4\} * \{3, 1\} = \{9, 6, 7, 14, 4\}$$

the upsampled version of which also yields the same interpolated output.   ■

Rational Sampling Rate Converters

There are different formats used in audio and video signals recording. The sampling frequency used for recording the signals varies. Therefore, it is often necessary to use rational sampling rate L/M converters to convert signals from one format to another. Let us say we want to increase the sampling rate of a signal $x(n)$ with sampling frequency f_s by a factor $L/M > 1$. The first stage of the interpolator is an upsampler by a factor L followed a digital anti-imaging filter with gain L and cutoff frequency $f_s/(2L)$. The first stage of the decimator is a digital anti-aliasing filter with gain 1 and cutoff frequency $f_s/(2M)$ followed by a downsampler by a factor M. These filters can be combined into one with a gain of L and cutoff frequency F_0 that is the minimum of $f_s/(2L)$ and $f_s/(2M)$. Figure 8.9 shows the block diagram of a sampling rate converter by a factor 3/2 with a combined lowpass filter. Figure 8.10a, b show, respectively, the input signal $x(n) = \cos(\frac{2\pi}{6}n)$

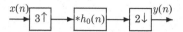

Fig. 8.9 Block diagram of a sampling rate converter by a rational factor 3/2 with a combined lowpass filter

and its DFT spectrum. Figure 8.10c, d show, respectively, the upsampled signal $x_u(n)$ by a factor of 3 and its replicated DFT spectrum. Due to upsampling, the power of the signal has been reduced by a factor of 3. This loss of the power has to be compensated in the filtering stage. Therefore, the gain of the filter is 3. As the minimum of $f_s/(2L)$ and $f_s/(2M)$ is $f_s/(2L)$, the lowpass filter cutoff frequency is $f_s/(2L)$. The filtered DFT spectrum and its IDFT $\cos(\frac{2\pi}{18}n)$ are shown, respectively, in Figure 8.10e, f. Figure 8.10g, h show, respectively, the decimated version by a factor of 2 of $\cos(\frac{2\pi}{18}n)$, which is the required sampling rate converted signal $\cos\left(\frac{2\pi}{9}n\right)$ by a factor of 3/2 with 9 samples in a cycle and its DFT spectrum.

Multistage Converters

If L or M or both are long, then the required lowpass filter becomes a narrowband filter making the converter difficult to implement. In order to make the design of the converter simpler, a set of cascaded converters can be used with factored L/M, where L and M are relatively small. That is,

$$\frac{L}{M} = \left(\frac{L_1}{M_1}\right)\left(\frac{L_2}{M_2}\right)\cdots\left(\frac{L_N}{M_N}\right)$$

Narrowband Filters

Narrowband filters have very small fractional bandwidths relatively. Typically, bandwidth of 1% is a narrow bandwidth. That is, the width of the passband or stopband is much smaller compared with the sampling frequency. For example, with the sampling frequency 10,000 Hz and the passband width 100 Hz, the filter is narrowband. Multirate converters use a set of lower-order filters, while the fixed-rate implementation uses one higher-order filter. High-order filters are difficult to design and implement in practice. In multirate implementation, the filtering task is carried out in three stages: a decimation stage, a filter stage, and an interpolation stage. The sampling rate is reduced by the decimator by a factor of M. This results in making the narrowband wider. The filtering problem is now easier. This filter stage produces the downsampled version of the desired signal. The interpolation stage upsamples the filter output to get back the desired signal at the sampling rate of the input signal. The number of stages of decimation and interpolation can be increased for improving the efficiency of implementation further. This way of implementing the narrowband filters can reduce the total filtering requirements much smaller.

Figure 8.11 illustrates narrowband filtering. Figure 8.11a shows one cycle of the input signal

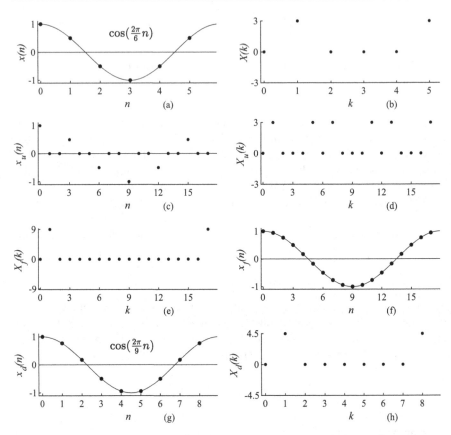

Fig. 8.10 (a) Input signal $x(n) = \cos(\frac{2\pi}{6}n)$ and (b) its DFT spectrum; (c) upsampled signal $x_u(n)$ by a factor of 3 and (d) its DFT spectrum; (e) filtered spectrum by a filter with gain 3 and cutoff frequency $f_s/(2L)$ and (f) its IDFT; (g) signal in (f) decimated by a factor of 2 (sampling rate converted signal $\cos(\frac{2\pi}{9}n)$ by a factor of 3/2 with 9 samples in a cycle) and (f) its DFT spectrum

$$\cos\left(\frac{2\pi}{64}n\right) + \cos\left(3\frac{2\pi}{64}n\right) + \cos\left(13\frac{2\pi}{64}n\right)$$

The sampling frequency f_s is $2\pi/64$ radians. The signal is composed of cosines waves of frequencies 1, 3, and 13 Hz. The filter is to retain only the 1 Hz component. A single optimal equiripple filter requires a filter of order 400. To get the same results by a multirate narrowband filter, we pass the input signal through a decimator with decimation factor $M = 4$. The filter required has a cutoff frequency $f_s/8$ with an order 100. The decimated signal is shown in Fig. 8.11b. Note that the filtering operation delays the signal. As the cutoff of frequency of the filter associated is less than that of 13 Hz, the highest frequency in the input signal gets eliminated in the decimation stage. Filtering operation can be affected by multiplying the DFTs of the input signal and the filter impulse response, which is of length 100, and find the

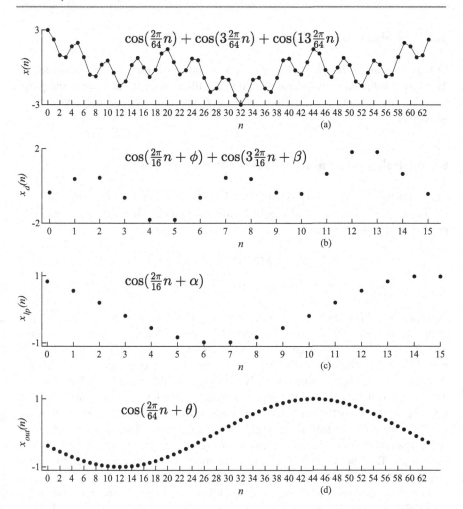

Fig. 8.11 (a) Input signal; (b) decimated signal; (c) lowpass filtered signal; (d) filtered signal

IDFT of the product. Note that the length 100 has to be increased to 128 by zero padding and then DFT of zero-padded impulse response and two cycles of the input signal must be computed for convolution.

Figure 8.11c shows the filtered signal by a filter of order 100. That filter should have sufficiently low cutoff frequency so that all the unwanted frequency components are adequately cutoff. The output of the filter is just the frequency component with frequency 1 Hz only as specified. However, this filtered signal has be upsampled with an upsampler with an upsampling factor $L = 4$ and gain 4. The filter used in interpolation is same as that used for decimation. Figure 8.11d shows the filtered signal.

8.3 Polyphase Decimator and Interpolator

In Figs. 8.4a and 8.7a, the filters work at at a higher sampling rate than necessary. This redundancy can be eliminated so that the filter can work at a lower sampling rate. In this section, we develop such filter structures, called polyphase filter structures.

8.3.1 Polyphase Implementation of the Decimator

Let us say the downsampling factor is 2 and a FIR filter with 4 coefficients. Only the even-indexed output values of the decimator is required. The governing equations producing the output with some consecutive even indexes are

$$y(8) = (h(0)x(8) + h(2)x(6)) + (h(1)x(7) + h(3)x(5))$$
$$y(10) = (h(0)x(10) + h(2)x(8)) + (h(1)x(9) + h(3)x(7))$$

Because the odd-indexed output values are not required, we can divide the input sequence into two subsequences, odd-indexed sequence and even-indexed sequence. The filter coefficients can also be divided into two sets, one set containing the even-indexed coefficients and the other set containing the odd-indexed coefficients. As odd-indexed input values convolve only with odd-indexed coefficients and even-indexed input values convolve only with even-indexed coefficients, we have partitioned the 4-coefficient filter into two subfilters, each one working its own input sequence of one-half of the original sampling rate f_s. The subfilters are called polyphase filters. Therefore, the filter structure consists of two filters with two input sequences. The outputs of the subfilters are added to produce the required output sequence.

The filter structure, with $M = 2$, is shown in Fig. 8.12. The leftmost delay unit delays the input sequence by one sampling interval. The top and bottom downsamplers output are, respectively, the even-indexed and odd-indexed input sequences of the original input sequence. These input subsequences entering into the respective subfilters produce two partial outputs, the sum of them becomes the required output sequence. This approach can be extended to any decimation factor M. There will be M input sequences and M subfilters. The following examples will help to understand the operation of the filter.

Example 8.6 Let the nonzero values of the input signal and the impulse response of the filter, with the starting index zero, be, respectively,

$$x(n) = \{2, 1, 3, 4, -1, 2, 4, 3, 1, 3, 2, 4\} \quad \text{and} \quad h(n) = \{1, -1, 4, 3, 2\}$$

Fig. 8.12 Polyphase
implementation of a
decimation filter with $M = 2$

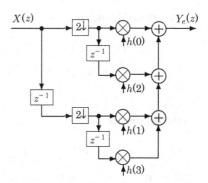

Find the every second, $M = 2$, convolution output of $x(n)$ and $h(n)$ directly and by
the polyphase approach.

Solution The complete output $y(n)$ of convolving $h(n)$ and $x(n)$ is

$$y(n) = x(n) * h(n) = \{2, -1, 10, 11, 14, 30, 16, 12, 18, 30, 20, 23, 15, 28, 16, 8\}$$

The decimated output, with $M = 2$, shown in boldface, is

$$\{2, 10, 14, 16, 18, 20, 15, 16\}$$

First, zero-pad the impulse response to make its length equal to an integral multiple
of $M = 2$ to get

$$h(n) = \{1, -1, 4, 3, 2, 0\}$$

By grouping the even- and odd-indexed values, we get

$$h1(n) = \{1, 4, 2\} \quad \text{and} \quad h2(n) = \{-1, 3, 0\}$$

The input data is grouped as

$$x1(n) = \{2, 3, -1, 4, 1, 2, 0\} \quad \text{and} \quad x2(n) = \{0, 1, 4, 2, 3, 3, 4\}$$

The initial value 0 in the second group is $x(-1)$.

The outputs of convolving the even- and odd-indexed coefficients of $h(n)$ with
those of $x(n)$ are, respectively,

$$\{1, 4, 2\} * \{2, 3, -1, 4, 1, 2, 0\} = \{2, 11, 15, 6, 15, 14, 10, 4, 0\}$$

$$\{-1, 3, 0\} * \{0, 1, 4, 2, 3, 3, 4\} = \{0, -1, -1, 10, 3, 6, 5, 12, 0\}$$

The decimated output is

$$\{2, 11, 15, 6, 15, 14, 10, 4, 0\} + \{0, -1, -1, 10, 3, 6, 5, 12, 0\}$$
$$= \{2, 10, 14, 16, 18, 20, 15, 16, 0\}$$

$M = 3$
Let the decimation factor be $M = 3$. Typical outputs produced are

$$y(6) = h(0)x(6) + h(1)x(5) + h(2)x(4) + (h(3)x(3) + h(4)x(2)) + h(5)x(1)$$
$$y(9) = h(0)x(9) + h(1)x(8) + h(2)x(7) + (h(3)x(6) + h(4)x(5)) + h(5)x(4)$$
$$y(12) = h(0)x(12) + h(1)x(11) + h(2)x(10) + (h(3)x(9) + h(4)x(8)) + h(5)x(7)$$

Due to decimation of the output, the data terms and impulse response terms get grouped. That is,

$$y(6) = (h(0)x(6) + h(3)x(3)) + (h(1)x(5) + h(4)x(2)) + (h(2)x(4) + h(5)x(1))$$
$$y(9) = (h(0)x(9) + h(3)x(6)) + (h(1)x(8) + h(4)x(5)) + (h(2)x(7) + h(5)x(4))$$
$$y(12) = (h(0)x(12) + h(3)x(9)) + (h(1)x(11) + h(4)x(8)) + (h(2)x(10) + h(5)x(7))$$

Example 8.7 Let the nonzero values of the input signal and the impulse response of the filter, with the starting index zero, be, respectively,

$$x(n) = \{2, 1, 3, 4, -1, 2, 4, 3, 1, 3, 2, 4\} \quad \text{and} \quad h(n) = \{1, -1, 4, 3, 2\}$$

Find the every third, $M = 3$, convolution output of $x(n)$ and $h(n)$ directly and by the polyphase approach.

Solution The complete output $y(n)$ of convolving $h(n)$ and $x(n)$ is

$$y(n) = x(n) * h(n) = \{\mathbf{2}, -1, 10, \mathbf{11}, 14, 30, \mathbf{16}, 12, 18, \mathbf{30}, 20, 23, \mathbf{15}, 28, 16, \mathbf{8}\}$$

The decimated output, with $M = 3$, shown in boldface, is

$$\{2, 11, 16, 30, 15, 8\}$$

By grouping impulse response values, we get

$$h1(n) = \{1, 3\}, \quad h2(n) = \{4, 0\}, \quad h3(n) = \{-1, 2\}$$

The input data is grouped as

$$x1(n) = \{2, 4, 4, 3, 0\}, \quad x2(n) = \{0, 1, -1, 3, 2\}, \quad x3(n) = \{0, 3, 2, 1, 4\}$$

The initial value 0 in the second group is $x(-1)$. The initial value 0 in the third group is $x(-2)$.

The outputs of convolving the respective groups of $h(n)$ and $x(n)$ are, respectively,

$$\{1, 3\} * \{2, 4, 4, 3, 0\} = \{2, 10, 16, 15, 9, 0, \}$$

$$\{4, 0\} * \{0, 1, -1, 3, 2\} = \{0, 4, -4, 12, 8, 0\}$$

$$\{-1, 2\} * \{0, 3, 2, 1, 4\} = \{0, -3, 4, 3, -2, 8\}$$

The output is

$$\{2, 10, 16, 15, 9, 0\} + \{0, 4, -4, 12, 8, 0\} + \{0, -3, 4, 3, -2, 8\} = \{2, 11, 16, 30, 15, 8\}$$

$M = 4$

Example 8.8 Let the nonzero values of the input signal and the impulse response of the filter, with the starting index zero, be, respectively,

$$x(n) = \{2, 1, 3, 4, -1, 2, 4, 3, 1, 3, 2, 4\} \quad \text{and} \quad h(n) = \{1, -1, 4, 3, 2, 1, 3, 2, 1\}$$

Find the every fourth, $M = 4$, convolution output of $x(n)$ and $h(n)$ directly and by the polyphase approach.

Solution The complete output $y(n)$ of convolving $h(n)$ and $x(n)$ is

$$y(n) = x(n) * h(n)$$
$$= \{\mathbf{2} - 1, 10, 11, \mathbf{14}, 32, 23, 22, \mathbf{35}, 48, 30, 35, \mathbf{33}, 48, 32, 24, \mathbf{17}, 19, 10, 4\}$$

The decimated output, with $M = 4$, shown in boldface, is

$$\{2, 14, 35, 33, 17\}$$

By grouping impulse response values, we get

$$h1(n) = \{1, 2, 1\}, \quad h2(n) = \{3, 2\}, \quad h3(n) = \{4, 3\}, \quad h4(n) = \{-1, 1\}$$

The input data is grouped as

$$x1(n) = \{2, -1, 1, 0\}, \quad x2(n) = \{0, 1, 2, 3\},$$
$$x3(n) = \{0, 3, 4, 2\}, \quad x4(n) = \{0, 4, 3, 4\}$$

The initial value 0 in the second group is $x(-1)$. The initial value 0 in the third group is $x(-2)$. The initial value 0 in the fourth group is $x(-3)$.

The outputs of convolving the respective groups of $h(n)$ and $x(n)$ are, respectively,

$$\{1, 2, 1\} * \{2, -1, 1, 0\} = \{2, 3, 1, 1, 1, 0\}$$

$$\{3, 2\} * \{0, 1, 2, 3\} = \{0, 3, 8, 13, 6, 0\}$$

$$\{4, 3\} * \{0, 3, 4, 2\} = \{0, 12, 25, 20, 6, 0\}$$

$$\{-1, 1\} * \{0, 4, 3, 4\} = \{0, -4, 1, -1, 4, 0\}$$

The output is

$$\{2, 3, 1, 1, 1, 0\} + \{0, 3, 8, 13, 6, 0\} + \{0, 12, 25, 20, 6, 0\} + \{0, -4, 1, -1, 4, 0\}$$
$$= \{2, 14, 35, 33, 17\}$$

8.3.2 Polyphase Implementation of the Interpolator

The basic definition of interpolation operation is that the input signal is upsampled first by a factor L and then the resulting image frequency components are filtered out to get the interpolated signal. The problem is that the filter works at a higher sampling rate than necessary in the straightforward implementation. In the polyphase implementation, the filters work at a lower rate. Let the input signal be

$$x(n) = \{x(0), x(1), x(2), \dots, \}$$

and it has to be interpolated by a factor of $L = 2$. By upsampling with $L = 2$, we get

$$x_u(n) = \{x(0), 0, x(1), 0, x(2), 0, \dots, \}$$

Let the FIR filter coefficients be

$$h(n) = \{h(0), h(1), h(2), h(3)\}$$

Then, the convolution of $x_u(n)$ and $h(n)$ yields typical outputs such as

$$y(2) = h(0)x(1) + h(1)0 \quad + h(2)x(0) + h(3)0$$
$$y(3) = h(0)0 \quad + h(1)x(1) + h(2)0 \quad + h(3)x(0)$$
$$y(4) = h(0)x(2) + h(1)0 \quad + h(2)x(1) + h(3)0$$
$$y(5) = h(0)0 \quad + h(1)x(2) + h(2)0 \quad + h(3)x(1)$$

Dropping the terms with zero inputs, we get

$$y(2) = h(0)x(1) + h(2)x(0)$$
$$y(3) = h(1)x(1) + h(3)x(0)$$
$$y(4) = h(0)x(2) + h(2)x(1)$$
$$y(5) = h(1)x(2) + h(3)x(1)$$

The even-indexed output samples are computed by convolving the input $x(n)$ with the subfilter consisting of the even-indexed values of the filter coefficients. The odd-indexed output samples are computed by convolving the input $x(n)$ with the subfilter consisting of the odd-indexed values of the filter coefficients. The block diagram of the polyphase implementation of the interpolation operation is shown in Fig. 8.13 with $L = 2$, which is just the transpose of the flowgraph of that of the decimation. The two subfilters constitute the first part. In the second part, the interleaving of the two partial outputs is carried out to get the interpolated signal. At each cycle, two outputs are produced. By upsampling and delaying the bottom subfilter output, the proper interleaved output is produced.

Example 8.9 Let the input signal and the impulse response of the filter be, respectively,

$$x(n) = \{2, 1, 3, 4\} \quad \text{and} \quad h(n) = \{1, 2, 1, 3\}$$

The interpolation factor is $L = 2$ and $x_u(n) = \{2, 0, 1, 0, 3, 0, 4, 0\}$. Find the convolution of $x_u(n)$ and $h(n)$ using the direct and polyphase method.

Solution The upsampled input is

$$x_u(n) = \{2, 0, 1, 0, 3, 0, 4, 0\}$$

Fig. 8.13 Polyphase implementation of an interpolation filter with $L = 2$

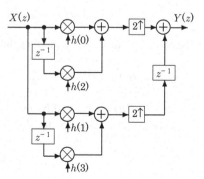

The convolution output of convolving $h(n)$ and $x_u(n)$ is

$$x_u(n) * h(n) = \{2, 4, 3, 8, 4, 9, 7, 17, 4, 12\}$$

The subfilters are

$$h1 = \{1, 1\} \quad \text{and} \quad h2 = \{2, 3\}$$

The convolution of $x(n)$ with the subfilters yields

$$y1 = \{2, 3, 4, 7, 4\} \quad \text{and} \quad y2 = \{4, 8, 9, 17, 12\}$$

The output $y(n)$ is obtained by interleaving the two partial outputs, which is the same that obtained by the direct method.

Example 8.10 Let the input signal and the impulse response of the filter be, respectively,

$$x(n) = \{2, 1, 3, 4\} \quad \text{and} \quad h(n) = \{1, -1, 4, 3, 2\}$$

The interpolation factor is $L = 3$ and $x_u(n) = \{2, 0, 0, 1, 0, 0, 3, 0, 0, 4, 0, 0\}$. Find the convolution of $x_u(n)$ and $h(n)$ using the direct and polyphase method.

Solution The upsampled input is

$$x_u(n) = \{2, 0, 0, 1, 0, 0, 3, 0, 0, 4, 0, 0\}$$

The convolution output of convolving $h(n)$ and $x_u(n)$ is

$$x_u(n) * h(n) = \{2, -2, 8, 7, 3, 4, 6, -1, 12, 13, 2, 16, 12, 8, 0\}$$

The subfilters are

$$h1 = \{1, 3\} \quad \text{and} \quad h2 = \{-1, 2\} \quad \text{and} \quad h3 = \{4, 0\}$$

The convolution of $x(n)$ with the subfilters yields

$$y1 = \{2, 7, 6, 13, 12\} \quad \text{and} \quad y2 = \{-2, 3, -1, 2, 8\} \quad \text{and} \quad y3 = \{8, 4, 12, 16, 0\}$$

The output $y(n)$ is obtained by interleaving the three sets of partial outputs, which is the same that obtained by the direct method.

8.4   Discrete Wavelet Transform

A major application of multirate signal processing is the Discrete Wavelet Transform (DWT). Fourier analysis is indispensable for signal and system analysis and design in science and engineering. Fourier analysis represents a signal in terms of sinusoidal basis functions. After using the Fourier analysis for a long time, it was found that, for certain applications, it is necessary to represent the signal in the time-frequency domain, rather than exclusively in the frequency domain. A modified version of the Fourier transform, called the short-time Fourier transform, was designed for this purpose. The length of the window limits the frequency resolution. A refined version of this transform is the wavelet transform. A set of basis functions was found by Haar before the use of the short-time Fourier transform itself. Several families of basis functions are used in the DWT. For this brief presentation, we use the Haar basis functions only, as it is the simplest and also practically useful.

In signal processing, signal representation to suit the required processing is the first step, as practical signals have arbitrary amplitude profile and it is difficult to process as such. The two most often used basis signals for representation are the impulse signal in the time domain and the sinusoidal signals in the frequency domain. The impulse signal is usually defined as the limit of a unit-area rectangular pulse, as its width tends to zero. If the pulse width is limited to a finite value, then its spectrum also becomes finite approximately. Therefore, any arbitrary signal can be represented by a set of suitable pulses. The pulses may be positive or negative and of appropriate height and width. Although pulses of other shapes are also used, we confine ourselves to the use of rectangular pulses for its simplicity to introduce the DWT. The basis functions of the DWT are called wavelets (short waves). Because of this, a small change affects only a few coefficients. A short pulse can be represented using few coefficients. The DWT representation of a signal also indicates the time of occurrence (time-frequency representation) of an event. Further, a signal can be represented into several resolutions, called multiresolution representation.

Fourier analysis

Time-domain $\leftrightarrow$ Frequency-domain

DWT

Time-domain $\leftrightarrow$ Time-frequency-domain

As the basis signals of the DWT are of finite nature, the correlation of any part of the signal with the basis signal can be determined by shifting the basis signal. In addition, the signal spectra are partitioned into unequal parts, thereby matching the features of various lengths of the signal to match with their spectral bandwidth requirements. These two features characterize the DWT and provide its advantages over other transforms in signal analysis. Mostly, the DWT is presented in three forms: (1) filter bank, (2) matrix formulation, and (3) polyphase matrix factorization.

The matrix DWT formulation is similar to that of the DFT. The representation of a signal is obtained by multiplying its samples by the transform matrix. The output vector is a function of time and frequency. Inherently, the DWT is suited to the analysis of nonstationary signals, whose spectrum changes at intervals during its duration.

8.4.1 Haar DWT

The 2-point 1-level DWT of the time-domain samples$\{x(0), x(1)\}$ is the same as that of the 2-point DFT.

$$\begin{bmatrix} X_\phi(0,0) \\ X_\psi(0,0) \end{bmatrix} = \frac{1}{\sqrt{2}} \begin{bmatrix} 1 & 1 \\ 1 & -1 \end{bmatrix} \begin{bmatrix} x(0) \\ x(1) \end{bmatrix} \quad \text{or} \quad X = W_{2,0} x \qquad (8.9)$$

where X, $W_{2,0}$, and x represent, respectively, the coefficient, transform, and input matrices, as in the case of the DFT. The subscript 2 indicates the input data length and 0 indicates the scale of decomposition. It is assumed that the length of the input sequence $x(n)$ is a power of 2. The letter ϕ in $X_\phi(j_0, k)$ indicates that it is an approximation coefficient obtained by some averaging of the input signal. The letter ψ in $X_\psi(j, k)$ indicates that it is a detail coefficient obtained by some differencing of the input signal. Haar DWT basis functions, with $N = 2$, are shown in Fig. 8.14a, b. As can be seen, scaled sums and differences of adjacent pairs of input values are formed and the sums form the top half of the output and the differences form the bottom half in computing the Haar DWT.

The 2-point 1-level inverse DWT (IDWT) of the coefficients $\{X_\phi(0, 0), X_\psi(0, 0)\}$ is defined as

$$\begin{bmatrix} x(0) \\ x(1) \end{bmatrix} = \frac{1}{\sqrt{2}} \begin{bmatrix} 1 & 1 \\ 1 & -1 \end{bmatrix} \begin{bmatrix} X_\phi(0,0) \\ X_\psi(0,0) \end{bmatrix} \quad \text{or} \quad x = W_{2,0}^{-1} X \qquad (8.10)$$

In computing the Haar IDWT, scaled sums and differences, of the values taken from the top half and the bottom half of the input, are formed and stored consecutively as adjacent pairs in the output.

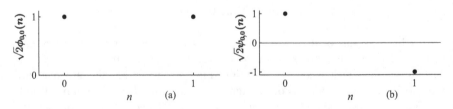

Fig. 8.14 Haar DWT basis functions with $N = 2$. (a) $\sqrt{2}\phi_{0,0}(n)$; (b) $\sqrt{2}\psi_{0,0}(n)$

The inverse transform matrix $W_{2,0}^{-1} = W_{2,0}^T$ is the inverse as well as the transpose of the forward transform matrix $W_{2,0}$. The forward and inverse transform matrices are real and orthogonal. That is, their product is an identity matrix.

Example 8.11 Using the Haar transform matrix, find the DWT of $x(n)$. Verify that $x(n)$ is reconstructed by computing the IDWT. Verify Parseval's theorem.

$$\{x(0) = 2, x(1) = -3\}$$

Solution

$$\begin{bmatrix} X_\phi(0,0) \\ X_\psi(0,0) \end{bmatrix} = \frac{1}{\sqrt{2}} \begin{bmatrix} 1 & 1 \\ 1 & -1 \end{bmatrix} \begin{bmatrix} 2 \\ -3 \end{bmatrix} = \begin{bmatrix} -\frac{1}{\sqrt{2}} \\ \frac{5}{\sqrt{2}} \end{bmatrix}$$

As in common with other transforms, the DWT output for an input is a set of coefficients that express the input in terms of the basis functions. Each basis function, during its existence, contributes to the value of the time-domain signal at each sample point. Therefore, multiplying each basis function by the corresponding DWT coefficient and summing the products get back the time-domain signal.

$$\left(\frac{1}{\sqrt{2}} \quad \frac{1}{\sqrt{2}} \right) \left(-\frac{1}{\sqrt{2}} \right) +$$
$$\left(\frac{1}{\sqrt{2}} \quad -\frac{1}{\sqrt{2}} \right) \quad \left(\frac{5}{\sqrt{2}} \right) =$$
$$\overline{\quad 2 \quad} \quad \overline{\quad -3 \quad}$$

Formally, the IDWT gets back the original input samples.

$$\begin{bmatrix} x(0) \\ x(1) \end{bmatrix} = \frac{1}{\sqrt{2}} \begin{bmatrix} 1 & 1 \\ 1 & -1 \end{bmatrix} \begin{bmatrix} -\frac{1}{\sqrt{2}} \\ \frac{5}{\sqrt{2}} \end{bmatrix} = \begin{bmatrix} 2 \\ -3 \end{bmatrix}$$

Parseval's theorem states that the sum of the squared-magnitude of a time-domain sequence (energy) equals the sum of the squared-magnitude of the corresponding transform coefficients. For the example sequence,

$$2^2 + (-3)^2 = 13 = \left(-\frac{1}{\sqrt{2}} \right)^2 + \left(\frac{5}{\sqrt{2}} \right)^2$$

∎

8.4.2 Two-Channel Filter Bank

The computation done in the last example can be implemented using the two-channel filter bank, shown in Fig. 8.15 in the frequency domain. There are two types of filter banks: (1) analysis filter banks and (2) synthesis filter banks. An analysis filter bank separates the signal into subband signals with the spectrum of each subband signal is a nonoverlapping portion of the spectrum of the input signal. In the synthesis filter bank, a set of subband signals are combined into a single signal. The DFT decomposes a N-point signal into N subband signals and the IDFT combines the N subband signals into one N-point signal. In other applications, the spectrum of the subband signals consists of a set of consecutive frequencies of the signal's spectrum, rather than one in the case of the DFT. The analysis and synthesis filter banks are located to the left and right of the dotted line in the figure. The reasons for the requirement of filter banks are that the low and high pass filters are not individually invertible, as their frequency response contains zero. In combination, inversion is possible despite the zeros. In order to invert a FIR filter with another FIR filter, the filter bank structure is required. In addition, practical filters are nonideal, there is aliasing in each filter bank and amplitude and phase distortion. The filter bank structure reconstructs the input signal with a constant gain and a constant delay, if its decomposed components are unaltered. For this to happen, the filters have to be designed carefully. Of course, the polyphase form of filters can be used for efficiency of implementation. There is lot of theory behind all these aspects, as presented in the specified reference at the end of the book.

Coming back to the numerical example presented earlier, let the lowpass filter impulse response be $\{h_l(0) = 1, h_l(1) = 1\}$. Let the highpass filter impulse response be $\{h_h(0) = 1, h_h(1) = -1\}$. We got the DWT coefficients by point-by-point multiplication of the data and the coefficients. To obtain the same output by convolution, we have to time-reverse one of the sequences. Convolving the input data $\{x(0) = 2, h(1) = -3\}$ with these sequences, the respective outputs are $\{x_l(-1) = 2, x_l(0) = -1, x_l(1) = -3\}$ and $\{x_h(-1) = -2, x_h(0) = 5, x_h(1) = -3\}$. The decimated output of these sequences by a factor of 2, starting with the convolution output with index 0 and scaling approximately, we get the DWT coefficients

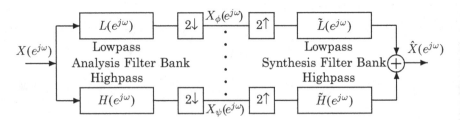

Fig. 8.15 Two-channel analysis and synthesis filter banks in the frequency domain

$$\left\{-\frac{1}{\sqrt{2}} \quad \text{and} \quad \frac{5}{\sqrt{2}},\right\}$$

which are the same as obtained using the matrix formulation earlier.

The impulse responses of the synthesis bank filters $\tilde{l}(n)$, (frequency response $\tilde{L}(e^{j\omega})$) analysis bank filters and $\tilde{h}(n)$ (frequency response $\tilde{H}(e^{j\omega})$) are related to those of the analysis bank filters.

$$\tilde{L}(e^{j\omega}) = H(e^{j(\omega+\pi)}) \quad \text{and} \quad \tilde{H}(e^{j\omega}) = -L(e^{j(\omega+\pi)})$$

Using these relations, the impulse response of the synthesis bank filter impulse responses are,

$$\tilde{l} = \{1, 1\}/\sqrt{2} \quad \text{and} \quad \tilde{h} = \{1, -1\}/\sqrt{2}$$

Upsampling the DWT coefficients, we get upsampled signal

$$\left\{-\frac{1}{\sqrt{2}}, 0\right\} \quad \text{and} \quad \left\{\frac{5}{\sqrt{2}}, 0\right\}$$

The sum of the output of the lowpass and highpass channels of the synthesis filter bank constitutes the reconstructed input $\hat{x}(n)$, which is the same as the input, as no processing was done on the input data.

The Haar DWT definition can be extended to any data length that is an integer power of 2. For example, the 1-level (scale 1) 4-point Haar DWT is

$$\begin{bmatrix} X_\phi(1,0) \\ X_\phi(1,1) \\ X_\psi(1,0) \\ X_\psi(1,1) \end{bmatrix} = \frac{1}{\sqrt{2}} \begin{bmatrix} 1 & 1 & 0 & 0 \\ 0 & 0 & 1 & 1 \\ 1 & -1 & 0 & 0 \\ 0 & 0 & 1 & -1 \end{bmatrix} \begin{bmatrix} x(0) \\ x(1) \\ x(2) \\ x(3) \end{bmatrix} \quad \text{or} \quad X = W_{4,1}x$$

The inverse DWT, with $W_{4,1}^{-1} = W_{4,1}^T$, is

$$\begin{bmatrix} x(0) \\ x(1) \\ x(2) \\ x(3) \end{bmatrix} = \frac{1}{\sqrt{2}} \begin{bmatrix} 1 & 0 & 1 & 0 \\ 1 & 0 & -1 & 0 \\ 0 & 1 & 0 & 1 \\ 0 & 1 & 0 & -1 \end{bmatrix} \begin{bmatrix} X_\phi(1,0) \\ X_\phi(1,1) \\ X_\psi(1,0) \\ X_\psi(1,1) \end{bmatrix} \quad \text{or} \quad x = W_{4,1}^{-1}X$$

Discontinuity Detection

Due to the finite and shifting ability, the DWT is inherently suitable for applications such as compression, denoising, and discontinuity detection. Consider the sinusoidal signal $x(n)$ with 32 samples shown in Fig. 8.16a. It has a discontinuity at the 23rd sample. The 1-level Haar DWT coefficients are shown in Fig. 8.16b, the 16 approximation coefficients first and the 16 detail coefficients second. The

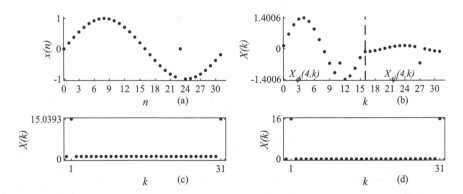

Fig. 8.16 (a) Input signal $x(n)$; (b) its DWT coefficients; (c) the DFT spectrum of $x(n)$; (d) the DFT of the input signal without the discontinuity

discontinuity is detected and its location also clearly marked. The 23rd sample in (a) corresponds to the 11th and $(11 + 16) = 27$th sample in (b). The DFT spectrum of the signal with and without the discontinuity are, shown in (c) and (d). While the two spectra are different, the location of the discontinuity cannot be detected due to the global nature of the DFT basis functions. Further, a discontinuity affects only the DWT coefficients in the neighborhood of the discontinuity, rather than all the coefficients as in the case of the DFT.

In this section, we have just presented what the DWT is in its simplest form. However, there is a lot of theory of the DWT, which requires a course in DWT to learn.

8.5   Summary

- A system that uses different sampling rates is called a multirate system.
- In addition to adders, multipliers, and delay units used in implementing single sampling rate digital systems, two more basic components, the upsampler and downsampler, are required in the implementation of multirate systems.
- In the frequency domain, the decimation operation retains only the lower frequency components of the input signal in the range $|\omega| \leq \pi/M$, where M is an integer, called the decimation factor. While a highpass filter removes the high frequency components, a downsampler reduces the sampling rate.
- In the time domain, the output of the downsampler retains every Mth sample, starting from the index $n = 0$, and discarding the rest. That is, $x_d = x(nM)$.
- Downsampling compresses the given signal in the time domain and expands its spectrum in the frequency domain.
- To change the sampling rate by a rational fraction, a cascade of an upsampler and a downsampler has to be used, with the condition that the factors must be relatively prime.

- A signal that is filtered first by a filter with cutoff frequency $f_s/(2M)$ and then downsampled by a factor of M is called a decimated signal by a factor of M.
- The sampling rate of a signal $x(n)$ with sampling rate f_s can be increased by a factor of L by inserting $L-1$ zeros, called upsampling, between successive values of the signal.
- The effect of zero-insertion is that the spectrum of $x_u(n)$ is L-fold periodic replication of that of $x(n)$.
- The basic definition of interpolation operation is that the input signal is upsampled first by a factor L and then the resulting image frequency components are filtered out to get the interpolated signal.
- The polyphase filter structures reduce the sampling rate of the filters in decimation and interpolation operations.
- It was found that, for certain applications, it is necessary to represent the signal in the time-frequency domain, rather than exclusively in the frequency domain, as in the Fourier analysis.
- The basis functions of the DWT are called wavelets (short waves). Because of this, a small change affects only a few coefficients. A short pulse can be represented using few coefficients. The DWT representation of a signal also indicates the time of occurrence (time-frequency representation) of an event. Further, a signal can be represented into several resolutions, called multiresolution representation.

Exercises

8.1 Given the nonzero samples of $x(n)$, find its DTFT $X(e^{j\omega})$. Express the DTFT $X_d(e^{j\omega})$ of $x(Mn)$ in terms of $X(e^{j\omega})$ using the formula

$$X_d(e^{j\omega}) = \frac{1}{M} \sum_{k=0}^{M-1} X(e^{j(\frac{\omega-2\pi k}{M})})$$

Verify that the inverse of $X_d(e^{j\omega})$ is the same as $x(Mn)$. Approximate the DTFT and inverse DTFT, respectively, by DFT and IDFT.

8.1.1

$$x(n) = \{x(0) = -1, x(1) = 4, x(2) = 3, x(3) = -3\}, \quad M = 1$$

8.1.2

$$x(n) = \{x(0) = -1, x(1) = 4, x(2) = 3, x(3) = -3\}, \quad M = 2$$

* **8.1.3**

$$x(n) = \{x(0) = -1, x(1) = 4, x(2) = 3, x(3) = -3\}, \quad M = 3$$

8.2 Given the samples over one period of a periodic signal $x(n)$ starting with index 0, find its DTFT $X(e^{j\omega})$. Filter $X(e^{j\omega})$ by an ideal lowpass filter with cutoff frequency π/M. Find the inverse of the filtered spectrum and downsample by a factor of M to get the decimated version $x_{dec}(n)$ of $x(n)$. Approximate the DTFT and inverse DTFT, respectively, by DFT and IDFT.

* 8.2.1

$$x(n) = \{2, 0.8660, -0.5, 0, 0.5, -0.8660, -2, -0.8660, 0.5, 0,$$
$$-0.5, 0.8660\}, \quad M = 4$$

8.2.2

$$x(n) = \{2, 1, 0, 1, 2, 1, 0, 1, 2, 1, 0, 1\}, \quad M = 4$$

8.2.3

$$x(n) = \{2, -0.1340, 1.5, -1, 0.5, -1.8660, 0, -1.8660,$$
$$0.5, -1, 1.5, -0.1340\}, \quad M = 2$$

8.3 Given the nonzero samples of $x(n)$, find its DTFT $X(e^{j\omega})$. Express the DTFT $X_u(e^{j\omega})$ of $x(n/L)$ in terms of $X(e^{j\omega})$ using the formula

$$X_u(e^{j\omega}) = X(e^{jL\omega}), \quad 0 < \omega < 2\pi$$

Verify that the inverse of $X_u(e^{j\omega})$ is the same as $x(n/L)$. Approximate the DTFT and inverse DTFT, respectively, by DFT and IDFT.

8.3.1

$$x(n) = \{2, 1, 3, 4\}, \quad L = 2$$

* 8.3.2

$$x(n) = \{2, 1, 3, 4\}, \quad L = 3$$

8.3.3

$$x(n) = \{2, 3\}, \quad L = 4$$

8.4 Given the samples over one period of a periodic signal $x(n)$ starting with index 0, find its DTFT $X(e^{j\omega})$. Express the DTFT $X_u(e^{j\omega})$ of $x(n/L)$ in terms of $X(e^{j\omega})$ using the formula

$$X_u(e^{j\omega}) = X(e^{jL\omega}), \quad 0 < \omega < 2\pi$$

Filter $X_u(e^{j\omega})$ by an ideal lowpass filter with cutoff frequency π/L. Find the inverse of the filtered spectrum multiplied by L to get the interpolated version $x_i(n)$ of $x(n)$. Approximate the DTFT and inverse DTFT, respectively, by DFT and IDFT.

8.4.1

$$x(n) = \{1, 0, -1, 0\}, \quad L = 2$$

* 8.4.2

$$x(n) = \{0, 1, 0, -1\}, \quad L = 3$$

8.5 Find the convolution of $x(n)$ and $h(n)$ in the time domain: (i) directly and (2) convolving $x(n)$ with $h_e(n)$ and $h_o(n)$.

8.5.1

$$x(0) = 1, x(1) = 0, x(2) = 3, x(3) = 0$$

$$h(0) = 3, h(1) = -2, h(2) = 1$$

8.5.2

$$x(0) = 1, x(1) = 0, x(2) = 1, x(3) = 0$$

$$h(0) = 3, h(1) = -2, h(2) = 1$$

8.5.3

$$x(0) = 4, x(1) = 0, x(2) = 1, x(3) = 0$$

$$h(0) = 3, h(1) = -2, h(2) = 1$$

8.6 Given a periodic signal $x(n)$, find its sampling rate converted version $y(n)$ by a factor L/M using DFT and IDFT.

8.6.1

$$x(n) = \cos\left(\frac{2\pi}{4}n\right), \quad L = 3, \quad M = 2$$

$$x(n) = \{1, 0, -1, 0\}$$

* 8.6.2

$$x(n) = \cos\left(\frac{2\pi}{6}n\right), \quad L = 2, \quad M = 3$$

8.7 Let the nonzero values of the input signal and the impulse response of the filter, with the starting index zero, be, respectively,

$$x(n) = \{3, 1, 2, 4, -3, 2, 1, 1, 2, 3, 2, 4\} \quad \text{and} \quad h(n) = \{2, -1, 3, 1, 2\}$$

Find the every second, $M = 2$, convolution output of $x(n)$ and $h(n)$ directly and by the polyphase approach.

* 8.8 Let the input signal and the impulse response of the filter be, respectively,

$$x(n) = \{2, 1, -3, 4\} \quad \text{and} \quad h(n) = \{2, 2, 3, 1\}$$

The interpolation factor is $L = 2$ and $x_u(n) = \{2, 0, 1, 0, -3, 0, 4, 0\}$. Find the convolution of $x_u(n)$ and $h(n)$ using the direct and polyphase method.

8.9 Using the Haar transform matrix, find the DWT of $x(n)$. Verify that $x(n)$ is reconstructed by computing the IDWT. Verify Parseval's theorem.

$$\{x(0) = 1, x(1) = 2\}$$

* 8.10 Using the Haar transform matrix, find the 1-level DWT of $x(n)$. Verify that $x(n)$ is reconstructed by computing the IDWT. Verify Parseval's theorem.

$$\{x(0) = 2, x(1) = 1, x(2) = 3, x(3) = 4\}$$

Fast Computation of the DFT

9

It is the availability of fast algorithms to compute the DFT that makes Fourier analysis indispensable in practical applications, in addition to its theoretical importance from its invention. In turn, the other versions of the Fourier analysis can be approximated adequately by the DFT. Although the algorithm was invented by Gauss in 1805, it is the widespread use of the digital systems that has given its importance in practical applications. The algorithm is based on the classical divide-and-conquer strategy of developing fast algorithms. A problem is divided into two smaller problems of half the size. Each smaller problem is solved separately, and the solution to the original problem is found by appropriately combining the solutions of the smaller problems. This process is continued recursively until the smaller problems reduce to trivial cases. Therefore, the DFT is never computed using its definition. While there are many variations of the algorithm, the order of computational complexity of all of them is the same in practice, $O(N \log_2 N)$ to compute a N-point DFT against $O(N^2)$ from its definition. It is this reduction in computational complexity by an order that has resulted in the widespread use of the Fourier analysis in practical applications of science and engineering. In this chapter, a particular variation of the algorithm, called the PM DFT algorithm, is presented. The algorithm is developed using the half-wave symmetry of periodic waveforms. This approach gives a better viewpoint of the algorithm than other approaches such as matrix factorization. The DFT is defined for any length. However, the practically most useful DFT algorithms are of length that is an integral power of 2. That is $N = 2^M$, where M is a positive integer. If necessary, zero padding can be employed to the sequences so that they satisfy this constraint.

Adapted from my book Fourier Analysis—A Signal Processing Approach, Springer, 2018 by permission.

9.1   Half-Wave Symmetry of Periodic Waveforms

Any practical signal can be decomposed into a set of sinusoidal components of even- and odd-indexed frequencies. The algorithm for fast computation of the DFT, which is of prime importance in practical applications, is based on separating the signal components into two sets: one set containing the odd-indexed frequency components and the other containing the even-indexed frequency components. A N-point of DFT becomes two $N/2$-point DFTs. This process is recursively continued until the subproblems become 1-point DFTs. A physical understanding of separating the even and odd frequency components can be obtained using the half-wave symmetry of periodic waveforms.

If a given periodic function $x(n)$ with period N satisfies the condition

$$x\left(n + \frac{N}{2}\right) = x(n),$$

then it is said to be even half-wave symmetric. The samples of the function over any half-period are the same as those in the succeeding half-period. In effect, the period is $N/2$. If the DFT is computed over the period N, then the odd-indexed DFT coefficients will be zero. That is, the function is composed of even-indexed frequency components only. If a given periodic function $x(n)$ with period N satisfies the condition

$$x\left(n + \frac{N}{2}\right) = -x(n),$$

then it is said to be odd half-wave symmetric. The samples of the function over any half-period are the negatives of those in the succeeding half-period. If the DFT is computed over the period N, then the even-indexed DFT coefficients will be zero. That is, the function is composed of odd-indexed frequency components only. It is due to the fact that any periodic function can be uniquely decomposed into even half-wave and odd half-wave symmetric components. If the even half-wave symmetric component is composed of the even-indexed frequency components, then the odd half-wave symmetric component must be composed of the odd-indexed frequency components. Therefore, if an arbitrary function is decomposed into its even half-wave and odd half-wave symmetric components, then we have divided the original problem of finding the N frequency coefficients into two problems, each of them being the determination of $N/2$ frequency coefficients. First, let us go through an example of decomposing a periodic function into its even half-wave and odd half-wave symmetric components.

An arbitrary periodic sequence $x(n)$ of period N can be expressed as the sum of its even and odd half-wave symmetric components $x_{eh}(n)$ and $x_{oh}(n)$, respectively, as

$$x(n) = x_{eh}(n) + x_{oh}(n) \tag{9.1}$$

in which

$$x_{eh}\left(n + \frac{N}{2}\right) = x_{eh}(n) \quad \text{and} \quad x_{oh}\left(n + \frac{N}{2}\right) = -x_{oh}(n)$$

Adding the two components, we get

$$x\left(n + \frac{N}{2}\right) = x_{eh}(n) - x_{oh}(n) \tag{9.2}$$

Solving for the components using Eqs. (9.1) and (9.2), we get

$$x_{eh}(n) = \frac{1}{2}\left(x(n) + x\left(n + \frac{N}{2}\right)\right) \quad \text{and} \quad x_{oh}(n) = \frac{1}{2}\left(x(n) - x\left(n + \frac{N}{2}\right)\right)$$

Figure 9.1a shows a periodic waveform of period $N = 16$

$$x(n) = \sin\left(\frac{2\pi}{16}n\right) + \cos\left(2\frac{2\pi}{16}n\right)$$

The first is the odd-indexed frequency component and the second is the even-indexed. Figure 9.1b shows the two components of the waveform $x(n)$. The odd-indexed component is shown by dots and the even-indexed component is shown by crosses. The samples of $x(n)$ are

$$\{1, 1.0898, 0.7071, 0.2168, 0, 0.2168, 0.7071, 1.0898,$$

$$1, 0.3244, -0.7071, -1.6310, -2, -1.6310, -0.7071, 0.3244\}$$

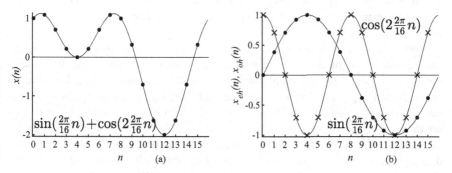

Fig. 9.1 (a) $x(n) = \sin\left(\frac{2\pi}{16}n\right) + \cos\left(2\frac{2\pi}{16}n\right)$; (b) The odd-indexed frequency component (dots) $\sin\left(\frac{2\pi}{16}n\right)$ and the even-indexed frequency component $\cos\left(2\frac{2\pi}{16}n\right)$ (crosses)

The samples of $x_{eh}(n)$ are

$$\{1, 0.7071, 0, -0.7071, -1, -0.7071, 0, 0.7071,$$
$$1, 0.7071, 0, -0.7071, -1, -0.7071, 0, 0.7071\}$$

The samples of $x_{oh}(n)$ are

$$\{0, 0.3827, 0.7071, 0.9239, 1, 0.9239, 0.7071, 0.3827,$$
$$0, -0.3827, -0.7071, -0.9239, -1, -0.9239, -0.7071, -0.3827\}$$

It can be verified that $x(n) = x_{eh}(n) + x_{oh}(n)$.
The DFT $X(k)$ of $x(n)$ is

$$\{0, -j8, 8, 0, \quad 0, 0, 0, 0, \quad 0, 0, 0, 0, \quad 0, 0, 8, j8\}$$

The DFT $X_{eh}(k)$ of $x_{eh}(n)$ is

$$\{0, 0, 8, 0, \quad 0, 0, 0, 0, \quad 0, 0, 0, 0, \quad 0, 0, 8, 0\}$$

The DFT $X_{oh}(k)$ of $x_{oh}(n)$ is

$$\{0, -j8, 0, 0, \quad 0, 0, 0, 0, \quad 0, 0, 0, 0, \quad 0, 0, 0, j8\}$$

It is obvious that $X(k) = X_{eh}(k) + X_{oh}(k)$. Due to the reduction in the number of frequency components, $x_{eh}(n)$ and $x_{oh}(n)$ can be represented by 8 samples. About the time index 8 in the middle of Fig. 9.1b, we can see the redundancy of data values of $x_{eh}(n)$ and $x_{oh}(n)$. The problem of computing the 16-point DFT of $x(n)$ has been reduced to two problems of 8-point DFTs of $x_{eh}(n)$ and $x_{oh}(n)$. This is the essence of all the DFT algorithms for sequence lengths that are an integral power of 2. While the sinusoidal waveforms are easy to visualize, as usual, the DFT problem using the equivalent complex exponential form is required for further description and practical use of the algorithms.

Recursively carrying out the even and odd half-wave symmetric components decomposition, along with frequency shifting and decimation of the frequency components, the frequency components are eventually isolated yielding their coefficients. In carrying out these operations, the coefficients are scaled and scrambled but remain unchanged otherwise. The decomposition of a waveform into its symmetric components is the principal operation in the algorithm, requiring repeated execution of add–subtract (plus–minus) operations. Since the frequency components are decomposed into smaller groups, this type of algorithms is named as the PM DIF DFT algorithms, where PM stands for plus–minus and DIF stands for decimation-in-frequency.

9.2 The PM DIF DFT Algorithm

A given waveform $x(n)$ can be expressed as the sum of N frequency components (e.g., with $N = 8$),

$$x(n) = X(0)e^{j0\frac{2\pi}{8}n} + X(1)e^{j1\frac{2\pi}{8}n} + X(2)e^{j2\frac{2\pi}{8}n} + X(3)e^{j3\frac{2\pi}{8}n}$$

$$+ X(4)e^{j4\frac{2\pi}{8}n} + X(5)e^{j5\frac{2\pi}{8}n} + X(6)e^{j6\frac{2\pi}{8}n} + X(7)e^{j7\frac{2\pi}{8}n}, \ n = 0, 1, \dots, 7$$

For the most compact algorithms, the input sequence $x(n)$ has to be expressed as 2-element vectors

$$\mathbf{a^0}(n) = \{a_0^0(n), a_1^0(n)\} = 2\{x_{eh}(n), x_{oh}(n), \ n = 0, 1, \dots, \frac{N}{2} - 1\}$$

The first and second elements of the vectors are, respectively, the scaled even and odd half-wave symmetric components $x_{eh}(n)$ and $x_{oh}(n)$ of $x(n)$. That is, the DFT expression is reformulated as

$$X(k) = \sum_{n=0}^{N-1} x(n)e^{-j\frac{2\pi}{N}kn}, \ k = 0, 1, \dots, N - 1$$

$$= \begin{cases} \displaystyle\sum_{n=0}^{(N/2)-1} \left(x(n) + x\left(n + \frac{N}{2}\right) \right) e^{-j\frac{2\pi}{N}kn} = \sum_{n=0}^{(N/2)-1} a_0^0(n)e^{-j\frac{2\pi}{N}kn}, \ k \text{ even} \\ \displaystyle\sum_{n=0}^{(N/2)-1} \left(x(n) - x\left(n + \frac{N}{2}\right) \right) e^{-j\frac{2\pi}{N}kn} = \sum_{n=0}^{(N/2)-1} a_1^0(n)e^{-j\frac{2\pi}{N}kn}, \ k \text{ odd} \end{cases}$$

The division by 2 in finding the symmetric components is deferred. For the example shown in Fig. 9.1,

$$2x_{eh}(n) = a_0^0(n) = 2\{1, 0.7071, 0, -0.7071, -1, -0.7071, 0, 0.7071\}$$

$$X(k) = \sum_{n=0}^{7} a_0^0(n)e^{-j\frac{2\pi}{16}kn}, \ k = 0, 2, 4, 6, 8, 10, 12, 14$$

$$= \sum_{n=0}^{7} a_0^0(n)e^{-j\frac{2\pi}{8}kn}, \ k = 0, 1, 2, 3, 4, 5, 6, 7 = \{0, 8, 0, 0, \ 0, 0, 0, 8\}$$

For the example shown in Fig. 9.1,

$$2x_{oh}(n) = a_1^0(n) = \{0, 0.3827, 0.7071, 0.9239, 1, 0.9239, 0.7071, 0.3827\}$$

$$X(k) = \sum_{n=0}^{7} a_1^0(n) e^{-j\frac{2\pi}{16}kn}, \quad k = 1, 3, 5, 7, 9, 11, 13, 15,$$

$$= \sum_{n=0}^{7} (e^{-j\frac{2\pi}{16}n} a_1^0(n)) e^{-j\frac{2\pi}{8}kn}, \quad k = 0, 1, 2, 3, 4, 5, 6, 7$$

$$= \{0, -j8, 0, 0, \ 0, 0, 0, j8\}$$

In order to reduce this computation to an 8-point DFT, we have to multiply the samples of $a_1^0(n)$ by the twiddle factor samples

$$\{e^{-j\frac{2\pi}{16}n} \ n = 0, 1, \ldots, 7\} = \{1, 0.9239 - j0.3827, 0.7071 - j0.7071,$$

$$0.3827 - j0.9239, \ j1, \ -0.3827 - j0.9239, \ -0.7071 - j0.7071, \ -0.9239 - j0.3827\}$$

This step results in frequency shifting of the DFT spectrum, and the odd-indexed coefficients become even-indexed. Therefore, the 16-point DFT becomes an 8-point DFT. The pointwise product of $e^{-j\frac{2\pi}{16}n}$ and $a_1^0(n)$ yields

$$\{0, 0.7071 - j0.2929, 1 - j1, 0.7071 - j1.7071, j2, -0.7071 - j1.7071,$$

$$-1 - j1, -0.7071 - j0.2929)$$

The 8-point DFT yields

$$\{-j8, 0, 0, 0, 0, 0, 0, j8\}$$

The recursive decomposition is continued until all the individual DFT coefficients are extracted, as shown in Fig. 9.3 for $N = 8$.

In terms of the frequency components,

$$\boldsymbol{a}^0(n) = \{a_0^0(n), a_1^0(n)\} = 2\{x_{eh}(n), x_{oh}(n)\}$$

$$= 2\left\{X(0)e^{j0\frac{2\pi}{8}n} + X(2)e^{j2\frac{2\pi}{8}n} + X(4)e^{j4\frac{2\pi}{8}n} + X(6)e^{j6\frac{2\pi}{8}n},\right.$$

$$\left. X(1)e^{j1\frac{2\pi}{8}n} + X(3)e^{j3\frac{2\pi}{8}n} + X(5)e^{j5\frac{2\pi}{8}n} + X(7)e^{j7\frac{2\pi}{8}n}\right\}, \quad n = 0, 1, 2, 3$$

The array of vectors $\boldsymbol{a}^0(n)$ is stored in the nodes at the beginning of the signal-flow graph of the algorithm shown in Fig. 9.2. Although a DFT algorithm can be expressed in other forms, the signal-flow graph is the most suitable form for its description. The repetitive nature of the basic computation is evident. The nodes, shown by discs, store a vector $\boldsymbol{a}^r(n)$. Arrows indicate the signal-flow path. The first elements of vectors stored in a pair of source nodes produce the vector for the sink node connected by an upward pointing arrow by add–subtract operation. Any integer

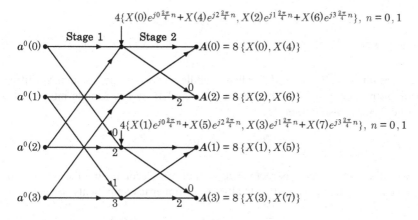

Fig. 9.2 The signal-flow graph of the PM DIF DFT algorithm, with $N = 8$. A twiddle factor $W_8^n = e^{-j\frac{2\pi}{8}n}$ is indicated only by its variable part of the exponent, n

value n near an arrow indicates that the value from a source node has to be multiplied by $e^{-j\frac{2\pi}{N}n}$ before the add–subtract operation. The second elements of vectors stored in a pair of source nodes produce the vector for the sink node connected by a downward pointing arrow. There are $\log_2 N - 1$ stages of the algorithm. The input nodes are source nodes and the output nodes are sink nodes. Rest of the nodes serve as both source and sink nodes.

Each of the two symmetric components has only four frequency components, and four samples are adequate to represent them. The even half-wave symmetric component is

$$2(X(0)e^{j0\frac{2\pi}{8}n} + X(2)e^{j2\frac{2\pi}{8}n} + X(4)e^{j4\frac{2\pi}{8}n} + X(6)e^{j6\frac{2\pi}{8}n})$$

$$= 2(X(0)e^{j0\frac{2\pi}{4}n} + X(2)e^{j1\frac{2\pi}{4}n} + X(4)e^{j2\frac{2\pi}{4}n} + X(6)e^{j3\frac{2\pi}{4}n}), \quad n = 0, 1, 2, 3$$

The set of frequency coefficients of this component is

$$2\{X(0), X(2), X(4), X(6)\}$$

The set of frequency coefficients of the odd half-wave symmetric component is

$$2\{X(1), X(3), X(5), X(7)\}$$

To reformulate this component as a 4-point DFT, it has to be multiplied by the exponential $e^{-j\frac{2\pi}{8}n}$ to get

$$2(X(1)e^{j1\frac{2\pi}{8}n} + X(3)e^{j3\frac{2\pi}{8}n} + X(5)e^{j5\frac{2\pi}{8}n} + X(7)e^{j7\frac{2\pi}{8}n})e^{-j\frac{2\pi}{8}n}$$

$$= 2(X(1)e^{j0\frac{2\pi}{8}n} + X(3)e^{j2\frac{2\pi}{8}n} + X(5)e^{j4\frac{2\pi}{8}n} + X(7)e^{j6\frac{2\pi}{8}n})$$

$$= 2(X(1)e^{j0\frac{2\pi}{4}n} + X(3)e^{j1\frac{2\pi}{4}n} + X(5)e^{j2\frac{2\pi}{4}n} + X(7)e^{j3\frac{2\pi}{4}n}), \; n = 0, 1, 2, 3$$

The multiplication by $e^{-j\frac{2\pi}{8}n}$ (called twiddle factor) is the frequency shifting operation necessary to shift the spectrum to the left by one sample interval. The twiddle factor is also written as

$$e^{-j\frac{2\pi}{8}n} = W_8^n, \quad W_8 = e^{-j\frac{2\pi}{8}}$$

Now, the same process is recursively continued. Decomposing the two 4-point waveforms into their even and odd half-wave symmetric components, we get

$$4\left\{X(0)e^{j0\frac{2\pi}{4}n} + X(4)e^{j2\frac{2\pi}{4}n}, X(2)e^{j1\frac{2\pi}{4}n} + X(6)e^{j3\frac{2\pi}{4}n}\right\}, \; n = 0, 1 \qquad (9.3)$$

$$4\left\{X(1)e^{j0\frac{2\pi}{4}n} + X(5)e^{j2\frac{2\pi}{4}n}, X(3)e^{j1\frac{2\pi}{4}n} + X(7)e^{j3\frac{2\pi}{4}n}\right\}, \; n = 0, 1 \qquad (9.4)$$

These vector arrays are stored in the middle nodes in the figure.

The even half-wave symmetric component of the waveform given by Eq. (9.3) can be expressed as a 2-point DFT

$$4(X(0)e^{j0\frac{2\pi}{4}n} + X(4)e^{j2\frac{2\pi}{4}n})$$

$$= 4(X(0)e^{j0\frac{2\pi}{2}n} + X(4)e^{j1\frac{2\pi}{2}n}), \; n = 0, 1$$

with frequency coefficients $4\{X(0), X(4)\}$. The coefficients

$$A(0) = \{A_0(0), A_1(0)\} = 8\{X(0), X(4)\}$$

are obtained by simply adding and subtracting the two sample values. These coefficients are stored in the top node at the end of the signal-flow graph in the figure.

The odd half-wave symmetric component of the waveform given by Eq. (9.3) is multiplied by the exponential $e^{-j\frac{2\pi}{8}(2n)} = e^{-j\frac{2\pi}{4}n}$ to get

$$4(X(2)e^{j1\frac{2\pi}{4}n} + X(6)e^{j3\frac{2\pi}{4}n})e^{-j\frac{2\pi}{4}n}$$

$$= 4(X(2)e^{j0\frac{2\pi}{4}n} + X(6)e^{j2\frac{2\pi}{4}n})$$

$$= 4(X(2)e^{j0\frac{2\pi}{2}n} + X(6)e^{j1\frac{2\pi}{2}n}), \; n = 0, 1$$

This is a 2-point DFT with frequency coefficients $4\{X(2), X(6)\}$. The coefficients

$$A(2) = \{A_0(2), A_1(2)\} = 8\{X(2), X(6)\}$$

are obtained by simply adding and subtracting the two sample values. These coefficients are stored in the second node from top at the end of the signal-flow graph in the figure.

The even half-wave symmetric component of the waveform defined by Eq. (9.4) can be expressed as

$$4(X(1)e^{j0\frac{2\pi}{4}n} + X(5)e^{j2\frac{2\pi}{4}n})$$

$$= 4(X(1)e^{j0\frac{2\pi}{2}n} + X(5)e^{j1\frac{2\pi}{2}n}), \ n = 0, 1$$

This is a 2-point DFT with frequency coefficients $4\{X(1), X(5)\}$. The coefficients

$$A(1) = \{A_0(1), A_1(1)\} = 8\{X(1), X(5)\}$$

are obtained by simply adding and subtracting the two sample values. These coefficients are stored in the third node from top at the end of the signal-flow graph shown in the figure.

The odd half-wave symmetric component of the waveform defined by Eq. (9.4) is multiplied by the exponential $e^{-j\frac{2\pi}{4}n}$ to get

$$4(X(3)e^{j1\frac{2\pi}{4}n} + X(7)e^{j3\frac{2\pi}{4}n})e^{-j\frac{2\pi}{4}(n)}$$

$$= 4(X(3)e^{j0\frac{2\pi}{4}n} + X(7)e^{j2\frac{2\pi}{4}n})$$

$$= 4(X(3)e^{j0\frac{2\pi}{2}n} + X(7)e^{j1\frac{2\pi}{2}n}), \ n = 0, 1$$

This is a 2-point DFT with frequency coefficients $4\{X(3), X(7)\}$. The coefficients

$$A(3) = \{A_0(3), A_1(3)\} = 8\{X(3), X(7)\}$$

are obtained by simply adding and subtracting the two sample values. These coefficients are stored in the fourth node from top at the end of the signal-flow graph shown in the figure.

The output vectors $\{A(0), A(1), A(2), A(3)\}$ are placed in the bit-reversed order. This order is obtained by reversing the order of bits of the binary number representation of the frequency indices. $\{0, 1, 2, 3\}$ is $\{00, 01, 10, 11\}$ in binary form. The bit-reversed $\{00, 10, 01, 11\}$ in binary form is $\{0, 2, 1, 3\}$ in decimal form. The bit-reversed order occurs at the output because of the repeated splitting of the frequency components into odd- and even-indexed frequency groups over the stages of the algorithm. Efficient algorithms are available to restore the natural order of the coefficients. In digital signal processing microprocessors, specialized instructions are available to carry out this task. The extraction of the coefficients, multiplied by 8, of $x(n) = \sin(\frac{2\pi}{8}n) + \cos(2\frac{2\pi}{8}n)$ is shown in Fig. 9.3. The input sequence $x(n)$ is

Input values stored in vector locations	Vector formation	Stage 1 output	Stage 2 output
$x(0) = 1$ $x(4) = 1$	$a_0(0) = 2$ $a_1(0) = 0$	0 4	$X(0) = A_0(0) = 0$ $X(4) = A_1(0) = 0$
$x(1) = \frac{1}{\sqrt{2}}$ $x(5) = -\frac{1}{\sqrt{2}}$	$a_0(2) = 0$ $a_1(2) = \sqrt{2}$	0 0	$X(2) = A_0(2) = 4$ $X(6) = A_1(2) = 4$
$x(2) = 0$ $x(6) = -2$	$a_0(1) = -2$ $a_1(1) = 2$	$-j2$ $j2$	$X(1) = A_0(1) = -j4$ $X(5) = A_1(1) = 0$
$x(3) = \frac{1}{\sqrt{2}}$ $x(7) = -\frac{1}{\sqrt{2}}$	$a_0(3) = 0$ $a_1(3) = \sqrt{2}$	$-j2$ 2	$X(3) = A_0(3) = 0$ $X(7) = A_1(3) = j4$

Fig. 9.3 The trace of the PM DIF DFT algorithm, with $N = 8$, in extracting the coefficients, scaled by 8, of $x(n) = \sin\left(\frac{2\pi}{8}n\right) + \cos\left(2\frac{2\pi}{8}n\right)$.

$$x(n) = \{1, 0.7071, 0, 0.7071, 1, -0.7071, -2, -0.7071\}$$

In the vector formation stage, we find the sum and difference of the two numbers stored in each group. For example, the first vector values are $1 \pm 1 = \{2, 0\}$. As given in the flow graph of the algorithm shown in Fig. 9.2, we find the stage 1 values using the values in the preceding stage.

$$(a_0(0) = 2) \pm (a_0(2) = -2) = \{0, 4\}$$

forms the first vector.

$$(a_0(1) = 0) \pm (a_0(3) = 0) = \{0, 0\}$$

forms the second vector.

$$a_1(0) = 0 \pm ((-j)a_1(2) = (-j)2) = \{-j2, j2\}$$

forms the third vector.

$$a_1(1) = \left(\frac{1}{\sqrt{2}} - j\frac{1}{\sqrt{2}}\right)\sqrt{2} \pm \left(a_1(3) = \left(-\frac{1}{\sqrt{2}} - j\frac{1}{\sqrt{2}}\right)(\sqrt{2})\right) = \{-j2, 2\}$$

forms the fourth vector.

As given in the flow graph of the algorithm shown in Fig. 9.2, we find the stage 2 values using the values in the preceding stage. Using the first two vectors, we get

$$0 \pm 0 = \{0, 0\} = \{X(0), X(4)\}$$

forms the first vector,

$$4 \pm 0 = \{4, 4\} = \{X(2), X(6)\}$$

forms the second vector,

$$-j2 \pm -j2 = \{-j4, 0\} = \{X(1), X(5)\}$$

forms the third vector, and

$$j2 \pm (-j)2 = \{0, j4\} = \{X(3), X(7)\}$$

forms the fourth vector. Similar to the operation of the binary search algorithm, the computation of the coefficients takes the correct path in the flow graph. The DFT of the input sequence $x(n)$ is

$$X(k) = \{0, -j4, 4, 0, 0, 0, 4, j4\}$$

Simple checks can be made on the accuracies of the DFT computation. The sum of the samples of $x(n)$ must be equal to $X(0)$. The sum of the samples of $(-1)^n x(n)$ must be equal to $X(4)$. The sum of the samples of $X(k)/8$ must be equal to $x(0)$. The sum of the samples of $(-1)^k X(k)/8$ must be equal to $x(4)$.

The number of complex multiplications and additions required for each stage are, respectively, $N/2$ and N, where N is the sequence length that is a power of 2. With $(\log_2 N) - 1$ stages and the initial vector formation requiring N complex additions, the computational complexity of the algorithm is $O(N \log_2 N)$ compared with that of $O(N^2)$ required for the direct computation from the DFT definition. Multiplication by twiddle factors of the forms $-j$ and $(1 - j1)/\sqrt{2}$ can be handled separately reducing the number of operations. If further speedup is required, two adjacent stages of the algorithm can be implemented together. This reduces the number of data transfers between the processor registers and the memory yielding significant reduction in the execution time of the algorithm. In case the number of stages is odd, one stage can be implemented separately.

The algorithm is so regular that one can easily get the signal-flow graph for any value of N that is an integral power of 2. The signal-flow graph of the algorithm is basically an interconnection of butterflies (a computational structure), shown in Fig. 9.4. The defining equations of a butterfly at the rth stage are given by

$$a_0^{(r+1)}(h) = a_0^{(r)}(h) + a_0^{(r)}(l)$$

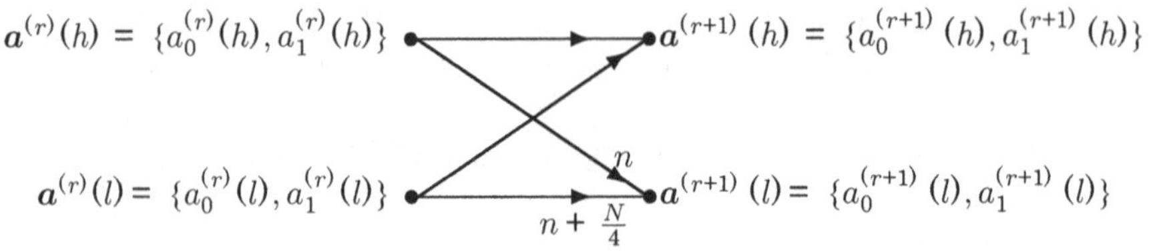

$$a^{(r)}(h) = \{a_0^{(r)}(h), a_1^{(r)}(h)\} \qquad a^{(r+1)}(h) = \{a_0^{(r+1)}(h), a_1^{(r+1)}(h)\}$$

$$a^{(r)}(l) = \{a_0^{(r)}(l), a_1^{(r)}(l)\} \qquad a^{(r+1)}(l) = \{a_0^{(r+1)}(l), a_1^{(r+1)}(l)\}$$

$$n$$
$$n + \frac{N}{4}$$

Fig. 9.4 The signal-flow graph of the butterfly of the PM DIF DFT algorithm, where $0 \le n < \frac{N}{4}$. A twiddle factor $W_N^n = e^{-j\frac{2\pi}{N}n}$ is indicated only by its variable part of the exponent, n

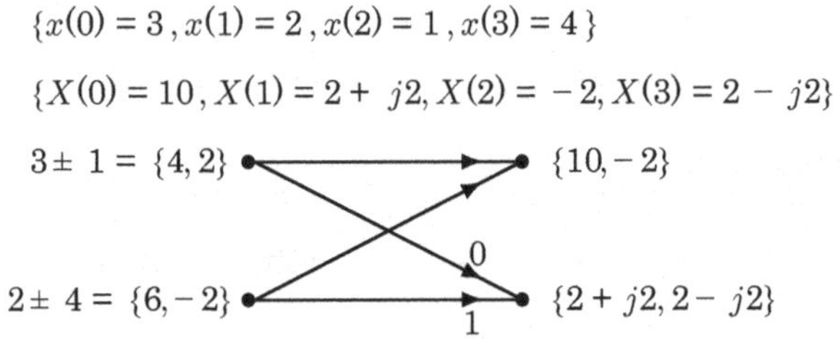

$$\{x(0) = 3\,, x(1) = 2\,, x(2) = 1\,, x(3) = 4\,\}$$

$$\{X(0) = 10, X(1) = 2 + j2, X(2) = -2, X(3) = 2 - j2\}$$

$$3 \pm 1 = \{4, 2\} \qquad \{10, -2\}$$

$$0$$

$$2 \pm 4 = \{6, -2\} \qquad \{2 + j2, 2 - j2\}$$
$$1$$

Fig. 9.5 The computation of a 4-point DFT by the PM DIF DFT algorithm. The twiddle factor $W_4^1 = e^{-j\frac{2\pi}{4}1}$ is indicated only by its exponent, 1

$$a_1^{(r+1)}(h) = a_0^{(r)}(h) - a_0^{(r)}(l)$$

$$a_0^{(r+1)}(l) = W_N^n a_1^{(r)}(h) + W_N^{n+\frac{N}{4}} a_1^{(r)}(l)$$

$$a_1^{(r+1)}(l) = W_N^n a_1^{(r)}(h) - W_N^{n+\frac{N}{4}} a_1^{(r)}(l),$$

where $W_N^n = e^{-j\frac{2\pi}{N}n}$. There are $(\log_2 N) - 1$ stages, each with $N/4$ butterflies. With $N = 8$, therefore, we see four butterflies in Fig. 9.2.

The computation of a 4-point DFT by the PM DIF DFT algorithm is shown in Fig. 9.5. The input is

$$x(n) = \{3, 2, 1, 4\}$$

The input vectors at the 2 nodes are, respectively,

$$3 \pm 1 = \{4, 2\} \quad \text{and} \quad 2 \pm 4 = \{6, -2\}$$

The output vectors at the 2 nodes are, respectively,

$$4 \pm 6 = \{10, -2\} \quad \text{and} \quad 2 \pm (-j)(-2) = \{2 + j2, 2 - j2\}$$

   The computation of the IDFT can be carried out by a similar algorithm, with the twiddle factors conjugated. Further, division by N is required. However, the DFT algorithm itself can be used to carry out the IDFT computation with the interchange of the real and imaginary parts of the input and output. At the end, division by N is required. Another method is to conjugate the input, compute its DFT, and conjugate the resulting output.

   The computation of a 4-point IDFT by the PM DIF DFT algorithm is as follows. The input $X(k)$, from the last example, and its conjugate are

$$X(k) = \{10, 2 + j2, -2, 2 - j2\} \quad \text{and} \quad X^*(k) = \{10, 2 - j2, -2, 2 + j2\}$$

The input vectors at the 2 nodes of the PM DIF DFT algorithm are, respectively,

$$10 \pm (-2) = \{8, 12\} \quad \text{and} \quad (2 - j2) \pm (2 + j2) = \{4, -j4\}$$

The output vectors at the 2 nodes are, respectively,

$$8 \pm 4 = \{12, 4\} \quad \text{and} \quad 12 \pm (-j)(-j4) = \{8, 16\}$$

Dividing by 4, we get

$$x(n) = \{3, 2, 1, 4\}$$

By interchanging the real and imaginary parts, we get

$$X(k) = \{j10, 2 + j2, -j2, -2 + j2\}$$

The input vectors at the 2 nodes of the PM DIF DFT algorithm are, respectively,

$$j10 \pm (-j2) = \{j8, j12\} \quad \text{and} \quad (2 + j2) \pm (-2 + j2) = \{j4, 4\}$$

The output vectors at the 2 nodes are, respectively,

$$j8 \pm j4 = \{j12, j4\} \quad \text{and} \quad j12 \pm (-j)(4) = \{j8, j16\}$$

By interchanging the real and imaginary parts and dividing by 4, we get

$$x(n) = \{3, 2, 1, 4\}$$

   The extraction of the coefficients, multiplied by 8, of the periodic waveform of period $N = 8$

Input values	Vector formation	Stage 1 output	Stage 2 output
$x(0)=1.5482$ $x(4)=-1.2802$	$a_0(0)=0.2679$ $a_1(0)=2.8284$	4 -3.4641	$X(0)=-8$ $X(4)=16$
$x(1)=-0.7929$ $x(5)=-6.2071$	$a_0(1)=-7.000$ $a_1(1)=5.4142$	-12 -2	$X(2)=-3.4641+j2$ $X(6)=-3.4641-j2$
$x(2)=2.2802$ $x(6)=1.4518$	$a_0(2)=3.7321$ $a_1(2)=0.8284$	$2.8284 - j0.8284$ $2.8284 + j0.8284$	$X(1)=5.6569-j5.6569$ $X(5)=j4$
$x(3)=-1.7929$ $x(7)=-3.2071$	$a_0(3)=-5.000$ $a_1(3)=1.4142$	$2.8284 - j4.8284$ $4.8284 - j2.8284$	$X(3)=-j4$ $X(7)=5.6569+j5.6569$

Fig. 9.6 The trace of the PM DIF DFT algorithm, with $N = 8$, in extracting the coefficients, scaled by 8, of $x(n)$

$$x(n) = -1 + 2\sin\left(\frac{2\pi}{8}n + \frac{\pi}{4}\right) - \cos\left(2\frac{2\pi}{8}n - \frac{\pi}{6}\right) + \sin\left(3\frac{2\pi}{8}n\right)$$

$$+ 2\cos\left(4\frac{2\pi}{8}n\right), \ n = 0, 1, \ldots, 7$$

is shown in Fig. 9.6. The 8 samples of $x(n)$ are shown in the first column. The vectors are shown in the second column. After 2 stages of processing as given in the flow graph in Fig. 9.2, the output transform values are shown in the last column. Since the sinusoidal components of the input waveform are given, we can find the transform values analytically. The DFT of the DC components (-1), with $N = 8$, is $X(0) = (-1)8 = -8$. The DFT of the frequency component with frequency index 4, with $N = 8$, is $X(4) = 2(8) = 16$. For the frequency component with frequency index 3, $X(3) = -j4$ and $X(5) = j4$. For the frequency component with frequency index 1, the phase shift becomes $-\pi/4$, if we express it in cosine form and

$$X(1) = 2(4)\left(\cos(\frac{\pi}{4}) - j\sin\left(\frac{\pi}{4}\right)\right) = 5.6569 - j5.6569 \text{ and } X(7)$$
$$= X^*(1) = 5.6569 + j5.6569$$

For the frequency component with frequency index 2,

Input values stored in vector locations	Vector formation	Stage 1 output	Stage 2 output
$x(0) = 3 + j1$ $x(4) = 4 + j2$	$a_0(0) = 7 + j3$ $a_1(0) = -1 - j1$	10 $4 + j6$	$X(0) = 10+j12$ $X(4) = 10-j12$
$x(1) = 1 + j2$ $x(5) = 1 + j4$	$a_0(2) = 2 + j6$ $a_1(2) = -j2$	12 4	$X(2) = 4+j2$ $X(6) = 4+j10$
$x(2) = 2 - j1$ $x(6) = 1 - j2$	$a_0(1) = 3 - j3$ $a_1(1) = 1 + j1$	$-j2$ -2	$X(1)=-1.4142-j3.4142$ $X(5)=1.4142-j0.5858$
$x(3) = -1 + j3$ $x(7) = -1 + j3$	$a_0(3) = -2 + j6$ $a_1(3) = 0 + j0$	$-1.4142-j1.4142$ $-1.4142-j1.4142$	$X(3)=-3.4142+j1.4142$ $X(7)=-0.5858-j1.4142$

Fig. 9.7 The trace of the PM DIF DFT algorithm, with $N = 8$, in extracting the coefficients, scaled by 8, of a complex-valued signal $x(n)$.

$$X(2) = -(4) \left(\cos(\frac{\pi}{6}) - j \sin \left(\frac{\pi}{6} \right) \right) = -3.4641 + j2 \text{ and } X(6)$$

$$= X^*(2) = -3.4641 - j2$$

While complex-valued signals do not appear naturally, it is a necessity in applications such as complex analysis, DFT algorithms, Hilbert transform, convolution of long real-valued signals (given later), etc. The extraction of the coefficients, multiplied by 8, of the periodic waveform of period $N = 8$

$$x(n) = \{3 + j1, 1 + j2, 2 - j1, -1 + j3, 4 + j2, 1 + j4, 1 - j2, -1 + j3\}$$

is shown in Fig. 9.7.

9.3   The PM DIT DFT Algorithm

We have given the physical explanation of the decomposition of waveforms in the DIF DFT algorithm. In a decimation-in-frequency (DIF) algorithm, the transform sequence, $X(k)$, is successively divided into smaller subsequences. For example, in the beginning of the first stage, the computation of a N-point DFT is decomposed into two problems: (1) computing the $(N/2)$ even-indexed $X(k)$ and (2) computing the $(N/2)$ odd-indexed $X(k)$. In a decimation-in-time (DIT) algorithm, the data sequence, $x(n)$, is successively divided into smaller subsequences. For example, in the beginning of the last stage, the computation of a N-point DFT is decomposed into two problems: (1) computing the $(N/2)$-point DFT of even-indexed $x(n)$ and (2) computing the $(N/2)$-point DFT of odd-indexed $x(n)$. The DIT DFT algorithm is based on zero padding, time shifting, and spectral redundancy. For understanding, the DIF DFT algorithms are easier. However, the DIT algorithms are used more often, as taking care of the data scrambling problem occurring at the beginning of the algorithm is relatively easier to deal with. The DIT DFT algorithms can be considered as the algorithms obtained by transposing the signal-flow graph of the corresponding DIF algorithms, that is by reversing the direction (signal flow) of all the arrows and interchanging the input and the output.

9.3.1 Basics of the PM DIT DFT Algorithm

The DIT algorithm is based on decomposing the data sequence recursively into smaller sequences. Consider the exponential $x(n) = e^{j\frac{2\pi}{8}n}$. The sample values over one period are

$$\left\{1, \frac{1}{\sqrt{2}} + j\frac{1}{\sqrt{2}}, j1, -\frac{1}{\sqrt{2}} + j\frac{1}{\sqrt{2}}, -1, -\frac{1}{\sqrt{2}} - j\frac{1}{\sqrt{2}}, -j1, \frac{1}{\sqrt{2}} - j\frac{1}{\sqrt{2}}\right\}$$

The samples of $x(n)$ can be expressed as the sum of the upsampled, by a factor of 2, even-indexed and odd-indexed components

$$\{1, 0, j1, 0, -1, 0, -j1, 0\} + \left\{0, \frac{1}{\sqrt{2}} + j\frac{1}{\sqrt{2}}, 0, -\frac{1}{\sqrt{2}} + j\frac{1}{\sqrt{2}}, 0, \right.$$

$$\left. -\frac{1}{\sqrt{2}} - j\frac{1}{\sqrt{2}}, 0, \frac{1}{\sqrt{2}} - j\frac{1}{\sqrt{2}}\right\}$$

The DFT of the even-indexed elements of $x(n)$ is

$$x_e(n) = \{1, j1, -1, -j1\} \leftrightarrow X_e(k) = \{0, 4, 0, 0\}$$

The DFT of the odd-indexed elements of $x(n)$ is

$$x_o(n) = \left\{\frac{1}{\sqrt{2}} + j\frac{1}{\sqrt{2}}, -\frac{1}{\sqrt{2}} + j\frac{1}{\sqrt{2}}, -\frac{1}{\sqrt{2}} - j\frac{1}{\sqrt{2}}, \frac{1}{\sqrt{2}} - j\frac{1}{\sqrt{2}}\right\}$$

$$\leftrightarrow X_o(k) = \{0, 2\sqrt{2} + j2\sqrt{2}, 0, 0\}$$

Due to the upsampling theorem, the DFTs of the upsampled sequences are the twofold repetition of $X_e(k)$ and $X_o(k)$, and we get

$$\{1, 0, j1, 0, -1, 0, -j1, 0\} \leftrightarrow \{0, 4, 0, 0, 0, 4, 0, 0\}$$

$$\left\{\frac{1}{\sqrt{2}} + j\frac{1}{\sqrt{2}}, 0. -\frac{1}{\sqrt{2}} + j\frac{1}{\sqrt{2}}, 0, -\frac{1}{\sqrt{2}} - j\frac{1}{\sqrt{2}}, 0, \frac{1}{\sqrt{2}} - j\frac{1}{\sqrt{2}}, 0\right\}$$

$$\leftrightarrow \{0, 2\sqrt{2} + j2\sqrt{2}, 0, 0, 0, 2\sqrt{2} + j2\sqrt{2}, 0, 0\}$$

Using the time-shift theorem, we get the DFT of upsampled and shifted $x_o(n)$ as

$$\left\{0, \frac{1}{\sqrt{2}} + j\frac{1}{\sqrt{2}}, 0, -\frac{1}{\sqrt{2}} + j\frac{1}{\sqrt{2}}, 0, -\frac{1}{\sqrt{2}} - j\frac{1}{\sqrt{2}}, 0, \frac{1}{\sqrt{2}} - j\frac{1}{\sqrt{2}}\right\}$$

$$\leftrightarrow \{0, 2\sqrt{2} + j2\sqrt{2}, 0, 0, 0, 2\sqrt{2} + j2\sqrt{2}, 0, 0\}e^{-j\frac{2\pi}{8}k}$$

$$= \{0, 4, 0, 0, 0, -4, 0, 0\}$$

Adding the two partial DFTs, we get the DFT of $x(n)$ as

$$X(k) = \{0, 4, 0, 0, 0, 4, 0, 0\} + \{0, 4, 0, 0, 0, -4, 0, 0\} = \{0, 8, 0, 0, 0, 0, 0, 0\}$$

The decomposition continues until the sequence lengths become 1, and the DFT of the data is itself. There are $\log_2 N$ stages for a sequence of length N. The computational complexity of each stage is of the order $O(N)$. Therefore, the computational complexity of computing a N-point DFT becomes $O(N \log_2 N)$.

The butterfly and the flow graph of the PM DIT DFT algorithm are the transpose of those of the corresponding DIF algorithms. The PM DIT DFT butterfly is shown in Fig. 9.8. The butterfly input–output relations at the rth stage are

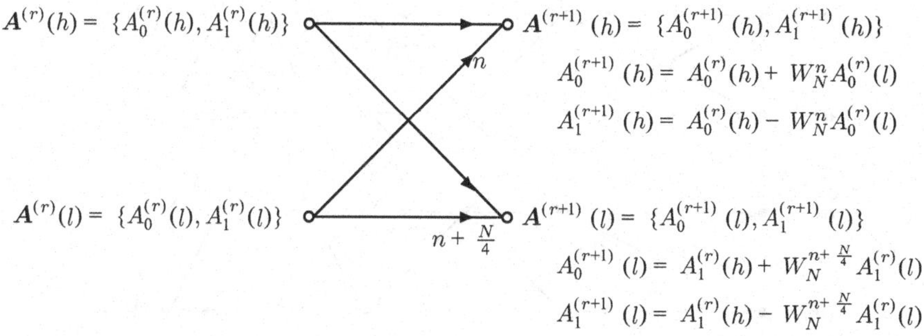

$A^{(r)}(h) = \{A_0^{(r)}(h), A_1^{(r)}(h)\}$ $A^{(r+1)}(h) = \{A_0^{(r+1)}(h), A_1^{(r+1)}(h)\}$

$$A_0^{(r+1)}(h) = A_0^{(r)}(h) + W_N^n A_0^{(r)}(l)$$

$$A_1^{(r+1)}(h) = A_0^{(r)}(h) - W_N^n A_0^{(r)}(l)$$

$A^{(r)}(l) = \{A_0^{(r)}(l), A_1^{(r)}(l)\}$ $A^{(r+1)}(l) = \{A_0^{(r+1)}(l), A_1^{(r+1)}(l)\}$

$$A_0^{(r+1)}(l) = A_1^{(r)}(h) + W_N^{n+\frac{N}{4}} A_1^{(r)}(l)$$

$$A_1^{(r+1)}(l) = A_1^{(r)}(h) - W_N^{n+\frac{N}{4}} A_1^{(r)}(l)$$

Fig. 9.8 The butterfly of the PM DIT DFT algorithm A twiddle factor $W_N^n = e^{-j\frac{2\pi}{N}n}$ is indicated only by its variable part of the exponent, n

$$A_0^{(r+1)}(h) = A_0^{(r)}(h) + W_N^n A_0^{(r)}(l)$$
$$A_1^{(r+1)}(h) = A_0^{(r)}(h) - W_N^n A_0^{(r)}(l)$$
$$A_0^{(r+1)}(l) = A_1^{(r)}(h) + W_N^{n+\frac{N}{4}} A_1^{(r)}(l)$$
$$A_1^{(r+1)}(l) = A_1^{(r)}(h) - W_N^{n+\frac{N}{4}} A_1^{(r)}(l),$$

where n is an integer whose value depends on the stage of computation r and the index h. The letter A is used to differentiate this butterfly from that of the DIF algorithm. The flow graph of the PM DIT DFT algorithm with $N = 16$ is shown in Fig. 9.9. The trace of the PM DIT DFT algorithm in extracting the coefficient scaled by 8 of the input $x(n)$, shown in Fig. 9.6, is shown in Fig. 9.10. The bit-reversal is carried out at the beginning in the DIT algorithm and at the end in the DIF algorithm. Different butterflies are used in the 2 algorithms, but the output is the same for the same input as expected. Both the butterflies and the flow graphs of the algorithms are the flow-graph transposes of each other. The computational complexity of both types of algorithms is the same. One type is more suitable for a particular application. While it is possible to draw the flow graph in other ways,

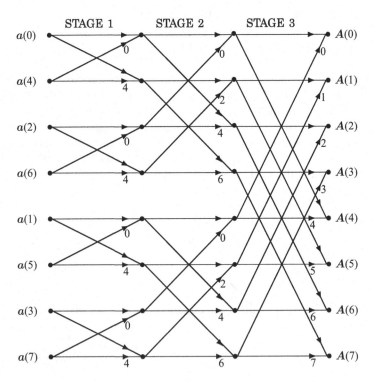

Fig. 9.9 The flow graph of the PM DIT DFT algorithm with $N = 16$. A twiddle factor $W_{16}^n = e^{-j\frac{2\pi}{16}n}$ is indicated only by its variable part of the exponent, n

Input values	Vector formation and swapping	Stage 1 output	Stage 2 output
$x(0)=1.5482$ $x(4)=-1.2802$	$a_0(0)=0.2679$ $a_1(0)=2.8284$	4 -3.4641	$X(0)=-8$ $X(4)=16$
$x(1)=-0.7929$ $x(5)=-6.2071$	$a_0(2)=3.7321$ $a_1(2)=0.8284$	$2.8284-j0.8284$ $2.8284+j0.8284$	$X(1)=5.6569-j5.6569$ $X(5)=j4$
$x(2)=2.2802$ $x(6)=1.4518$	$a_0(1)=-7.000$ $a_1(1)=5.4142$	-12 -2	$X(2)=-3.4641+j2$ $X(6)=-3.4641-j2$
$x(3)=-1.7929$ $x(7)=-3.2071$	$a_0(3)=-5.000$ $a_1(3)=1.4142$	$5.4142-j1.4142$ $5.4142+j1.4142$	$X(3)=-j4$ $X(7)=5.6569+j5.6569$

Fig. 9.10 The trace of the PM DIT DFT algorithm, with $N=8$ in extracting the coefficient scaled by 8 of $x(n)$.

the usual ones are for the DIT DFT algorithm to place the input in the bit-reversed order with the output in the normal order and vice versa for the DIF DFT algorithm.

9.4   Efficient Computation of the DFT of Real Data

It is the reduction of the computational complexity from $O(N^2)$ to $O(N \log_2 N)$ that is most important in most of the applications. However, if it is essential, the computational complexity and the storage requirements can be further reduced by a factor of about 2 for computing the DFT of real data. Two algorithms are commonly used to compute the DFT of real-valued data. One algorithm computes the DFT of a single real-valued data. Another algorithm computes the DFTs of two real-valued data sets simultaneously.

9.4.1   Computing Two DFTs of Real Data Simultaneously

In this approach, we pack two real data sets in a linear manner into one complex data set of the same length. Let $a(n)$ and $b(n)$ be the two real-valued data sets each of length N. Let the respective DFTs be $A(k)$ and $B(k)$. We form the complex data $c(n)$ such that its real and imaginary parts are, respectively, $a(n)$ and $b(n)$. Let the DFT of $c(n)$ is $C(k)$. Then, using the linearity property of the DFT,

$$c(n) = a(n) + jb(n) \leftrightarrow C(k) = A(k) + jB(k)$$

Fig. 9.11 The flowchart of
the DFT algorithm for
computing the DFT of two
real-valued data sets
simultaneously

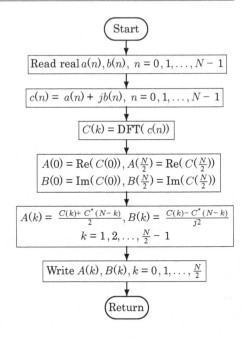

Since the DFT of real-valued data is conjugate-symmetric, $A(N - k) = A^*(k)$ and $B(N - k) = B^*(k)$. Then,

$$C(N - k) = A^*(k) + jB^*(k) \quad \text{and} \quad C^*(N - k) = A(k) - jB(k)$$

Solving for $A(k)$ and $B(k)$ from the last two equations, we get

$$A(k) = \frac{C(k) + C^*(N - k)}{2} \quad \text{and} \quad B(k) = \frac{C(k) - C^*(N - k)}{j2} \tag{9.6}$$

The flowchart of the algorithm is shown in Fig. 9.11. The two real input data sets, each of length N, are read into real arrays $a(n)$ and $b(n)$. The complex data array $c(n) = a(n) + jb(n)$ is formed, where j is the imaginary unit, $\sqrt{-1}$. The DFT of $c(n)$, $C(k)$, is computed. The DFTs $A(k)$ and $B(k)$ of $a(n)$ and $b(n)$ are separated from $C(k)$ using Eq. (9.6). The values $A(0)$, $B(0)$, $A(N/2)$, and $B(N/2)$ are readily taken from $C(k)$, since their values are real. Only half of the values of $A(k)$ and $B(k)$ are computed, since the other half is its complex conjugate.

The IDFTs of two DFTs, $A(k)$ and $B(k)$ of real-valued data $a(n)$ and $b(n)$, can be computed simultaneously. The flowchart of the algorithm is shown in Fig. 9.12. As the DFTs are conjugate-symmetric, only one half of the values of $A(k)$ and $B(k)$ are read. The values $A(0)$, $B(0)$, $A(N/2)$, and $B(N/2)$ form the real imaginary parts of $C(0)$ and $C(N/2)$, since their values are real. The second half of $C(k)$ is found using $C(N - k) = A^*(k) + jB^*(k)$. As the IDFT of $C(k)$ is $c(n) = a(n) + jb(n)$, the real and imaginary parts of $c(n)$ are, respectively, $a(n)$ and $b(n)$.

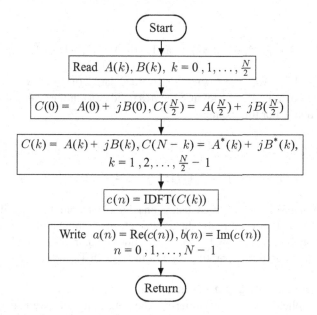

Fig. 9.12 The flowchart of the IDFT algorithm for computing the IDFT of the transform of two real-valued data sets simultaneously

Example 9.1 Compute the DFT of the sequences $a(n) = \{2, 1, 4, 3\}$ and $b(n) = \{1, 2, 2, 3\}$ using a DFT algorithm for complex data of the same length. Verify that the IDFT of the complex data formed gets back $a(n)$ and $b(n)$.

Solution Packing the sequences into a complex data, we get

$$c(n) = a(n) + jb(n) = \{2 + j1, 1 + j2, 4 + j2, 3 + j3\}$$

Taking the DFT of $c(n)$, we get

$$C(k) = \{10 + j8, -3 + j1, 2 - j2, -1 - j3\}$$

Using Eq. (9.6), we get

$$A(0) = \text{Re}(C(0)) = 10, \quad A(2) = \text{Re}(C(2)) = 2$$

$$A(1) = \frac{(-3 + j1) + (-1 + j3)}{2} = -2 + j2, \quad A(3) = A^*(1) = -2 - j2$$

$$B(0) = \text{Im}(C(0)) = 8, \quad B(2) = \text{Im}(C(2)) = -2$$

$$B(1) = \frac{(-3 + j1) - (-1 + j3)}{j2} = -1 + j1, \quad B(3) = B^*(1) = -1 - j1$$

Therefore,

$$A(k) = \{10, -2 + j2, 2, -2 - j2\}, \quad \text{and} \quad B(k) = \{8, -1 + j1, -2, -1 - j1\}$$

Given two DFTs, $A(k)$ and $B(k)$, of real-valued data, $a(n), b(n)$, we form $A(k) + jB(k)$ and compute its IDFT yielding the $a(n)$ and $b(n)$ in the real and imaginary parts, respectively. For this example,

$$A(k) + jB(k) = \{10 + j8, -3 + j1, 2 - j2, -1 - j3\}$$

and its IDFT is $a(n) + jb(n) = \{2 + j1, 1 + j2, 4 + j2, 3 + j3\}$.   ∎

Since two DFTs are computed using one DFT for complex-valued data, for computing the DFT of each real-valued data set, the computational complexity is one-half of that of the algorithm for complex-valued data. In addition, $N-2$ addition operations are required in separating each of the two DFTs.

Linear Filtering of Long Data Sequences
In practical applications of linear filtering, the data sequences are real-valued and often very long compared with the impulse response. Typically, convolution is carried out over sections of the input using methods, such as the overlap-save method. As there are large number of sections, two sections can be packed into a complex data and their transform is computed. The DFT of the complex data is multiplied by the DFT of the impulse response, which is computed only once. The IDFT of the product yields the convolution of the two input sequences in the real and imaginary parts. One advantage of this procedure is that no separation and combination of the DFTs are required.

9.4.2   DFT of a Single Real Data Set

The computation of the DFT of a single real data set is referred as RDFT. The inverse DFT of the transform of real data is referred as RIDFT. Since the DFT of real data is conjugate-symmetric, only one-half of the number of butterflies in each stage of the DFT algorithm for complex-valued data is necessary. The redundant butterflies can be easily eliminated in the DIT algorithm for complex-valued data in deriving the corresponding RDFT algorithm. Similarly, the RIDFT algorithm can be derived from the corresponding DIF algorithm.

The PM DIT RDFT Butterfly
The equations characterizing the input–output relation of the PM DIT RDFT butterfly, shown in Fig. 9.13, are

Fig. 9.13 The butterfly of the PM DIT RDFT algorithm

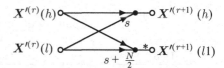

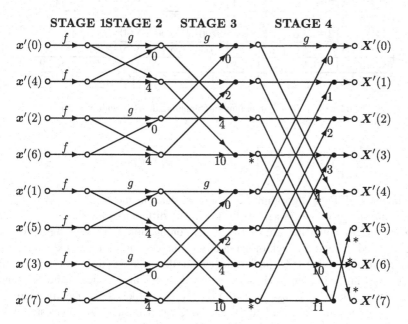

Fig. 9.14 The flow graph of the PM DIT RDFT algorithm with $N = 16$

$$X'^{(r+1)}(h) = X'^{(r)}(h) + W_N^s X'^{(r)}(l)$$
$$X'^{(r+1)}(l1) = (X'^{(r)}(h) - W_N^s X'^{(r)}(l))^*$$

where s is an integer whose value depends on the stage of the computation r and the index h. The butterfly for real data is essentially the same as that for complex data with some differences. Only half of the butterflies are used in each group of butterflies in each stage. Since a butterfly produces one output at the upper half and another output at the lower half of a group of butterflies in each stage, the lower half output has to be conjugated and stored in a memory location in the upper half.

The flow graph of the PM DIT RDFT algorithm with $N = 16$ is shown in Fig. 9.14. The trace of the RDFT algorithm for computing the DFT with $N = 16$ is shown in Fig. 9.15. The data storage scheme is

$$x'(n) = \left\{ x(n), x(n + \frac{N}{2}) \right\}, \quad n = 0, 1, \ldots, \frac{N}{2} - 1$$

Input		Stage 1 Output		Stage 2 Output	Stage 3 Output		Stage 4 Output	
$x(0)$=1	$x(8)$=0	1	1	8 + 6	16.00	0.00	$X(0)$= 41	$X(16)$= -9
$x(1)$=3	$x(9)$=2	7	-1	$1+j1$	$-0.41-j3.24$		$X(1)$=1.8162$-j7.8620$	
$x(2)$=2	$x(10)$=0	2	2	8 -4	$-6+j4$		$X(2)$= $-6.71+j4.71$	
$x(3)$=5	$x(11)$=2	6	4	$2-j4$	$2.41-j5.24$		$X(3)$=2.2557$-j3.3292$	
$x(4)$=3	$x(12)$=4	5	1	$11-1$	25.00	-3.00	$X(4)$= $j3.00$	
$x(5)$=4	$x(13)$=2	6	2	$1-j2$	$3.83-j3.41$		$X(5)$=2.5727$+j7.1561$	
$x(6)$=5	$x(14)$=1	7	3	14 0	-1		$X(6)$= $-5.29-j3.29$	
$x(7)$=3	$x(15)$=4	7	-1	$3+j1$	$-1.83+j0.59$		$X(7)$= $-2.6447-j1.3768$	

Fig. 9.15 The trace of the RDFT algorithm for computing the DFT with $N = 16$

$$X'(0) = \left\{ X(0), X\left(\frac{N}{2}\right) \right\} \quad \text{and} \quad X'(k) = X(k), \quad k = 1, \ldots, \frac{N}{2} - 1$$

As in the DIT DFT algorithms for complex data, in the last stage, two 8-point DFTs are merged to form a 16-point DFT. The differences are that: (1) only half the number of butterflies are used and (2) conjugation and swapping operations are required. Due to the lack of symmetry of the data at the input and output end, special butterflies are used. The first stage butterflies, denoted by f, compute 2-point DFTs. The input and output data are stored as shown in the trace of the algorithm. In addition, the input is also placed in bit-reversed order. Subsequently, there is a g butterfly for each group of butterflies. The g butterflies use the values stored in a pair of nodes. Both the nodes have two real values. The sum and difference of the values stored in the first locations of the top and bottom nodes, respectively, form the output for the first and second locations of the top node. For example, in the g butterfly of the first stage, the output values are $1 + 7 = 8$ and $1 - 7 = -6$. The output value at the bottom node is the sum of the second value of the top node and the product of the second value at the bottom node multiplied by $-j$. For example, $1 + (-j)(-1) = 1 + j1$ is stored in the second node.

The input and output data are stored as shown in the trace of the algorithm. The input and output for the rest of the butterflies are complex, and the computation is as shown in Fig. 9.13.

Let the input 16-point real data sequence be

$$\{1, 3, 2, 5, 3, 4, 5, 3, 0, 2, 0, 2, 4, 2, 1, 4\}$$

The first 9 values of the DFT $X(k)$ are

$$\{X(k), k = 0, 1, \ldots, 8\} = \{41, 1.8162 - j7.8620, -6.7071 + j4.7071, 2.2557$$

$$-j3.3292,$$

$$j3, 2.5727 + j7.1561, -5.2929 - j3.2929, -2.6447 - j1.3768, -9\}$$

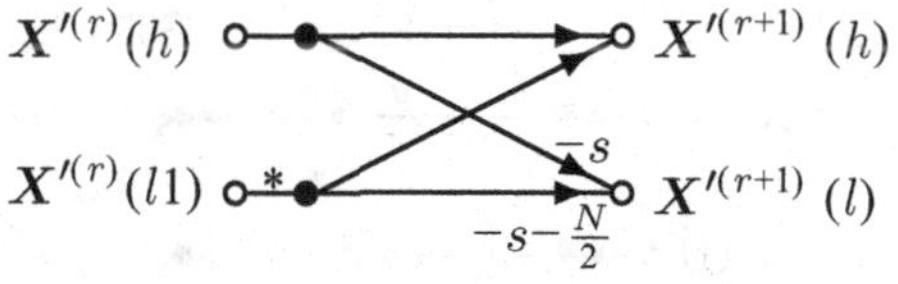

Fig. 9.16 RIDFT butterfly

The other values can be obtained using the conjugate symmetry of the DFT of real data. The even-indexed values $xe(n)$ of $x(n)$ are

$$xe(n) = \{1, 2, 3, 5, 0, 0, 4, 1\}$$

The DFT $X(k)$ of $xe(n)$ is

$$\{16, -0.4142 - j3.2426, -6 + j4, 2.4142 - j5.2426, 0, 2.4142 + j5.2426,$$
$$-6 - j4, -0.4142 + j3.2426\}$$

The odd-indexed values $xo(n)$ of $x(n)$ are

$$xo(n) = \{3, 5, 4, 3, 2, 2, 2, 4\}$$

The DFT $X(k)$ of $xo(n)$ is

$$\{25, 3.8284 - j3.4142, -1, -1.8284 + j0.5858, -3, -1.8284 - j0.5858,$$
$$-1, 3.8284 + j3.4142\}$$

The last stage output is the first $(N/2) + 1 = 9$ DFT coefficients. These coefficients are obtained by merging the 5 DFT coefficients of the two 8-point DFTs. The whole computation is similar to complex-valued algorithms with few differences.

The PM DIF RIDFT Butterfly

The RIDFT algorithm is derived from the corresponding DIF DFT algorithm for complex-valued data. Only half of the butterflies of the complex-valued algorithm are necessary. The RIDFT butterfly is shown in Fig. 9.16. It is similar to that of the algorithm for complex-valued data, except that the input data at the bottom node is read from the first half of the DFT coefficients and conjugated. In addition, the twiddle factors are conjugated compared with those of the RDFT algorithm. The equations governing the butterfly are

$$X'^{(r+1)}(h) = X'^{(r)}(h) + (X'^{(r)}(l1))^*$$
$$X'^{(r+1)}(l) = W_N^{-s}(X'^{(r)}(h) - (X'^{(r)}(l1))^*)$$

The flow graph of the RIDFT algorithm with $N = 16$ is shown in Fig. 9.17. The f butterflies compute a 2-point IDFT, without the division operation by 2, of the of

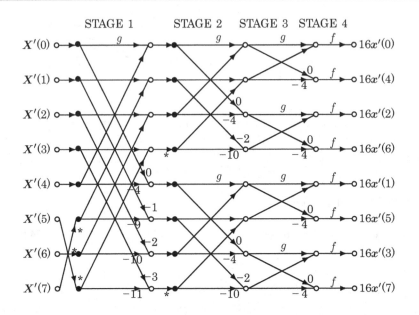

Fig. 9.17 The flow graph of the RIDFT algorithm with $N = 16$

Input		Stage 1 Output		Stage 2 Output		Stage 3 Output		Stage 4 Output	
$X(0)= 41$	$X(16)= -9$	32	0.00	32	-24	8	8	$x(0)=16$	$x(8)=0$
$X(1)=1.8162- j7.8620$		$-0.83-j6.49$		$4+j4$		56	-8	$x(4)=48$	$x(12)=64$
$X(2)= -6.71 + j4.71$		$-12+j8$		32	-16	16	16	$x(2)=32$	$x(10)=0$
$X(3)=2.2557- j3.3292$		$4.83-j10.49$		$8-j16$		48	32	$x(6)=80$	$x(14)=16$
$X(4)= j3.00$		50	-6.00	44	-4	40	8	$x(1)=48$	$x(9)=32$
$X(5)=2.5727+j7.1561$		$7.66-j6.83$		$4-j8$		48	16	$x(5)=64$	$x(13)=32$
$X(6)= -5.29- j3.29$		-2		56	0	56	24	$x(3)=80$	$x(11)=32$
$X(7)= -2.6447- j1.3768$		$-3.66+j1.17$		$12+j4$		56	-8	$x(7)=48$	$x(15)=64$

Fig. 9.18 The trace of the RIDFT algorithm

the two real values stored in each node. The g butterflies use the values stored in a pair of nodes. The top node has two real values. The sum and difference of these values are stored in the first location of the top and bottom nodes, respectively. For example, in the g butterfly of the first stage, the output values are $41 - 9 = 32$ and $41 + 9 = 50$. Let the complex value stored in the bottom node is $a + jb$. Then, $2a$ and $-2b$ are stored in the second location of the top and bottom nodes, respectively. For the first g butterfly, $a + jb = 0 + j3$. Therefore, 0 and -6 are stored.

The trace of the RIDFT algorithm is shown in Fig. 9.18. The input DFT coefficients are the output in Fig. 9.15. The output values have to be divided by 16.

9.5   Summary

- Fourier analysis is indispensable in the analysis of linear systems and signals in practice as well, due to the availability of fast algorithms to approximate the Fourier spectrum.
- The computational complexity of computing the 1-D N-point DFT, using fast algorithms, is $O(N \log_2 N)$ for N a power of 2, whereas that of the implementation from the definition is $O(N^2)$.
- The fast algorithms reduce the inherent redundancy of operations in the definition of the DFT. While there are different approaches in designing the DFT algorithms, the power of 2 algorithms based on the classical divide-and-conquer strategy of developing fast algorithms yield the practically most useful DFT algorithms.
- While there are alternate ways of deriving the power of 2 algorithms (such as matrix decomposition), the best physical appreciation of the algorithms is obtained using the half-wave symmetry of periodic waveforms.
- The sum and difference of the samples of the first half and second half of the period of a periodic waveform decompose it into two components. One component is composed of the even-indexed frequency components and the other is composed of the odd-indexed frequency components. This decomposition decomposes a N-point DFT into two $N/2$-point DFTs. Recursive decomposition yields the final algorithm.
- There are $\log_2 N$ stages, each stage with a computational complexity of $O(N)$. Therefore, the computational complexity becomes $O(N \log_2 N)$ for a N-point DFT.
- The decomposition can be carried out by decomposing the frequency components into two sets. One set consists of the even-indexed frequency components, and the other consists of the odd-indexed frequency components. The DFTs of the two sets are computed and combined to obtain the whole DFT. These types of algorithms are called the decimation-in-frequency (DIF) algorithms.
- The decomposition can also be carried out by decomposing the time-domain samples into two sets. One set consists of the even-indexed samples and the other consists of the odd-indexed samples. The DFTs of the two sets are computed and merged to obtain the whole DFT. These types of algorithms are called the decimation-in-time (DIT) algorithms.
- As usual, Fourier analysis, in theoretical analysis, is almost always carried out using the complex exponential as the basis waveform. This results in a factor-of-2 redundancy in computing the DFT of real-valued data.
- This redundancy can be reduced by computing the DFT of two sets of real-valued data at the same time.
- An alternative method is to reduce the redundancy in the flow graph of the algorithms for complex data.

Exercises

9.1 Given the samples of a waveform $x(n)$, find the samples of its even half-wave symmetric and odd half-wave symmetric components. Verify that the sum of the samples of the two components adds up to the samples of $x(n)$. Compute the DFT of $x(n)$ and its components. Verify that the DFT of the even half-wave symmetric component consists of zero-valued odd-indexed spectral values. Verify that the DFT of the odd half-wave symmetric component consists of zero-valued even-indexed spectral values.

* **9.1.1** $x(n) = \{\check{0}, 1, 2, 3\}$.

9.1.2 $x(n) = \{\check{0}, 1, 0, 1\}$.

9.1.3 $x(n) = \{\check{1}, 3, -1, -3\}$.

9.1.4 $x(n) = \{\check{2}, 1, 3, 4\}$.

9.1.5 $x(n) = \{\check{3}, 1, 2, 4\}$.

9.2 Given a waveform $x(n)$ with period 8, find the samples of the waveform. (a) Give the trace of the PM DIF DFT algorithm in computing its DFT $X(k)$. (b) Give the trace of the PM DIT DFT algorithm in computing its DFT $X(k)$. Verify that both are the same. In both cases, find the IDFT of $X(k)$ using the same DFT algorithms, give the trace and verify that the samples of the input $x(n)$ are obtained.

9.2.1 $x(n) = -2e^{-j\left(\frac{2\pi}{8}n + \frac{\pi}{6}\right)}$.

* **9.2.2** $x(n) = -e^{-j\left(\frac{2\pi}{8}6n - \frac{\pi}{3}\right)}$.

9.2.3 $x(n) = e^{j\left(\frac{2\pi}{8}0n - \frac{\pi}{3}\right)} + e^{j\left(\frac{2\pi}{8}4n - \frac{\pi}{6}\right)}$.

9.2.4 $x(n) = -e^{j\left(\frac{2\pi}{8}n + \frac{\pi}{4}\right)} + e^{j\left(\frac{2\pi}{8}2n + \frac{\pi}{6}\right)}$.

9.2.5 $x(n) = 3e^{j\left(\frac{2\pi}{8}6n - \frac{\pi}{4}\right)} + e^{j\left(\frac{2\pi}{8}7n + \frac{\pi}{3}\right)}$.

9.3 Given two waveforms $x(n)$ and $y(n)$ with period 8, find the samples of the waveforms. Use the PM DIT DFT algorithm for complex data to find their DFTs using the algorithm for computing the DFTs of two real data sets simultaneously. Compute the IDFTs of the DFTs using a DFT algorithm for complex data sets simultaneously. Verify the DFT $X(k)$ and $Y(k)$ by expressing $x(n)$ and $y(n)$ into its complex exponential components.

9.3.1 $x(n) = \cos\left(\frac{2\pi}{8}1n + \frac{\pi}{6}\right)$ and $y(n) = \cos\left(\frac{2\pi}{8}3n - \frac{\pi}{6}\right)$.

9.3.2 $x(n) = \cos\left(\frac{2\pi}{8}2n - \frac{\pi}{6}\right)$ and $y(n) = \cos\left(\frac{2\pi}{8}7n - \frac{\pi}{3}\right)$.

* **9.3.3** $x(n) = \cos\left(\frac{2\pi}{8}3n + \frac{\pi}{4}\right)$ and $y(n) = \cos\left(\frac{2\pi}{8}5n - \frac{\pi}{3}\right)$.

9.3.4 $x(n) = \cos\left(\frac{2\pi}{8}0n + \frac{\pi}{6}\right)$ and $y(n) = \cos\left(\frac{2\pi}{8}6n + \frac{\pi}{3}\right)$.

9.3.5 $x(n) = \cos\left(\frac{2\pi}{8}6n + \frac{\pi}{4}\right)$ and $y(n) = \cos\left(\frac{2\pi}{8}5n - \frac{\pi}{3}\right)$.

9.4 Given a waveform $x(n)$ with period 8, find the samples of the waveform. Find its DFT $X(k)$ using the PM RDFT algorithm. Find the IDFT of $X(k)$ using the PM RIDFT algorithm to get back the samples of $x(n)$. Verify the DFT $X(k)$ by expressing $x(n)$ into its complex exponential components.

9.4.1 $\cos\left(\frac{2\pi}{8} 0n + \frac{\pi}{6}\right)$.

9.4.2 $\cos\left(\frac{2\pi}{8} 1n - \frac{\pi}{6}\right)$.

9.4.3 $\cos\left(\frac{2\pi}{8} 2n + \frac{\pi}{4}\right)$.

* **9.4.4** $\cos\left(\frac{2\pi}{8} 3n - \frac{\pi}{4}\right)$.

9.4.5 $\cos\left(\frac{2\pi}{8} 4n + \frac{\pi}{3}\right)$.

Effects of Finite Wordlength

10

Several signals encountered in practice are of random nature. Their future values are unpredictable and represented by some averages such as correlation in the time domain and power spectrum in the frequency domain. Practical signals are usually of continuous nature, and it is sampled with an adequate sampling interval to convert them into discrete signal. And, for processing them by digital devices, they are quantized. Quantization is the process of rounding off or assigning one of a finite set of possible values to the infinite-valued discrete samples. Digital filters are very efficient in practice, provided that adequate wordlength is available. That is, the quantization level is sufficiently small. Wordlength here means the number of binary bits used to the representation of signal samples and filter coefficients. In practice, only finite-length precision is available and infinite precision is unnecessary. In digital processing, it is usually assumed that, initially, infinite wordlength is available in the analysis and design, and then the wordlength effect is analyzed to determine the adequate finite wordlength for a particular problem to make the analysis simpler. Provided the quantization errors are small compared with the data of interest, the difficult quantization analysis problem can be adequately simplified using statistical methods. One problem with finite wordlength is that the values may have to be truncated, making them imprecise. Any value that requires more bits than available has to be truncated or rounded to represent them in the given wordlength with reduced accuracy. Another problem is the possibility of overflow due to addition operation. Further, the major problem is the truncation or rounding the product with $2b$ bits resulting due to the multiplication of two b-bit numbers. The analysis of quantized systems is difficult and the ultimate solution is to simulate them.

10.1   Quantization Noise Model

The effect of arithmetic noise is analyzed using a linear noise model by placing an additive noise source at the output of the multipliers and after quantization. Both the sampling and quantization problems are similar. Naturally occurring signals have their sampling interval and quantization level tending to zero. But, both physical devices and human beings cannot have such a resolution. Therefore, finite wordlength is sufficient, and that finite value has to be determined for each problem.

10.1.1   2's Complement Number System

Digital devices, hardware or software, mostly use the 2's complement number system. It is a binary number system that can represent both positive and negative numbers. Table 10.1 shows the number system with $b = 4$ bits to represent numbers from 0.875 to -1. In the unsigned binary number system, the most significant bit (MSB) is the most weighted bit in the number. In the 2's complement number system, the leftmost bit is reserved for indicating the sign of the number. The least significant bit (LSB) is usually the rightmost bit in a binary data string. The LSB is the least weighted bit in the number. As there are only 3 bits left, the quantization step is $2^{-3} = 0.125$. For positive numbers, the bits representing the decimal number N is given by

$$N = \sum_{i=0}^{b-2} x_i 2^{i-b+1},$$

where b is the number of bits in the binary number x. Let the positive 4-bit signed binary number be

$$\{x_3, x_2, x_1, x_0\}, \quad = \{0, 1, 1, 0\}$$

Then, its value in the decimal system, assuming that the binary point is placed after the MSB, is

$$N = x_2 2^{-1} + x_1 2^{-2} + x_0 2^{-3} \quad \text{and} \quad (1)2^{-1} + (1)2^{-2} + (0)2^{-3} = 0.750,$$

as shown in the left-side table.

For negative numbers, the bits representing the decimal number N are given by

$$N = -x_{b-1} + \sum_{i=0}^{b-2} x_i 2^{i-b+1}$$

Table 10.1 The 2's complement representation, using four bits, of the decimal numbers from 0.875 to −1, positive numbers on the left and negative numbers on the right

Decimal	2's complement $x_3x_2x_1x_0$	Decimal	2's complement $x_3x_2x_1x_0$
0.000	0.000	−1.000	1.000
0.125	0.001	−0.875	1.001
0.250	0.010	−0.750	1.010
0.375	0.011	−0.625	1.011
0.500	0.100	−0.500	1.100
0.625	0101	−0.375	1.101
0.750	0.110	−0.250	1.110
0.875	0.111	−0.125	1.111

Table 10.2 Conversion of positive fractional decimal number 0.625 to 2's complement representation with 4 bits

Decimal	2's complement $x_3x_2x_1x_0$
$0.625 \times 2 = 1 + 0.250$	0.1
$0.250 \times 2 = 0 + 0.500$	0.10
$0.500 \times 2 = 1 + 0$	0.101

That is, subtract 1, the MSB, from the number obtained by the summation. For example, let the binary number be {1, 1, 1, 0}. This is a negative number as the MSB is 1. Its decimal equivalent is $-1 + 0.750 = -0.250$, as shown in the right-side table. To find the 2's complement of a binary number, simply complement the number, add 1 to the LSB, and ignore the carry. For example, the 2's complement of 0110 is $1001 + 0001 = 1010$ and -0.750 in decimal. Another way is to keep the first 1 bit from the LSB and complement the rest of the bits to its left. We keep the bits same up to the first 1 from the right in 0110 and complement the last two bits to get 1010 as before.

Table 10.2 shows the conversion of positive fractional decimal number 0.625 to 2's complement representation with 4 bits. The procedure is to keep multiplying the resulting fractionl part after multiplying it by 2. The integer part gives the bits from left to right after the decimal point.

10.1.2 Uniform Probability Density Function

In the finite wordlength representation of an infinite-precision number, there is always some uncertainty about the actual amplitude of that number. The corresponding signal due to this uncertainty is called as quantization noise. It has to be ensured that the power of this noise is sufficiently small by providing adequate wordlength so that a stable system remains stable and the signal-to-noise ratio (SNR) is high.

Assume that we use rounding and the quantization errors are uniformly distributed from $-2^{-b}/2$ to $2^{-b}/2$, where the quantization step is 2^{-b}. Note that, for quantization error analysis, the number of bits b does not include the sign bit. Then, the mean of the noise is zero. Let $q = 2^{-b}$. The probability density function of the

Fig. 10.1 The uniform probability density function of the quantization error due to rounding

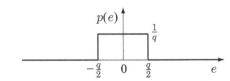

Table 10.3 Errors in DFT computation due to quantization with 4 bits 2's complement representation of data with rounding

		$x(n)$	$X(k)$
1	Unquantized	{0.7238, −0.2924}	{0.4314, 1.0162}
2	Unquantized	{0.7238, −0.2524}	{0.4714, 0.9762}
3	Quantized	{0.75, −0.25}	{0.5, 1}

quantization error is shown in Fig. 10.1. As the mean is zero, the variance (power) reduces to that of the signal power. The variance σ_q^2, assuming the error is uniformly distributed with mean zero, is given as

$$\sigma_q^2 = \frac{1}{q} \int_{-q/2}^{q/2} e^2 de = \frac{q^2}{12} = \frac{2^{-2b}}{12}$$

The variance, for uniform density function, is equal to the square of the width of the density function divided by 12. Each multiplier in the discrete system model will be replaced by a noise source with power σ_q^2 and a multiplier with infinite wordlength.

10.2   Quantization Effects in DFT Computation

Not only the order of speedup of the power of 2 DFT algorithms in computing the DFT but its reduction of the quantization effects is also effective in its immense practical importance. The major problems with finite worldlength are the arithmetic overflow, round-off errors, and twiddle factor quantization. The fast DFT algorithms are an interconnection of butterflies, as shown in the last chapter. The magnitude of the twiddle factors is 1. Therefore, if the input $|x(n)| < 1$ and scaled by 0.5 at each stage, then there will be no overflow. Consider the computation of 2-point DFTs with 4 bits, shown in Table 10.3. Note that computing the 2-point DFT is just finding the sum and difference of the two given data values. Assume rounding and 2's complement representation with 4 bits. The inputs to the 2-point DFT, in the first case, are $x(0) = 0.7238, x(1) = -0.2924$, the DFTs of which are $X(0) = 0.4314, X(1) = 1.0162$, as shown in the first row of Table 10.3. The inputs to the 2-point DFT, in the second case, are $x(0) = 0.7238, x(1) = -0.2524$, the DFTs of which are $X(0) = 0.4714, X(1) = 0.9762$, as shown in the second row of Table 10.3. After quantization, the input values become $x(0) = 0.75, x(1) = -0.25$, as shown in the third row of Table 10.3. Computing the DFT, we get $X(0) = 0.5, X(1) = 1$. The rounding of these numbers is shown in Fig. 10.2.

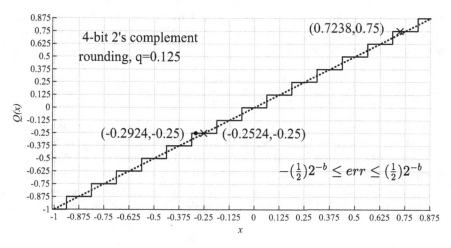

Fig. 10.2 4-bit 2's complement quantizer input–output characteristic with rounding

Figure 10.2 shows the 4-bit 2's complement quantizer input–output characteristic with rounding. In rounding a number, it is assigned to the nearest quantization level. The error is bounded by one-half of the quantization step for practical purposes, $err = \pm0.0625$ with 4-bit representation. It is assumed that a number that is exactly halfway between two quantization levels is assigned to the nearest higher level.

In the second case of computing the 2-point DFT, which has the same quantized representation as that of the first case, the output is the same as in the first case. That is, due to quantization, the uniqueness of the DFT is lost. Further, the output is also different. The uniqueness and correctness can be exactly restored with infinite bit representation. As that is not possible in practice, adequate finite wordlength has to be selected so that the computed DFT values are adequately accurate.

10.2.1   Quantization Errors in the Direct Computation of the DFT

We consider the quantization in the computation of DFT directly from its definition. We derive the SNR as a figure of merit. The DFT computation is usually carried out assuming that the input samples are complex-valued. In computing a single DFT coefficient, N complex multiplications are required. Since each such multiplication requires 4 real multiplications and modeled as 4 real multiplications with no noise in addition with 4 additive noise sources, there are $4N$ noise sources. Assume that both the real and imaginary parts of the samples are represented using b bits.

The variance of quantization errors from $4N$ multiplication operations is

$$\sigma_q^2 = 4\sigma_{q1}^2 = 4N\frac{2^{-2b}}{12} = N\frac{2^{-2b}}{3}$$

The variance is proportional to the length of the DFT, N. Let us increase the length N fourfold to $4N$. Let the variance be the same. Then, we have to increase the number of bits by 1. Then,

$$4N\frac{2^{-2b}}{3} = 2^2 N\frac{2^{-2(b+1)}}{3} = N\frac{2^{-2b}}{3}$$

As DFT coefficients are upscaled by its length N, overflow due to addition will occur in the computation. Therefore, the input has to be scaled to keep it in the range $|x(n)| < 1$, which can be ensured by dividing all the input samples by N. Remember that the magnitude of the twiddle factors is less than or equal to unity. Unfortunately, this extremely severe scaling decreases the effective precision of the samples low and decreases the SNR drastically. Let the complex-valued input samples be

$$x(n), n = 0, 1, \ldots, N - 1,$$

which is an uncorrelated set of values with uniform magnitude probability density $-\frac{1}{N}$ to $\frac{1}{N}$. Therefore, overflow is eliminated in its DFT computation with the DFT values $|X(k)| < 1$. The power of the complex input values is given by

$$\sigma_x^2 = \frac{N}{2} \int_{-\frac{1}{N}}^{\frac{1}{N}} |x|^2 dx = \frac{1}{3N^2}$$

This result can also be arrived, since the power is the width of the density function squared divided by 12. That is,

$$\sigma_x^2 = \frac{\left(\frac{2}{N}\right)^2}{12} = \frac{1}{3N^2}$$

Since there is a gain by N from input $x(n)$ to $X(k)$, the output signal power is given by

$$N\sigma_x^2 = \frac{1}{3N}$$

Consequently, SNR is given by

$$\text{SNR} = \frac{N\sigma_x^2}{\sigma_q^2} = \frac{1}{3N}\frac{3}{N2^{-2b}} = \frac{2^{2b}}{N^2}$$

SNR is inversely proportional to N^2. To maintain the same noise power, one more bit is required for each time the length of the DFT is increased by a factor of two.

10.2.2 Quantization Errors in Computation of the DFT Using Power of 2 Algorithms

In the fast power of 2 DFT algorithms, the computation is carried out over $\log_2 N$ stages, with sharing as many partial results as possible. Exactly $N - 1$ complex multiplications are required to compute a single DFT coefficient, while direct computation requires N complex multiplications. This computation over stages enables to carry out distributed scaling, which scales the noise produced by the preceding stage by the current stage along with the partial results. This is the reason for the reduction of round-off noise in fast DFT algorithms. Out of the $\log_2 N$ stages of computation, some stages have only trivial multiplications by -1 and $-j$, which do not produce round-off noise. We derive the SNR ratio assuming that all stages have nontrivial multiplications. Therefore, the results will be conservative by a small factor. The order of reduction of the noise between single stage scaling and distributed scaling is, however, more important.

The output of each stage of a fast algorithm increases by no more than by a constant factor 2. Therefore, by scaling each stage input by a factor of 2, we avoid the possibility of overflow. Each such scaling reduces the noise power by a factor of 4. In each of the succeeding stages, the noise power is further reduced power by a factor of 4. The number of multiplications required to compute a single DFT coefficient in the first stage is $N/2$, in the second stage $N/4$, and so on over the $\log_2 N$ stages. Therefore, the total variance due to the quantization errors at the output of the algorithm, with

$$\sigma_q^2 = \frac{2^{-2b}}{3} \quad \text{and} \quad N = 2^n,$$

is

$$\sigma_{qt}^2 = \sigma_q^2 \left(\frac{N}{2} \left(\frac{1}{4} \right)^{n-1} + \frac{N}{4} \left(\frac{1}{4} \right)^{n-2} + \cdots + 1 \right)$$

$$= \sigma_q^2 \left(\left(\frac{1}{2} \right)^{n-1} + \left(\frac{1}{2} \right)^{n-2} + \cdots + \frac{1}{2} + 1 \right)$$

$$= \sigma_q^2 \left(\frac{1 - (\frac{1}{2})^n}{1 - \frac{1}{2}} \right) \simeq \frac{2}{3} 2^{-2b}$$

where $(\frac{1}{2})^n$ becomes negligible as n increases. The input signal power remains the same and is given by

$$N\sigma_x^2 = \frac{1}{3N}$$

The SNR becomes

$$\mathrm{SNR} \ = \ \frac{N\sigma_x^2}{\sigma_{qt}^2} = \frac{2^{2b}}{2N}$$

SNR is inversely proportional to N compared to N^2 when all the scaling was carried out at the input. For each quadrupling of N, the wordlength has to be increased by 1 bit to maintain the same SNR.

Example 10.1 Let the number of input samples to the DFT be $N = 512$ with the desired SNR be 40 dB. Determine the number of bits required: (i) with all the scaling carried out at the input and (ii) with distributed scaling in fast DFT algorithms.

Solution

(i) $10\log_{10}(X) = 40, \quad \log_{10}(X) = 4, \quad X = 10^4 = 2^{13.2877} = 2^{(2b-18)},$

$$b = (13.2877 + 18)/2 = 15.6439$$

Rounding, we get $b = 16$.

(ii) $10\log_{10}(X) = 40, \quad \log_{10}(X) = 4, \quad X = 10^4 = 2^{13.2877} = 2^{(2b-10)},$

$$b = (13.2877 + 10)/2 = 11.6439$$

Rounding, we get $b = 12$.   ∎

10.3 Quantization Effects in Digital Filters

Of necessity, we have to implement digital filters using fixed-point digital systems. Consequently, the filter performance deteriorates from the desired one. There are four aspects we have to look into. The structure of the filter has a significant effect on its sensitivity to coefficient quantization. Effects of rounding or truncation of signal samples have a random noise source inserted at the point of quantization. Appropriate scaling to prevent overflows has to be taken care of. The quantization effects introduce nonlinearities in the linear system, which may result in oscillations, called limit cycles even with zero or constant input. Remember that no new frequencies other than the input frequency components can exist in a linear system. Due to quantization, some of the effects are: (1) uniqueness of output may be lost; (2) true output may be lost; (3) appearance of limit cycles; and (4) loss of stability in IIR systems. The wordlength has to be sufficient so that the system response is acceptable.

10.3.1 Coefficients Quantization

Quantized coefficients of the filter will be different from the infinite-valued coefficients depending on the quantization step. Therefore, the poles and zeros and, consequently, the frequency response will be different from the desired one. Stability of IIR filters may be affected due to the quantization of the coefficients. The coefficient sensitivity of a filter depends on the distance between its poles. The poles of higher-order filters tend to occur in clusters. Therefore, in order to increase the distance between poles, it is a common practice to decompose higher-order filters into second-order sections connected in cascade or parallel. The distance between pair of poles is usually longer. There are special filter structures that are designed for low coefficient sensitivity. Poles with locations close to the unit-circle affect the frequency response more. Further, the design of the filter coefficients may limit the order of the filter, which results in decomposition of longer filters into a set of smaller ones.

Consider the IIR filter transfer function

$$H(z) = \frac{b_0 + b_1 z^{-1} + \cdots + b_m z^{-1}}{1 + a_1 z^{-1} + \cdots + a_n z^{-1}}$$

with infinite-precision coefficients. With the memory for the storage of the coefficient fixed to N bits, the transfer function, after quantization, becomes

$$H_q(z) = \frac{Q_N(b_0) + Q_N(b_1)z^{-1} + \cdots + Q_N(b_m)z^{-m}}{1 + Q_N(a_1)z^{-1} + \cdots + Q_N(a_n)z^{-n}}$$

The quantization effects are more in higher-order filters, if implemented in direct form. Therefore, they are usually decomposed into first and second-order sections and implemented in parallel or cascade connection. In the cascade connection, $H(z)$ is decomposed into

$$H(z) = C\, H_1(z) H_2(z) \cdots H_M(z),$$

where C is a constant. In the parallel connection, $H(z)$ is decomposed into

$$H(z) = K + \sum_{n=1}^{N} H_n(z),$$

where K is a constant.

Figure 10.3a shows the magnitude of the frequency response of the 6th order IIR Butterworth bandpass filter with infinite, 16-bit and 8-bit quantized coefficients. The response with 16-bits is almost identical to that with infinite-precision coefficients. With 8 bits to represent the coefficients (response shown in a dotted line), the deviation of the responses is quite evident. Figure 10.3b shows the magnitude of

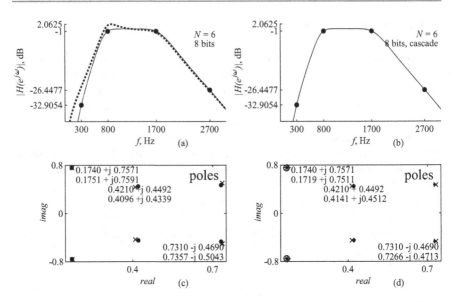

Fig. 10.3 (a) The magnitude of the frequency response of the 6th order IIR Butterworth bandpass filter with infinite, 16-bit and 8-bit quantized coefficients; (b) the response with 8-bit quantized coefficients, when implemented as a cascade of 3 second-order sections; (c) the pole locations with infinite-precision and 8-bit quantized coefficients in direct implementation; (d) the pole locations with infinite-precision and 8-bit quantized coefficients in cascade implementation

the frequency response of the 6th order IIR Butterworth bandpass filter with 8-bit quantized coefficients, when implemented as a cascade of 3 second-order sections. The response with 8 bits, this time, is almost identical with that implemented with infinite-precision coefficients. The point is that, with proper filter structures, fixed-point implementation systems are capable of giving adequate response with relatively low wordlength. Figure 10.3c shows the pole locations with infinite-precision and 8-bit quantized coefficients in direct implementation. The poles have moved, as shown with crosses. Figure 10.3d shows the pole locations with infinite-precision and 8-bit quantized coefficients in cascade implementation. The poles have moved by a smaller distance.

Figure 10.4a shows the magnitude of the frequency response of the 8th order FIR optimal equiripple lowpass filter with infinite and 8-bit quantized coefficients, which are almost the same. Figure 10.4b shows the response with 6-bit quantized coefficients and the response deviates from the desired one. Figure 10.4c shows the zero locations with infinite-precision and 8-bit quantized coefficients in direct implementation, the two sets of the coefficients are almost identical. Figure 10.4d shows the zero locations with infinite-precision and 6-bit quantized coefficients in direct implementation, with more deviation.

The use of finite wordlength makes the filter nonlinear to some extent, and the frequency response gets deviated from the desired one. As analyzing nonlinear systems is very difficult, an approximate linear model is used to find the finite

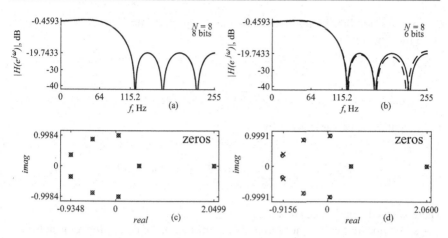

Fig. 10.4 (a) The magnitude of the frequency response of the 8th order FIR optimal equiripple lowpass filter with infinite and 8-bit quantized coefficients; (b) the response with 6-bit quantized coefficients; (c) the zero locations with infinite-precision and 8-bit quantized coefficients in direct implementation; (d) the zero locations with infinite-precision and 6-bit quantized coefficients in direct implementation

wordlength effects. The quantization errors are represented by statistical models. As the errors are more severe in fixed-point realization, the analysis of quantization errors in such systems only is usually carried out.

10.3.2 Round-off Noise

Let a first-order system is characterized by the difference equation

$$y(n) = x(n) + a\,y(n-1), \quad |a| < 1$$

with no noise sources. With noise sources, we get an additional equation

$$q(n) = e(n) + a\,q(n-1)$$

where $q(n)$ is the quantization noise with the average of $e(n)$ is zero. Consequently, the impulse response is $h(n) = a^n u(n)$ due to noise model alone. The finite wordlength effects make the linear system into nonlinear one, the analysis of which is extremely difficult. In such cases, one of the often used methods to analyze systems is to make them linear with some assumptions. While the results are indicative of the effects of finite wordlength, the best analysis of finite wordlength effects is the simulation of the filter.

The round-off noise produced in the multiplication is a random signal, and it is assumed to be uncorrelated with the input signal and additive. Further, the noise is assumed to be a stationary white noise sequence, $e(n)$, with its autocorrelation

$\sigma_q^2 \delta(n)$. Considering the noise alone as the input signal, the output of the filter is given by convolution as

$$y(n) = \sum_{k=0}^{n} h(k)e(n-k)$$

The output noise power is given by

$$\sigma_{qo}^2(n) = \sum_{k=0}^{n} \sum_{l=0}^{n} h(k)h(l)E(e(n-k)e(n-l))$$

where E stands for "expected value of."

As its autocorrelation is nonzero only at index zero, the noise product is nonzero only when $l = k$. Therefore,

$$\sigma_{qo}^2(n) = \sigma_q^2 \sum_{k=0}^{n} (h(k))^2 = \frac{2^{-2b}}{12} \sum_{k=0}^{n} (h(k))^2 \tag{10.1}$$

The impulse response, $h(n)$, of stable systems tends to zero as $k \to \infty$. Consequently, the steady-state value of the noise power is σ_{qo}^2. For the first-order filter, the output noise power is

$$\sigma_{qo}^2 = \frac{2^{-2b}}{12} \sum_{k=0}^{\infty} a^{2k} = \frac{2^{-2b}}{12} \frac{1}{1-a^2}$$

Therefore, the power gain of the filter is $\frac{1}{1-a^2}$. Equivalently, in the frequency domain, with

$$a^n u(n) \leftrightarrow \frac{1}{1 - ae^{-j\omega}},$$

the average output noise power is given by

$$\sigma_{qo}^2 = \frac{\sigma_q^2}{2\pi} \int_{-\pi}^{\pi} |H(e^{j\omega})|^2 d\omega = \frac{2^{-2b}}{12} \frac{1}{2\pi} \int_{-\pi}^{\pi} |H(e^{j\omega})|^2 d\omega \tag{10.2}$$

Example 10.2 The difference equation

$$y(n) = x(n) + 0.9y(n-1)$$

characterizes an IIR filter, and it is realized with a wordlength of 4 bits. Using the time- and frequency-domain relations, find the output round-off noise power, due to quantization. With a wordlength of 5 bits, what is the noise power? Repeat the problem, if the difference equation governing the filter is

$$y(n) = x(n) + 0.7y(n - 1)$$

Solution In the time domain, the output noise power, with $a = 0.9$, is

$$\frac{2^{-2b}}{12}\frac{1}{1 - a^2} = \frac{2^{-2(4)}}{12}\frac{1}{1 - (0.9)^2} = 0.0017$$

up to a precision of four decimal places. With 5 bits, we get

$$\frac{2^{-2b}}{12}\frac{1}{1 - a^2} = \frac{2^{-2(5)}}{12}\frac{1}{1 - (0.9)^2} = 0.0004$$

As the wordlength is increased by 1 bit, the noise power reduces by 4 times. With $a = 0.7$ and 4 bits,

$$\frac{2^{-2b}}{12}\frac{1}{1 - a^2} = \frac{2^{-2(4)}}{12}\frac{1}{1 - (0.7)^2} = 0.00063828$$

With $a = 0.7$ and 5 bits,

$$\frac{2^{-2b}}{12}\frac{1}{1 - a^2} = \frac{2^{-2(5)}}{12}\frac{1}{1 - (0.7)^2} = 0.00015957$$

The noise power is more with the pole located close to the unit-circle.

As the impulse response, in practice, has arbitrary amplitude profile, the convolution sum in Eq. (10.1) can be evaluated by summing the terms until the difference between two consecutive partial sums becomes negligible. For example, the first noise power 0.0069 is obtained in about 40 iterations. As usual, the use of DFT makes the evaluation of Eq. (10.2) and, hence, the output noise power, much faster due to fast algorithms to compute the DFT. With sufficient number of samples, DFT can approximate the DTFT adequately for practical purposes, as shown in an earlier chapter. The DTFT of the impulse response 0.9^n of the filter is

$$H(e^{j\omega}) = \frac{e^{j\omega}}{e^{j\omega} - 0.9}$$

The numerator coefficient is 1. The denominator coefficients are $\{1, -0.9\}$. Let us use 4 samples to approximate Eq. (10.2). With 3 zeros appended, the DFT of the numerator coefficient is

$$\{1, 0, 0, 0\} \leftrightarrow \{1, 1, 1, 1\}$$

With 2 zeros appended, the DFT of the denominator coefficients is

$$\{1, -0.9, 0, 0\} \leftrightarrow \{0.1, 1 + j0.9, 1.9, 1 - j0.9\}$$

Dividing the numerator DFT pointwise by the denominator DFT, we get

$$\{10, 0.5525 - j0.4972, 0.5263, 0.5525 + j0.4972\},$$

the magnitude and its square of which are

$$\{10, 0.7433, 0.5263, 0.7433\} \text{ and } \{100, 0.5525, 0.2770, 0.5525\}$$

The sum of the second set of coefficients is 101.382. As the DTFT spectrum is continuous, the area enclosed by the spectrum is required. Since we took 4 samples, the frequency increment of the DFT spectrum is $(2\pi/4)$. Multiplying this value with $\frac{\sigma_q^2}{2\pi}$, we get $\frac{\sigma_q^2}{4}$. Therefore, the output noise power is approximated as $\frac{\sigma_q^2}{4} 101.382 = \frac{2^{-2(4)}}{12} \frac{101.382}{4} = 0.0083$, which is very inaccurate due to the longer sampling interval compared with the expected value 0.0017. By increasing the number of samples, we get a better approximation. With $N = 16$ and 64, we get, respectively, $\{0.0025, 0.0017\}$. The DFT has fast algorithms for its computation. ∎

Example 10.3 Two filters governed by the difference equations

$$y(n) = 0.8x(n) + 0.9y(n-1) \quad \text{and} \quad y(n) = x(n) + 0.7y(n-1)$$

are connected in cascade in that order and are realized with a wordlength of 4 bits. Find the output round-off noise power, due to quantization.

Solution The model for the output round-off noise power analysis of the cascade filter is shown in Fig. 10.5. The noise sources $e_1(n)$, $e_2(n)$, and $e_3(n)$ are due to the multipliers with coefficients 0.8, 0.9, and 0.7. The noise power is given by

$$\sigma_{qo}^2 = \sigma_q^2 \sum_{k=0}^{\infty} (h(k))^2 = \frac{2^{-2b}}{12} \sum_{k=0}^{\infty} (h(k))^2$$

For the first stage, the noise power is

$$\sigma_{qo}^2 = \sigma_q^2 \frac{1}{1 - 0.9^2} = \sigma_q^2 5.2632,$$

where

$$\sigma_q^2 = \frac{2^{-2b}}{12}$$

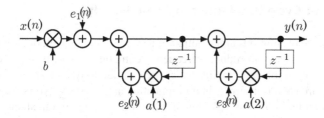

Fig. 10.5 Modeling the filter for the round-off error analysis due to quantization.

and $b = 4$. For the second stage, the noise power is

$$\sigma_{qo}^2 = \sigma_q^2 \frac{1}{1 - 0.7^2} = \sigma_q^2 1.9608$$

The impulse response of the cascade stages is the convolution of the individual stages, which is

$$h(n) = (4.5(0.9)^n - 3.5(0.7)^n)u(n)$$

The corresponding noise power is

$$\sigma_{qo}^2 = \sigma_q^2 \left(\frac{4.5^2}{1 - 0.9^2} + \frac{3.5^2}{1 - 0.7^2} - \frac{4.5(3.5)}{1 - 0.9(0.7)} \right) = \sigma_q^2 88.031$$

The total output noise power is

$$\sigma_q^2(88.031 + 88.031 + 1.9608) = \sigma_{qo}^2 = \sigma_q^2 178.0228$$

Note that noise sources $e_1(n)$ and $e_2(n)$ are affected by the impulse response of the combined system, whereas $e_3(n)$ is affected only by the impulse response of the second stage with coefficient 0.7. ∎

In the cascade interconnection, the noise generated in any section is affected by succeeding sections. Therefore, ordering the sections makes a difference in the value of the output noise power. A reasonable strategy is to place the sections in decreasing gain, so that the noise produced in the earlier sections is not amplified too much. In the parallel interconnection, the total noise generated is simply the linear combination of each of the individual sections.

10.3.3 Limit-Cycle Oscillations in Recursive Filters

Limit cycles are oscillations at the output of a system, for constant or zero input. As any linear system cannot produce a frequency component other than present in the input, these oscillations are nonlinearities created due to quantization effects. The main causes are round-off errors in multiplication and overflow errors in addition.

Another problem in the IIR digital filter due to signal quantization is the production of sustained oscillations, called limit cycles, even after the input signal is reduced to zero level. Due to rounding, the situation could occur when the magnitude of the rounded output value is equal to the input value. That is, the input and output differ by less than one-half of the quantization step. Limit cycles are not possible in FIR filters, since oscillations cannot be sustained without feedback.

Consider the filter governed by the difference equation

$$y(n) = x(n) - 0.9y(n - 1)$$

Let the input be $x(n) = 10\delta(n)$ with the initial condition $y(-1) = 0$. Let the multiplier output is rounded to the nearest integer. By iteration, the first few values of the output of the filter are computed as

$$y(0) = 10$$
$$y(1) = -0.9(10) = -9$$
$$y(2) = -0.9(-9) = 8.1 = 8$$
$$y(3) = -0.9(8) = -7.2 = -7$$
$$y(4) = -0.9(-7) = 6.3 = 6$$
$$y(5) = -0.9(6) = -5.4 = -5$$
$$y(6) = -0.9(-5) = 4.5 = 5$$
$$y(7) = -0.9(5) = -4.5 = -5$$
$$y(8) = -0.9(-5) = 4.5 = 5$$

The first few values of the output with infinite precision are

$$\{10, -9, 8.1, -7.296.561, -5.9049, 5.3144, -4.783, 4.3047, -3.8742\},$$

which is approaching zero as $n \to \infty$. With finite precision, the output oscillates between 5 and -5 after few iterations without any input. Figure 10.6a shows the outputs in the two cases. The range -5 to 5 is called the dead-band. During the limit-cycle effect, the pole is changed from -0.9 to -1 due to signal quantization. Limit cycles can occur due to the nonzero initial condition alone, even if the input is zero.

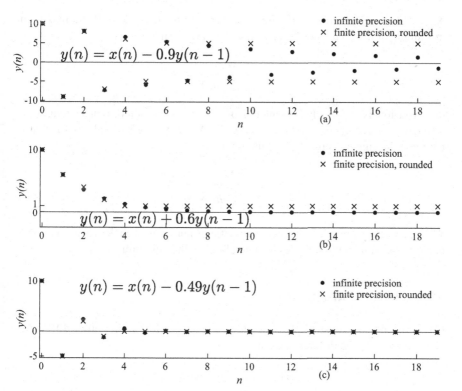

Fig. 10.6 (a) Limit-cycle oscillations in IIR digital filters with the initial condition zero; (b) limit-cycle oscillations in IIR digital filters with the initial condition zero and pole located far-off from the unit-circle; (c) limit-cycle oscillations in IIR digital filters with the initial condition zero and pole at $z = -0.49$

The use of truncation instead of rounding creates some problem and is not used. With the pole located far-off from the unit-circle, the dead-band is reduced, as shown in Fig. 10.6b. In this case, the pole is positive (0.6) and the period of oscillation is 1 and the output becomes a constant. With increasing the wordlength, the amplitude of the oscillation can be reduced, and with sufficiently longer wordlength, limit cycles will be absent. With the magnitude of the poles of the filter is less than 0.5, the problem is avoided, as shown in Fig. 10.6c.

10.3.4 Scaling

To ensure a high SNR, the internal signals in the stages of a filter should be kept as high as possible without the possibility of overflow. This requires appropriate scaling of the values of the signals at various stages of a filter. For stable systems, the maximum value of the output cannot exceed the input value multiplied by the factor

$\sum_{n=0}^{\infty} |h(n)|$, where $h(n)$ is the impulse response of the system. If the maximum magnitude of the input, say M, is scaled to be less than

$$\frac{M}{\sum_{m=0}^{\infty} |h(m)|},$$

then overflow is effectively prevented. However, the reduction in signal strength can result in a corresponding reduction in SNR. A more effective scale factor that is based on the signal energy and commonly used is

$$\frac{M}{\sqrt{\sum_{m=0}^{\infty} |h(m)|^2}}$$

With this scaling, overflow will occur sometimes but a higher SNR is achieved. SNR deteriorates more with the poles located closer to the unit-circle.

10.4   Lattice Structure

Consider the lattice structure, shown in Fig. 10.7, for a first-order FIR filter characterized by the difference equation

$$y(n) = x(n) + b(1)x(n-1)$$

It is similar to the well-known butterfly computational structure repeatedly used in fast DFT algorithms. In lattice realization of digital filters, a cascade of m stages of this structure produces the same output as that of the FIR filter characterized by a mth order difference equation. In the first stage of the cascade, both the inputs are connected to the input $x(n)$. The output from the top node of the last stage is the filter output $y(n)$. The cascade is characterized by the following set of recursive equations:

$$r_0(n) = s_0(n) = x(n)$$
$$r_m(n) = r_{m-1}(n) + K_m s_{m-1}(n-1)$$
$$s_m(n) = K_m r_{m-1}(n) + s_{m-1}(n-1), \quad m = 1, 2, \ldots, M$$

and the output of the Mth order FIR filter is $y(n) = r_M(n)$.

Consider the lattice structure, shown in Fig. 10.8, for a second-order FIR filter characterized by the difference equation

$$y(n) = x(n) + b(1)x(n-1) + b(2)x(n-2)$$

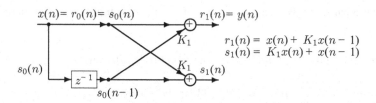

Fig. 10.7 Lattice structure of a first-order FIR filter

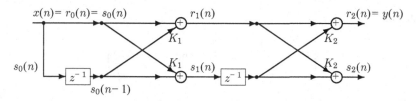

Fig. 10.8 Lattice structure of a second-order FIR filter

The coefficient of $x(n)$, $b(0)$, is assumed to be unity for mathematical convenience. The output $y(n)$ can be obtained from a two-stage lattice realization shown in Fig. 10.8. The output of the first stage is

$$r_1(n) = x(n) + K_1 x(n-1)$$
$$s_1(n) = K_1 x(n) + x(n-1)$$

The output of the second stage is

$$r_2(n) = r_1(n) + K_2 s_1(n-1)$$
$$s_2(n) = K_2 r_1(n) + s_1(n-1)$$

Solving for $y(n) = r_2(n)$, we get

$$y(n) = r_2(n) = x(n) + K_1(1 + K_2)x(n-1) + K_2 x(n-2)$$

Comparing the difference equation of the direct form of the filter, we get

$$b(1) = K_1(1 + K_2), \ b(2) = K_2 \quad \text{or} \quad K_1 = \frac{b(1)}{1 + b(2)}, \ K_2 = b(2)$$

The bottom output $s_2(n)$ for a second-order filter is given by

$$s_2(n) = K_2 x(n) + K_1(1 + K_2)x(n-1) + x(n-2)$$
$$= b(2)x(n) + b(1)x(n-1) + x(n-2)$$

The coefficients, in this case, are the same as for the top output but appear in reverse order.

Example 10.5 Given the difference of a second-order FIR filter

$$y(n) = x(n) + 3x(n-1) - 2x(n-2),$$

find the corresponding lattice structure coefficients.

Solution

$$\left\{ K_1 = \frac{3}{1-2} = -3, \quad \text{and} \quad K_2 = -2 \right\}$$

■

The lattice structures are less sensitive to errors due to finite wordlength implementation of digital filters. It has got many variations including IIR version. It is more often used in speech processing.

Example 10.6 Given the difference of a second-order FIR filter

$$y(n) = x(n) + \frac{5}{3}x(n-1) + x\frac{4}{3}(n-2),$$

find the corresponding response of direct and lattice realizations with infinite- and finite precision coefficients.

Solution The lattice structure coefficients are

$$\left\{ K_1 = \frac{5/3}{1+4/3} = \frac{5}{7} = 0.7143, \quad \text{and} \quad K_2 = \frac{4}{3} = 1.3333 \right\}$$

Now, the coefficients are truncated to one digit after the decimal point.

$$\{ K_1 = 0.7, \quad \text{and} \quad K_2 = 1.3 \}$$

The direct-form realization coefficients become

$$\{ K_1 = 1.6, \quad \text{and} \quad K_2 = 1.3 \}$$

The first few values of the response of the second-order filter using infinite- and finite precision coefficients in direct and lattice realizations are shown in Fig. 10.9a. The difference between the responses in the direct and lattice realizations with

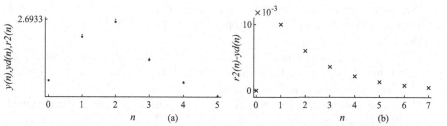

Fig. 10.9 (**a**) The response of a second-order filter using infinite- and finite precision coefficients in direct and lattice realizations; (**b**) the difference between the responses in the direct and lattice realizations with truncated coefficients

truncated coefficients is shown in Fig. 10.9b. The response of the lattice filter is more closer to the ideal response shown by dots in Fig. 10.9a. ∎

10.5 Summary

- Several signals encountered in practice are of random nature. Their future values are unpredictable and represented by some averages such as correlation coefficient in the time domain and power spectrum in the frequency domain.
- Wordlength here means the number of binary bits used to the representation of signal samples and filter coefficients.
- Provided that the quantization errors are small compared with the data of interest, the difficult quantization analysis problem can be adequately simplified using statistical methods.
- The precise analysis of quantized systems is difficult, and the ultimate solution is to simulate them.
- The effect of arithmetic noise is analyzed using a linear noise model by placing an additive noise source at the output of the multipliers and after quantization.
- Digital devices, hardware or software, mostly use the 2's complement number system. It is a binary number system that can represent both positive and negative numbers.
- In the finite wordlength representation of an infinite-precision number, there is always some uncertainty about the actual amplitude of that number. The corresponding signal due to this uncertainty is called as quantization noise.
- Not only the order of speedup of the power of 2 DFT algorithms in computing the DFT but its reduction of the quantization effects is also effective in its immense practical importance.
- In rounding a number, it is assigned to the nearest quantization level. The error is bounded by one-half of the quantization step for practical purposes.

- For each quadrupling of the length N, the wordlength has to be increased by 1 bit to maintain the same SNR by appropriate scaling of the input in computing the DFT.
- Due to quantization, some of the effects in digital filters are: (1) uniqueness of output may be lost; (2) true output may be lost; (3) appearance of limit cycles; and (4) loss of stability in IIR systems.
- Quantized coefficients of the filter will be different from the infinite-valued coefficients depending on the quantization step. Therefore, the poles and zeros and, consequently, the frequency response will be different from the desired one. Stability of IIR filters may be affected due to the quantization of the coefficients.
- While the results of analysis are indicative of the effects of finite wordlength in digital filters, the best analysis of finite wordlength effects is the simulation of the filter.
- In the cascade interconnection, the noise generated in any section is affected by succeeding sections. Therefore, ordering the sections makes a difference in the value of the output noise power. A reasonable strategy is to place the sections in decreasing gain, so that the noise produced in the earlier sections is not amplified too much. In the parallel interconnection, the total noise generated is simply the linear combination of each of the individual sections.
- Limit cycles are oscillations at the output of a system, for constant or zero input. As any linear system cannot produce a frequency component other than present in the input, these oscillations are nonlinearities created due to quantization effects. The main causes are round-off errors in multiplication and overflow errors in addition.
- To ensure a high SNR, the internal signals in the stages of a filter should be kept as high as possible without the possibility of overflow. This requires appropriate scaling of the values of the signals at various stages of a filter.
- In lattice realization of digital filters, a cascade of m stages of this structure produces the same output as that of the FIR filter characterized by a mth-order difference equation.
- The lattice structures are less sensitive to errors due to finite wordlength implementation of digital filters.

Exercises

10.1 Determine the 2's complement 4-bit representation B of the decimal number x with rounding. Determine also the 2's complement representation of $-x$. Add the two representations and verify that the sum of the two numbers is zero.

10.1.1 $x = 0.6321$.

* **10.1.2** $x = -0.0728$.

10.1.3 $x = 0.8321$.

10.2 A white noise signal has amplitudes uniformly distributed between -1 and 1. Find the total signal variance.

* **10.3** Let the number of input samples to the DFT be $N = 256$ with the desired SNR 30 dB. Determine the number of bits required: (1) with all the scaling carried out at the input and (2) with distributed scaling in fast DFT algorithms.

10.4 The difference equation

$$y(n) = x(n) + 0.8y(n-1)$$

characterizes an IIR filter, and it is realized with a wordlength of 4 bits. Using the time- and frequency-domain relations, find the output round-off noise power, due to quantization. Use 256-point DFT to compute the noise power in the frequency domain. Verify that the noise power is the same in both the domains.

* **10.5** Two filters governed by the difference equations

$$y(n) = 0.6x(n) + 0.8y(n-1) \quad \text{and} \quad y(n) = x(n) + 0.7y(n-1)$$

are connected in cascade in that order and are realized with a wordlength of 4 bits. Find the output round-off noise power, due to quantization.

10.6 Consider the filter governed by the difference equation

$$y(n) = x(n) - 0.85y(n-1)$$

Let the input be $x(n) = 10\delta(n)$ with the initial condition $y(-1) = 0$. Let the multiplier output is rounded to the nearest integer. By iteration, find the first 10 values of the output of the filter. List also the first 10 output values with infinite precision. Is there limit-cycle oscillation?

* **10.7** Given the difference of a second-order FIR filter

$$y(n) = x(n) + x(n-1) - 2x(n-2),$$

find the corresponding lattice structure coefficients.

Transform Pairs and Properties

A

See Tables A.1, A.2, A.3, A.4, A.5, and A.6.

Table A.1 DFT pairs

$x(n)$, period $= N$	$X(k)$, period $= N$
$\delta(n)$	1
1	$N\delta(k)$
$e^{j(\frac{2\pi}{N}mn)}$	$N\delta(k-m)$
$\cos(\frac{2\pi}{N}mn)$	$\frac{N}{2}(\delta(k-m) + \delta(k-(N-m)))$
$\sin(\frac{2\pi}{N}mn)$	$\frac{N}{2}(-j\delta(k-m) + j\delta(k-(N-m)))$
$x(n) = \begin{cases} 1 & \text{for } n = 0, 1, \ldots, L-1 \\ 0 & \text{for } n = L, L+1, \ldots, N-1 \end{cases}$	$e^{(-j\frac{\pi}{N}(L-1)k)}\frac{\sin(\frac{\pi}{N}kL)}{\sin(\frac{\pi}{N}k)}$

© The Author(s), under exclusive license to Springer Nature Switzerland AG 2021
D. Sundararajan, *Digital Signal Processing*,
https://doi.org/10.1007/978-3-030-62368-5

Table A.2 DFT properties

Property	$x(n), h(n)$, period $= N$	$X(k), H(k)$, period $= N$
Linearity	$ax(n) + bh(n)$	$aX(k) + bH(k)$
Duality	$\frac{1}{N} X(N \mp n)$	$x(N \pm k)$
Time shifting	$x(n \pm m)$	$e^{\pm j \frac{2\pi}{N} mk} X(k)$
Frequency shifting	$e^{\mp j \frac{2\pi}{N} mn} x(n)$	$X(k \pm m)$
Time convolution	$\sum_{m=0}^{N-1} x(m) h(n - m)$	$X(k) H(k)$
Frequency convolution	$x(n) h(n)$	$\frac{1}{N} \sum_{m=0}^{N-1} X(m) H(k - m)$
Time expansion	$h(mn) =$ $\begin{cases} x(n) \text{ for } n = 0, 1, \dots, N-1 \\ 0 \quad \text{otherwise} \end{cases}$ where m is any positive integer	$H(k) = X(k \bmod N),$ $k = 0, 1, \dots, mN - 1$
Time reversal	$x(N - n)$	$X(N - k)$
Conjugation	$x^*(N \pm n)$	$X^*(N \mp k)$
Parseval's theorem	$\sum_{n=0}^{N-1} |x(n)|^2$	$\frac{1}{N} \sum_{k=0}^{N-1} |X(k)|^2$

Table A.3 DTFT pairs

$x(n)$	$X(e^{j\omega})$, Period$= 2\pi$
$\begin{cases} 1 \text{ for } -N \le n \le N \\ 0 \text{ otherwise} \end{cases}$	$\dfrac{\sin(\omega \frac{(2N+1)}{2})}{\sin(\frac{\omega}{2})}$
$\frac{\sin(an)}{\pi n}, \ 0 < a \le \pi$	$\begin{cases} 1 \text{ for } |\omega| < a \\ 0 \text{ for } a < |\omega| \le \pi \end{cases}$
$a^n u(n), \ |a| < 1$	$\frac{1}{1 - ae^{-j\omega}}$
$(n + 1) a^n u(n), \ |a| < 1$	$\frac{1}{(1 - ae^{-j\omega})^2}$
$a^{|n|}, \ |a| < 1$	$\frac{1 - a^2}{1 - 2a\cos(\omega) + a^2}$
$a^n \sin(\omega_0 n) u(n), \ |a| < 1$	$\frac{(a)e^{-j\omega} \sin(\omega_0)}{1 - 2(a)e^{-j\omega} \cos(\omega_0) + (a)^2 e^{-j2\omega}}$
$a^n \cos(\omega_0 n) u(n), \ |a| < 1$	$\frac{1 - (a)e^{-j\omega} \cos(\omega_0)}{1 - 2(a)e^{-j\omega} \cos(\omega_0) + (a)^2 e^{-j2\omega}}$
$\delta(n)$	1
$\sum_{k=-\infty}^{\infty} \delta(n - kN)$	$\frac{2\pi}{N} \sum_{k=-\infty}^{\infty} \delta(\omega - \frac{2\pi}{N} k)$
$u(n)$	$\pi \delta(\omega) + \frac{1}{1 - e^{-j\omega}}$
1	$2\pi \delta(\omega)$
$e^{j\omega_0 n}$	$2\pi \delta(\omega - \omega_0)$
$\cos(\omega_0 n)$	$\pi(\delta(\omega + \omega_0) + \delta(\omega - \omega_0))$
$\sin(\omega_0 n)$	$j\pi(\delta(\omega + \omega_0) - \delta(\omega - \omega_0))$

Table A.4 DTFT properties

Property	$x(n), h(n)$	$X(e^{j\omega}), H(e^{j\omega})$
Linearity	$ax(n) + bh(n)$	$aX(e^{j\omega}) + bH(e^{j\omega})$
Time shifting	$x(n \pm n_0)$	$e^{\pm j\omega n_0} X(e^{j\omega})$
Frequency shifting	$x(n)e^{\pm j\omega_0 n}$	$X(e^{j(\omega \mp \omega_0)})$
Time convolution	$\sum_{m=-\infty}^{\infty} x(m)h(n-m)$	$X(e^{j\omega})H(e^{j\omega})$
Frequency convolution	$x(n)h(n)$	$\frac{1}{2\pi}\int_0^{2\pi} X(e^{jv})H(e^{j(\omega-v)})dv$
Time expansion	$h(n)$	$H(e^{j\omega}) = X(e^{ja\omega})$
	$h(an) = x(n),\ a > 0$ is an integer	
	and $h(n) = 0$ zero otherwise	
Time reversal	$x(-n)$	$X(e^{-j\omega})$
Conjugation	$x^*(\pm n)$	$X^*(e^{\mp j\omega})$
Difference	$x(n) - x(n-1)$	$(1 - e^{-j\omega})X(e^{j\omega})$
Summation	$\sum_{l=-\infty}^{n} x(l)$	$\frac{X(e^{j\omega})}{(1-e^{-j\omega})} + \pi X(e^{j0})\delta(\omega)$
Frequency differentiation	$(n)^m x(n)$	$(j)^m \frac{d^m X(e^{j\omega})}{d\omega^m}$
Parseval's theorem	$\sum_{n=-\infty}^{\infty} |x(n)|^2$	$\frac{1}{2\pi}\int_0^{2\pi} |X(e^{j\omega})|^2 d\omega$
Conjugate symmetry	$x(n)$ real	$X(e^{j\omega}) = X^*(e^{-j\omega})$
Even symmetry	$x(n)$ real and even	$X(e^{j\omega})$ real and even
Odd symmetry	$x(n)$ real and odd	$X(e^{j\omega})$ imaginary and odd

Table A.5 z-transform pairs

$x(n)$	$X(z)$	ROC
$\delta(n)$	1	$|z| \geq 0$
$\delta(n-p),\ p > 0$	z^{-p}	$|z| > 0$
$u(n)$	$\frac{z}{z-1}$	$|z| > 1$
$a^n u(n)$	$\frac{z}{z-a}$	$|z| > |a|$
$na^n u(n)$	$\frac{az}{(z-a)^2}$	$|z| > |a|$
$nu(n)$	$\frac{z}{(z-1)^2}$	$|z| > |1|$
$\cos(\omega_0 n)u(n)$	$\frac{z(z-\cos(\omega_0))}{z^2 - 2z\cos(\omega_0)+1}$	$|z| > 1$
$\sin(\omega_0 n)u(n)$	$\frac{z\sin(\omega_0)}{z^2 - 2z\cos(\omega_0)+1}$	$|z| > 1$
$a^n \cos(\omega_0 n)u(n)$	$\frac{z(z-a\cos(\omega_0))}{z^2 - 2az\cos(\omega_0)+a^2}$	$|z| > |a|$
$a^n \sin(\omega_0 n)u(n)$	$\frac{az\sin(\omega_0)}{z^2 - 2az\cos(\omega_0)+a^2}$	$|z| > |a|$

Table A.6 z-transform properties

Property	$x(n)u(n), h(n)u(n)$	$X(z), H(z)$
Linearity	$ax(n)u(n) + bh(n)u(n)$	$aX(z) + bH(z)$
Left shift	$x(n+m)u(n), \ m > 0$	$z^m X(z) - z^m \sum_{n=0}^{m-1} x(n)z^{-n}$
Right shift	$x(n-m)u(n), \ m > 0$	$z^{-m} X(z) + z^{-m} \sum_{n=1}^{m} x(-n)z^n$
Multiplication by a^n	$a^n x(n)u(n)$	$X(\frac{z}{a})$
Time convolution	$x(n)u(n) * h(n)u(n)$	$X(z)H(z)$
Summation	$\sum_{m=0}^{n} x(m)$	$\frac{z}{z-1} X(z)$
Multiplication by n	$nx(n)u(n)$	$-z \frac{dX(z)}{dz}$
Initial value	$x(0)$	$\lim_{z \to \infty} X(z)$
Final value	$\lim_{n \to \infty} x(n)$	$\lim_{z \to 1}((z-1)X(z))$
		ROC of $(z-1)X(z)$
		includes the unit circle

Answers to Selected Exercises

Chapter 1

1.1.2. Samples of the digital signal with 3-digit decimal precision for the fractional part are

$$\{0.809, 0.156, -0.588, -0.988, -0.809, -0.156, 0.588, 0.988\}$$

1.2.1. Period 5. The samples values of $x(n)$ are

$$\{0, 0.9511, 0.5878, -0.5878, -0.9511, 0, 0.9511, 0.5878,$$
$$- 0.5878, -0.9511, 0, 0.9511\}$$
$$x(78) = x(3) = -0.5878.$$

1.3.5. Even signal.

$$x(-3) = 0, x(-2) = 0.4135, x(-1) = 0.8270, x(0) = 1,$$
$$x(1) = 0.8270, x(2) = 0.4135, x(3) = 0$$

1.4.4. The samples of $x(n)$ are

$$x(n) = \{0, 0, 1, -\hat{0}.8, 0.6400, -0.5120, 0.4096\}$$
$$xe(n) = \{0.2048, -0.2560, 0.8200, -\hat{0}.8, 0.8200, -0.2560, 0.2048\}$$
$$xo(n) = \{-0.2048, 0.2560, 0.1800, \hat{0}, -0.1800, -0.2560, 0.2048\}$$

1.5.5. Average power 4 is finite.
1.6.2.

$$x(n) = 3\delta(n) + \delta(n + 11) - 4\delta(n - 12) + 3\delta(n - 5)$$

© The Author(s), under exclusive license to Springer Nature Switzerland AG 2021
D. Sundararajan, *Digital Signal Processing*,
https://doi.org/10.1007/978-3-030-62368-5

1.7.2.

$$x(n) = u(n+1) + u(n-2) - 7u(n-4) + 8u(n-5) - 3u(n-6)$$

1.8.1.

$$x(n) = 2\cos\left(\frac{-\pi}{6}\right)\cos\left(\frac{2\pi}{8}n\right) - 2\sin\left(\frac{-\pi}{6}\right)\sin\left(\frac{2\pi}{8}n\right)$$

$$= \sqrt{3}\cos\left(\frac{2\pi}{8}n\right) + \sin\left(\frac{2\pi}{8}n\right)$$

The samples of the sinusoid $x(n)$ are

$$\{1.7321, 1.9319, 1, -0.5176, -1.7321, -1.9319, -1, 0.5176\}$$

1.9.1. $z(n) = 1\cos(\frac{2\pi}{6}n + \frac{\pi}{3})$. The samples of the sinusoid $x(n)$ are

$$\{0.5000, -0.5000, -1.0000, -0.5000, 0.5000, 1.0000\}$$

The samples of the sinusoid $y(n)$ are

$$\{1.0000, -1.0000, -2.0000, -1.0000, 1.0000, 2.0000\}$$

The samples of the sinusoid $z(n) = x(n) + y(n)$ are

$$\{0.5000, -0.5000, -1.0000, -0.5000, 0.5000, 1.0000\}$$

1.10.3.

$$xc(n) = (e^{j(\frac{2\pi}{8}n+\pi)} + e^{-j(\frac{2\pi}{8}n+\pi)})$$

The samples of the sinusoid $x(n)$ and $xc(n)$ are the same.

$$\{-2.0000, -1.4142, -0.0000, 1.4142, 2.0000, 1.4142, 0.0000, -1.4142\}$$

1.11.3. The samples of the sinusoid $x(n)$ are

$$\{0.8660 - 0.8660, -0.0000\}$$

The samples of the shifted sinusoid $x(n-2)$ are

$$\{-0.8660, -0.0000, 0.8660\}$$

The samples of the shifted and scaled sinusoid $x((1/3)n - 2)$ are

$$\{-0.8660, 0, 0, 0, 0, 0, 0.8660, 0, 0\}$$

1.12.1. The samples of the waveform $x(n)$ are

$$\{\hat{1}, -2, 3, 4\}$$

The samples of the waveform $x(n-2)$ are

$$\{\hat{0}, 0, 1, -2, 3, 4\}$$

The samples of the shifted and scaled waveform $x(2n-2)$ are

$$\{\hat{0}, 1, 3\}$$

Chapter 2

2.1.3.

$$b = \frac{\tau T_s}{\tau T_s + 1} \quad \text{and} \quad a = -\frac{1}{\tau T_s + 1}$$

$$y_{zi}(n) = 2(-a)^{(n+1)}$$

$$y_{zs}(n) = b\left(\frac{1 - (-a)^{(n+1)}}{1 - (-a)}\right), \quad (-a) \neq 1, \quad n = 0, 1, 2, \ldots$$

The total solution of the continuous system is

$$y(t) = (1 + e^{-\tau t})u(t)$$

$$y(n) = bx(n) + (-a)y(n-1), \quad n = 0, 1, 2 \ldots$$

With $T_s = 0.002$, $\tau = 2$, $a = -0.9960$, and $b = 0.0040$.
The first few exact samples of the complete response are

$$\{2, 1.9960, 1.9920, 1.9881, 1.9841, 1.9802, 1.9763, 1.9724, 1.9685, 1.9646\}$$

The approximate samples of the complete response by iteration of the difference equation are

$$\{1.9960, 1.9920, 1.9881, 1.9842, 1.9802, 1.9763, 1.9724,$$
$$1.9686, , 1.9647, 1.9609\}$$

The approximate samples of the zero-input response are

{1.9920, 1.9841, 1.9762, 1.9683, 1.9605, 1.9527, 1.9449,

1.9371, 1.9294, 1.9217}

The approximate samples of the zero-state response are

{0.0040, 0.0080, 0.0119, 0.0158, 0.0198, 0.0237,

0.0276, 0.0314, 0.0353, 0.0391}

With $T_s = 0.02$, $\tau = 2$, $a = -0.9615$, and $b = 0.0385$.
The first few exact samples of the complete response are

{2, 1.9608, 1.9231, 1.8869, 1.8521, 1.8187, 1.7866, 1.7558, 1.7261, 1.6977}

The approximate samples of the complete response by iteration of the difference equation are

{1.9615, 1.9246, 1.8890, 1.8548, 1.8219, 1.7903, 1.7599,

1.7307, 1.7026, 1.6756}

The approximate samples of the zero-input response are

{1.9231, 1.8491, 1.7780, 1.7096, 1.6439, 1.5806, 1.5198,

1.4614, 1.4052, 1.3511}

The approximate samples of the zero-state response are

{0.0385, 0.0754, 0.1110, 0.1452, 0.1781, 0.2097, 0.2401,

0.2693, 0.2974, 0.3244}

2.2.4.

$$y = \{2, -3, 9, 0, 23, \overset{\wedge}{-7}, -12\}$$

2.3.2.

$$y(n) = (6(0.6)^{(n-3)} - 5(0.5)^{(n-3)})u(n-3)$$

The first four values of the convolution output are

$$0, 0, 0, 1, 1.1000, 0.9100, 0.6710$$

2.4.3. The input sequence is

$\{0.8660, 0.9659, 0.5000, -0.2588, -0.8660, -0.9659, -0.5000,$

$0.2588, 0.8660, 0.9659\}$

By iteration, the first few values of the total response of the system are

$3.1321, 0.6054, 1.5421, -1.0971, -1.2229, -1.9419,$

$-0.6066, 0.4423, 1.6813, 1.6210$

The zero-state response values are

$1.7321, 1.5854, 0.8561, -0.6169, -1.5590, -1.7066,$

$-0.7713, 0.5576, 1.6006, 1.6775$

The zero-input response values are

$1.4000, -0.9800, 0.6860, -0.4802, 0.3361, -0.2353, 0.1647,$

$-0.1153, 0.0807, -0.0565$

The impulse response values are

$2, -0.4000, 0.2800, -0.1960, 0.1372, -0.0960,$

$0. - 0.0471, 0.0329, -0.0231$

The impulse response is

$$h(n) = 1.4286\delta(n) + 0.5714(-0.7)^n u(n), \ n = 0, 1, 2, \ldots$$

The zero-state response is found using convolution as

$$y(n) = \mathrm{Re}\left(e^{-j\frac{\pi}{6}}\left(1.4286e^{j(\frac{2\pi}{8}n)} + 0.5714(-0.7)^n\right.\right.$$

$$\left.\left. \times \frac{(-1.4286e^{j(\frac{2\pi}{8})})^{n+1} - 1}{(-1.4286e^{j(\frac{2\pi}{8})}) - 1}\right)\right)$$

The zero-input response is

$$-2(-0.7)^{(n+1)}$$

The sum of zero-input and zero-state responses is the total output response.

2.5.4. Due to the absolute operator in the past output term, the system is nonlinear.

$$xc = 2xa + 3xb = \{1\hat{3}, 4, 6, -4\}$$

$$ya = \{\hat{2}, -2.4000, 1.3200, 3.0760\}$$

$$yb = \{\hat{3}, -0.1000, -0.0700, -4.0490\}$$

$$(2ya + 3yb) = \{1\hat{3}, -5.1000, 2.4300, -5.9950\}$$

$$yc = \{1\hat{2}, -5.1000, 2.4300, -5.7010\}$$

As the outputs $(2ya(n) + 3yb(n))$ and $yc(n)$ are different, the system is nonlinear.

2.6.4. The system is time variant, since the index of the input is time variant.

$$y = \{\hat{2}, 3, 3, 2\}$$

$$yd = \{\hat{0}, 0, 3, 2, 2, 8\}$$

2.7.4. System is noncausal.

2.8.1. The first few values of the zero-state response are

$$1, 1.3403 + j0.8415, 0.6561 + j1.5825, -0.4651 + j1.4071,$$

$$-1.0257 + j0.3689, -0.5369 - j0.6638$$

The steady-state response of the system to the input e^{j1n} is $1.1355e^{j(1n-0.8701)}$

2.9.5.

$$\sum_{n=1}^{\infty} h(n) = \sum_{n=1}^{\infty} \frac{1}{n^2} = 1 + \frac{1}{4} + \frac{1}{9} + \cdots = \frac{\pi^2}{6}$$

is absolutely convergent. Therefore, the system is stable.

2.11. The impulse response of the second-order system from the parallel configuration is

$$hp1(n) + hp2(n) = -5(-0.7)^{(n)}u(n) + (5)(-0.5^n)u(n), \quad 0, 1, 2, \ldots$$

The impulse response of the second-order system from the cascade configuration is

$$hc1(n) * hc2(n) = (-0.7)^{(n-1)}u(n-1) * (-0.5^n)u(n), \quad 0, 1, 2, \ldots$$

The first four values of the impulse response of the single equivalent system from either expression are

$$\{0, 1.0000, -1.2000, 1.0900\}$$

Chapter 3

3.1.4. The DFT $X(k)$ of $x(n)$ is

$$\{9 + j1, 5 + j1, -1 + j3, -9 - j1\}$$

The energy is 50.
The least squares error is $\{29.5, 30.32, 30.32\}$.

3.2.2. Samples of $x(n)$ from index 0 are $\{-0.2679, 3, -3.7321, 5\}$.
The DFT coefficients $X(k)$ are $\{4, 3.4641 + j2, -12, 3.4641 - j2\}$.

3.3.3.

$$y_c(n) = \{0, \hat{3}, 0, 3\}$$

$$y_l(n) = \{2, -1, -3, 3, -2, \hat{4}, 3\}$$

3.4.2.

$$y_{xh_c}(n) = \{-3, \hat{9}, -14, 2\}$$

$$y_{xh_l}(n) = \{\hat{5}, -14, 11, -6, 4, -9, 3\}$$

Signal power is 30.

3.5.1.

$$P_{xy}(k) = \{12, 6 + j14, -8, 6 - j14\}$$

3.6.3.

$$xe(n) = \{0.2048, -0.2560, 0.8200, -0.8000, 0.8200, -0.2560, 0.2048\}$$

$$xo(n) = \{-0.2048, 0.2560, 0.1800, \hat{0}, -0.1800, -0.2560, 0.2048\}$$

3.7.3.

$$\{2, 1, 4 + j2, -3\} \rightarrow \{4 + j2, -2 - j6, 8 + j2, -2 + j2\} \rightarrow \{8, -12, 16 + j8, 4\}$$

3.8.1.

$$x_r(n) = \sqrt{3}\cos(\frac{2\pi}{8}n) + \sin(\frac{2\pi}{8}n)$$

3.9.3.

$$z(n) = 0.2611\cos(\frac{2\pi}{6}n - 0.6545).$$

3.10.1.

$$xe(n) = \{2, 0, 0, 1.5, \quad 2, 0, 0, 1.5\}$$
$$xo(n) = \{0, 1, -3, 2.5, \quad 0, -1, 3, -2.5\}$$

Chapter 4

4.2.2. One period of the DTFT $X(e^{j\omega})$ is

$$\{X\left(e^{j0}\right) = 8\left(\frac{2\pi}{4}\right)\delta(\omega), X\left(e^{j\frac{2\pi}{4}}\right) = (-2+j4)\left(\frac{2\pi}{4}\right)\delta\left(\omega - \frac{2\pi}{4}\right),$$

$$X\left(e^{j2\frac{2\pi}{4}}\right) = 4\left(\frac{2\pi}{4}\right)\delta\left(\omega - 2\frac{2\pi}{4}\right),$$

$$X\left(e^{j3\frac{2\pi}{4}}\right) = (-2-j4)\left(\frac{2\pi}{4}\right)\delta\left(\omega - 3\frac{2\pi}{4}\right)\}$$

4.3.1. The DFT of $x(n)$ is $\{X(k), k = 0, 1, 2, 3\} = \{8, 5 - j1, -2, 5 + j1\}$. The samples of $X(e^{j\omega})$

$$X(e^{j\omega}) = 4 + 3e^{-j\omega} - e^{-j2\omega} + 2e^{-j3\omega}$$

at $\omega = 0, \frac{\pi}{2}, \pi, \frac{3\pi}{2}$ are also the same.

4.4.3.

$$y(n) = ((4)0.8^{(n)} - (3)0.6^{(n)})u(n)$$

The first 4 values of $y(n)$ are

$$\{1, \quad 1.4000, \quad 1.4800, \quad 1.4000\}$$

4.5.2.

$$x(n)h(n) = \{3, -2, 3, 16, 2\}$$

4.6.3. Taking the inverse DTFT, we get

$$s(n) = (5 - 4(0.8)^n)u(n)$$

The first few samples of $s(n)$, $n = 0, 1, 2, \ldots$ are

$$\{1, 1.8, 2.44, 2.952, 3.3616, 3.6893, 3.9514, 4.1611,$$
$$4.3289, 4.4631, 4.5705\}$$

4.9.

$$y(n) = (-110.25)\left(\frac{4}{5}\right)^n + n(14.1667)\left(\frac{2}{3}\right)^n$$
$$+(111.25)\left(\frac{2}{3}\right)^n, \ n = 0, 1, \ldots$$

The first four values of the sequence $y(n)$ are

$$\{y(0) = 1, \quad y(1) = 0.1333, \quad y(2) = -2.2267, \quad y(3) = -4.5961\}$$

4.10.2. The analytic signal is

$$x_a(n) = x(n) + jx_H(n) = \{3 + j1.5, -1 + j0.5, 2 - j1.5, 2 - j0.5\}\}$$

The DFT of $x_a(n)$ is

$$X_H(k) = \{6, 2 + j6, 4, 0\}$$

a one-sided spectrum.

Chapter 5

5.1.2.

$$X(z) = 2(-1z^{-1} - 2z^{-3}).$$

5.2.3.

$$\{x(7) = 1, x(9) = 2\}.$$

5.3.1.

$$y(n) = \{\hat{3}, 8, -8, -8\}$$

5.4.3.

$$X(z) = \left(\frac{z}{(z-1)^2}\right)$$

5.5.3.

$$X_0(z) = \frac{3z+2}{z}$$

$$X(z) = \left(\frac{z^2}{z^2-1}\right) X_0(z), \quad |z| > 1$$

5.6.1

$$y(n) = \big((0.0924)(0.6956)^n + (2)(0.6107)(1.6956)^n$$

$$\times \cos(2.4896n + (0.7332))\big) u(n)$$

The first four values of the sequence $y(n)$ are

$$\{y(0) = 1, \quad y(1) = -2, \quad y(2) = 3, \quad y(3) = -2\}$$

5.9.

$$y(n) = \left((-0.7420 + j1.9127)(e^{j\frac{2\pi}{8}})^n + (1.2420 - j1.0467)\left(\frac{5}{6}\right)^n\right.$$

$$\left. + \frac{5}{2}\left(\frac{5}{6}\right)^n\right), \ n = 0, 1, \dots$$

Taking the real part the sequence of $y(n)$, we get the first four values of $y(n)$ are

$$\{y(0) = 3, \quad y(1) = 1.2412, \quad y(2) = 0.6859, \quad y(3) = 1.3377\}$$

Chapter 6

6.1. The shifted impulse response values, with a precision of four digits after the decimal point, are

$$\{h_s(n), n = 0, 1, \dots, 18\}$$

$$= \{-0.0161, 0.0234, 0.0449, 0.0164, -0.0450, -0.0757,$$

$$-0.0166, 0.1288, 0.2836, 0.3500, 0.2836, 0.1288, -0.0166,$$

$$-0.0757, -0.0450, 0.0164, 0.0449, 0.0234, -0.0161\}$$

6.3. The impulse response values are

$\{h_{sw}(n), n = 0, 1, \ldots, 21\}$

$= \{0.0039, -0.0002, -0.0117, -0.0082, 0.0198, 0.0311,$

$\quad -0.0181, -0.0765, -0.0183, 0.1889, 0.3889, 0.3889, 0.1889,$

$\quad -0.0183, -0.0765 - 0.0181, 0.0311, 0.0198,$

$\quad -0.0082, -0.0117, -0.0002, 0.0039\}$

6.5. The impulse response values, with a precision of four digits after the decimal point, are

$\{h_{BP}(n), n = 0, 1, \ldots, 22\}$

$= \{0.0027, -0.0000, 0.0063, 0.0056, -0.0269,$

$\quad -0.0174, 0.0000, -0.0553, 0.0771, 0.2669, -0.0574, 0.6,$

$\quad -0.0574, 0.2669, 0.0771, -0.0553, 0.0000, -0.0174,$

$\quad -0.0269, 0.0056, 0.0063, -0.0000, 0.0027\}$

6.7. The 21 shifted and widowed impulse response values, with a precision of four digits after the decimal point, of the differentiator are

$\{h_{diff}(n), n = 0, 1, \ldots, 20\}$

$= \{-0.0080, 0.0114, -0.0210, 0.0385, -0.0663, 0.1080,$

$\quad -0.1705, 0.2701, -0.4561, 0.9775, 0, -0.9775, 0.4561,$

$\quad -0.2701, 0.1705, -0.1080, 0.0663,$

$\quad -0.0385, 0.0210, -0.0114, 0.0080\}$

6.9. The 21 impulse response values, with a precision of four digits after the decimal point, are

$\{0.0014, -0.0175, -0.0270, -0.0195, 0.0133, 0.0507, 0.0511,$

$\quad - 0.0169, -0.1415, -0.2641, 0.6848, -0.2641, -0.1415,$

$\quad - 0.0169, 0.0511, 0.0507, 0.0133, -0.0195, -0.0270, -0.0175, 0.0014\}$

$$\delta_p = 0.0138 \quad \text{and} \quad \delta_s = 0.0551$$

6.11. The 19 impulse response values, with a precision of four digits after the decimal point, are

$$\{-0.0000, -0.0805, 0.0000, 0.0315, 0.0000, 0.1451, 0.0000,$$

$$0.2666, 0.0000, 0.3189, 0.0000, 0.2666, 0.0000, 0.1451,$$

$$0.0000, 0.0315, 0.0000, -0.0805, -0.0000\}$$

$$\delta_p = 0.0493 \qquad \text{and} \qquad \delta_s = 0.1480$$

6.13. The 21 impulse response values, with a precision of four digits after the decimal point, are

$$\{-0.0000, -0.0044, -0.0000, -0.0204, -0.0000 - 0.0623, -0.0000,$$

$$- 0.1656, -0.0000, -0.6196, 0, 0.6196, 0.0000, 0.1656, 0.0000, 0.0623,$$

$$0.0000, 0.0204, 0.0000, 0.0044, 0.0000\}$$

$$\delta_p = 5.9583e - 04$$

Chapter 7

7.2.

$$H(z) = \frac{0.5867(z^2 - 2z + 1)}{(z^2 - 1.1389z + 0.4943)}$$

7.4.

$$H(z) = \frac{0.4321(z^2 - 1.3478z + 1)}{(z^2 - 0.5824z - 0.1358)}$$

7.6.

$$H(z) = \frac{0.1730(z^2 - 2z + 1)(z^2 - 2z + 1)(z - 1)}{(z^2 - 0.7206z + 0.5982)(z^2 - 1.3935z + 0.9097)(z + 0.2772)}$$

7.8.

$$H(z) = \frac{0.1813(z^2 - 1.3478z + 1)(z^2 - 1.3478z + 1)}{(z^2 + 0.7896z + 0.4919)(z^2 - 1.7995z + 0.8420)}$$

Chapter 8

8.1.3.

$$X_d(k) = \{-4, -2.5 + j2.5981, 0.5$$
$$+ j2.5981, 2, 0.5 - j2.5981, -2.5 - j2.5981\}$$

The 2-point of IDFT of $X_d(0) = -4, \quad X_d(3) = 2$

$$x_d(0) = -1, \quad x_d(1) = -3$$

8.2.1.

$$x_{dec}(n) = \{1, -0.5, -0.5\}$$

8.3.2.

$$x_u(n) = \{2, 0, 0, 1, 0, 0, 3, 0, 0, 4, 0, 0\}$$

8.4.2.

$$x_i(n) = \{0, 0.5, 0.8660, 1, 0.8660, 0.5, 0, -0.5, -0.8660,$$
$$- 1, -0.8660, -0.5\}$$

8.6.2.

$$x(n) = \{1, 0, -1, 0\}$$

$$X(k) = \{0, 2, 0, 2\}$$

8.8.

$$\{4, 4, 8, 4, -3, -5, -1, 5, 12, 4, 0\}$$

8.10. The DWT coefficients are

$$\{2.1213, \quad 4.9497, \quad 0.7071, \quad -0.7071\}$$

The energy is 30.

Chapter 9

9.1.1.

$$xe(n) = \{1, 2, 1, 2\}, xo(n) = \{-1, -1, 1, 1\}$$
$$X(k) = \{6, -2 + j2, -2, -2 - j2\},$$
$$Xe(k) = \{6, 0, -2, 0\}, Xo(k) = \{0, -2 + j2, 0, -2 - j2\}$$

9.2.2.

$$x(n) = \{-0.5000 - j0.8660, 0.8660 - j0.5000, 0.5000 + j0.8660,$$
$$- 0.8660 + j0.5000 - 0.5000 - j0.8660,$$
$$0.8660 - j0.5000, 0.5000 + j0.8660, -0.8660 + j0.5000\}$$
$$X(k) = \{0, 0, -4.0000 - j6.9282, 0, 0, 0, 0, 0\}$$

9.3.3.

$$x(n) = \{0.7071, -1.0000, 0.7071, 0.0000, -0.7071,$$
$$1.0000, -0.7071, 0.0000\}$$
$$y(n) = \{0.5000, -0.9659, 0.8660, -0.2588, -0.5000,$$
$$0.9659, -0.8660, 0.2588\}$$
$$x(n) + jy(n) = \{0.7071 + j0.5000, -1.0000 - j0.9659, 0.7071$$
$$+ j0.8660, 0.0000 - j0.2588,$$
$$- 0.7071 - j0.5000, 1.0000 + j0.9659,$$
$$- 0.7071 - j0.8660, 0.0000 + j0.2588\}$$
$$X(k) + jY(0, 0, 0, -0.6357 + j4.8284, 0,$$
$$6.2925 - j0.8284, 0, 0\}$$
$$X(k) = \{0, 0, 0, 2.8284 + j2.8284, 0, 2.8284 - j2.8284, 0, 0\}$$
$$Y(k) = \{0, 0, 0, 2 + j3.4641, 0, 2 - j3.4641, 0, 0\}$$

9.4.4.

$$x(n) = \{0.7071, 0.0000, -0.7071, 1.0000, -0.7071,$$
$$- 0.0000, 0.7071, -1.0000\}$$
$$X(k) = \{0, 0, 0, 2.8284 - j2.8284, 0, 2.8284 + j2.8284, 0, 0\}$$

Chapter 10

10.1.2.

$$B = 0.0001$$
$$B^* = 1.1111$$

10.3.

$$b = 13 \text{ and } b = 10$$

10.5.

$$\sigma_{qo}^2 = \sigma_q^2 295.1277$$

10.7.

$$\{K_1 = \frac{1}{1-2} = -1, \quad \text{and} \quad K_2 = -2\}$$

References

1. Mitra, S. K. (2011). *Digital signal processing - a computer-based approach*. New York: McGraw Hill.
2. Proakis, J. G., & Manolakis, D. G. (2007). *Digital signal processing - principles, algorithms, and applications*. Englewood Cliffs: Prentice-Hall.
3. Sundararajan, D. (2003). *Digital signal processing - theory and practice*. Singapore: World Scientific.
4. Sundararajan, D. (2008). *Signals and systems – a practical approach*. Singapore: Wiley.
5. Sundararajan, D. (2015). *Discrete wavelet transform, a signal processing approach*. Singapore: Wiley.
6. Sundararajan, D. (2018). *Fourier analysis - a signal processing approach*. Singapore: Springer.
7. The Mathworks. (2021). *Matlab signal processing tool box user's guide*. Novi: The Mathworks, Inc.

Index

Printed in the United States
by Baker & Taylor Publisher Services